Modern Birkhäuser Classics

Many of the original research and survey monographs, as well as textbooks, in pure and applied mathematics published by Birkhäuser in recent decades have been groundbreaking and have come to be regarded as foundational to the subject. Through the MBC Series, a select number of these modern classics, entirely uncorrected, are being re-released in paperback (and as eBooks) to ensure that these treasures remain accessible to new generations of students, scholars, and researchers.

T0214489

Wavelets Made Easy

Yves Nievergelt

Reprint of the 2001 Edition

 Birkhäuser

Yves Nievergelt
Department of Mathematics
Eastern Washington University
Cheney, WA
USA

ISBN 978-1-4614-6005-3   ISBN 978-1-4614-6006-0 (eBook)
DOI 10.1007/978-1-4614-6006-0
Springer New York Heidelberg Dordrecht London

Library of Congress Control Number: 2012952313

© Springer Science+Business Media New York 2013
This work is subject to copyright. All rights are reserved by the Publisher, whether the whole or part of the material is concerned, specifically the rights of translation, reprinting, reuse of illustrations, recitation, broadcasting, reproduction on microfilms or in any other physical way, and transmission or information storage and retrieval, electronic adaptation, computer software, or by similar or dissimilar methodology now known or hereafter developed. Exempted from this legal reservation are brief excerpts in connection with reviews or scholarly analysis or material supplied specifically for the purpose of being entered and executed on a computer system, for exclusive use by the purchaser of the work. Duplication of this publication or parts thereof is permitted only under the provisions of the Copyright Law of the Publisher's location, in its current version, and permission for use must always be obtained from Springer. Permissions for use may be obtained through RightsLink at the Copyright Clearance Center. Violations are liable to prosecution under the respective Copyright Law.
The use of general descriptive names, registered names, trademarks, service marks, etc. in this publication does not imply, even in the absence of a specific statement, that such names are exempt from the relevant protective laws and regulations and therefore free for general use.
While the advice and information in this book are believed to be true and accurate at the date of publication, neither the authors nor the editors nor the publisher can accept any legal responsibility for any errors or omissions that may be made. The publisher makes no warranty, express or implied, with respect to the material contained herein.

Printed on acid-free paper

Springer is part of Springer Science+Business Media (www.birkhauser-science.com)

Yves Nievergelt

Wavelets Made Easy

Birkhäuser
Boston • Basel • Berlin

Yves Nievergelt
Department of Mathematics
Eastern Washington University
Cheney, WA 99004-2431
USA

Library of Congress Cataloging-in-Publication Data

Nievergelt, Yves.
    Wavelets made easy / Yves Nievergelt.
        p. cm.
    Includes bibliographical references and index.
    ISBN 0-8176-4061-4 (acid-free paper) – ISBN
3-7643-4061-4   (acid-free paper)
    1. Wavelets (Mathematics) I. Title.
QA403.3.N54 1999
515'.2433–dc21   98-29994
                                                                       CIP

AMS Subject Classifications: 42

Printed on acid-free paper
©1999 Birkhäuser Boston
©2001 Birkhäuser Boston, 2nd printing
    with corrections

Birkhäuser

All rights reserved. This work may not be translated or copied in whole or in part without the written permission of the publisher (Birkhäuser Boston, c/o Springer-Verlag New York, Inc., 175 Fifth Avenue, New York, NY 10010, USA), except for brief excerpts in connection with reviews or scholarly analysis. Use in connection with any form of information storage and retrieval, electronic adaptation, computer software, or by similar or dissimilar methodology now known or hereafter developed is forbidden.
The use of general descriptive names, trade names, trademarks, etc., in this publication, even if the former are not especially identified, is not to be taken as a sign that such names, as understood by the Trade Marks and Merchandise Marks Act, may accordingly be used freely by anyone.

ISBN 0-8176-4061-4   SPIN 10787109
ISBN 3-7643-4061-4

Formatted from the author's TEX files by Integre Technical Publishing Company, Inc., Albuquerque, NM.
Printed and bound by R.R. Donnelley and Sons, Harrisonburg, VA.
Printed in the United States of America.

9 8 7 6 5 4 3 2

Contents

Preface

This book explains the nature and computation of mathematical wavelets, which provide a framework and methods for the analysis and the synthesis of signals, images, and other arrays of data. The material presented here addresses the audience of engineers, financiers, scientists, and students looking for explanations of wavelets at the undergraduate level. It requires only a working knowledge or memories of a first course in linear algebra and calculus. The first part of the book answers the following two questions:

What are wavelets? Wavelets extend Fourier analysis.

How are wavelets computed? Fast transforms compute them.

To show the practical significance of wavelets, the book also provides transitions into several applications: analysis (detection of crashes, edges, or other events), compression (reduction of storage), smoothing (attenuation of noise), and synthesis (reconstruction after compression or other modification). Such applications include one-dimensional signals (sounds or other time-series), two-dimensional arrays (pictures or maps), and three-dimensional data (spatial diffusion). The applications demonstrated here do not constitute recipes for real implementations, but aim only at clarifying and strengthening the understanding of the mathematics of wavelets.

The second part of the book explains orthogonal projections, discrete and fast Fourier transforms, and Fourier series, as a preparation for the third part and as an answer to the following question:

How are wavelets related to other methods of signal analysis?

The third part of the book invokes occasional results from advanced calculus and focuses on the following question, which provides a transition into the theory and research on the subject:

How are wavelets designed? (Designs use Fourier transforms.)

More details appear in the chapter summaries on the following page. The material has been taught in various forms for a decade in an undergraduate course at Eastern Washington University, to engineers and students majoring in mathematics or computer science. I thank them for their patience in reading through several drafts.

YVES NIEVERGELT

Eastern Washington University
Cheney, WA

Outline

Part A, which can be read before or after Part B, provides an immediate and very basic introduction to wavelets.

Chapter 1 gives a first elementary yet rigorous explanation of the nature of mathematical wavelets, in particular, Alfred Haar's wavelets, without either calculus or linear algebra.

Chapter 2 presents multidimensional wavelets and some applications, with one use of matrix algebra but without calculus.

Chapter 3 introduces computational features of Ingrid Daubechies' wavelets with one and more dimensions, with some matrix algebra but without calculus. Chapter 3 also aims at justifying the need for some clarification of Daubechies wavelets through theory, which will be the subject of Parts B and C.

Part B presents the mathematical context in which wavelets arose: Joseph Fourier's analysis of signals and functions.

Chapter 4 reviews the topics from linear algebra that explain how wavelets approximate functions: linear spaces, inner products, norms, orthogonal projections, and least-squares regression.

Chapter 5 focuses on the discrete fast Fourier transform of James W. Cooley and John W. Tukey, which provides a framework simpler than Daubechies wavelets to explain fast transforms.

Chapter 6 treats Fourier series, which demonstrate least-squares approximations of functions within a framework simpler than, but similar to, that of Daubechies wavelets in Chapter 8.

Part C explains Fourier transforms and their use in wavelet design.

Chapter 7 presents the Fourier transform and its inverse on the real line, in the plane, and in space. This is the essential concept for the design and the mathematical foundations of wavelets.

Chapter 8 explains how Ingrid Daubechies applied the Fourier Transform to design wavelets. The explanations also show how the Fourier Transform applies to the design of other wavelets.

Chapter 9 shows how accurately wavelets can approximate signals.

PART A

Algorithms for Wavelet Transforms

CHAPTER 1

Haar's Simple Wavelets

1.0 INTRODUCTION

This chapter explains the nature of the simplest wavelets and an algorithm to compute a fast wavelet transform. Such wavelets have been called "Haar's wavelets" since Haar's publication in 1910 (reference [19] in the bibliography). To analyze and synthesize a signal—which can be any array of data—in terms of simple wavelets, this chapter employs shifts and dilations of mathematical functions, but does *not* involve either calculus or linear algebra.

The first step in applying wavelets to any signal or physical phenomenon consists in representing the signal under consideration by a mathematical function, as in Figure 1.1(a). The usefulness of mathematical functions lies in their efficiency and versatility in representing various types of signals or phenomena. For instance, the horizontal axis in Figure 1.1(a)–(c) may correspond to time ($r = t$), while the vertical axis may correspond to the intensity of a signal ($s = f(r)$), for example, a sound; the values $s = f(r) = f(t)$ measure the sound at each time t at a fixed location. Alternatively, the horizontal axis may correspond to a spatial dimension ($r = x$), and then the values $s = f(r) = f(x)$ measure the intensity of the sound at each location x at a common time. Similarly, the same function f may represent the intensity of light along a cross section of an image.

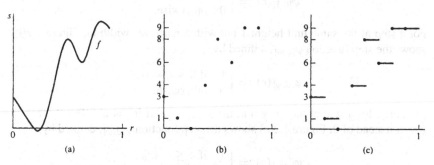

Figure 1.1 (a) Signal. (b) Sample. (c) Approximation.

In any event, because the same type of mathematical function f can represent many types of signals or phenomena, the same type of analysis or synthesis of f, in terms of wavelets or otherwise, will apply to all the signals or phenomena represented by f.

1.1   SIMPLE APPROXIMATION

Because practical measurements of real phenomena require time and resources, they provide not all values but only a finite sequence of values, called a **sample,** of the function representing the phenomenon under consideration, as in Figure 1.1(b). Therefore, the first step in the analysis of a signal with wavelets consists in approximating its function by means of the sample alone. One of the simplest methods of approximation uses a horizontal stair step extended through each sample point, as in Figure 1.1(c). The resulting steps form a new function, denoted here by $\tilde{f}$ and called a **simple function** or **step function,** which approximates the sampled function f. Although approximations more accurate than simple steps exist, they demand more sophisticated mathematics, so this chapter restricts itself to simple steps. A precise notation will prove useful to indicate the location of such steps. (The following notation is consistent with Y. Meyer's books on wavelets [31, p. 94].)

Definition 1.1   For all numbers u and w, the notation $[u, w[$ represents the **interval** of all numbers from u *included* to w *excluded*:

$$[u, w[= \{r : u \le r < w\}. \qquad \qquad \square$$

(The symbol $\square$ marks the end of a definition or other formal unit.)

The analysis of the approximating function $\tilde{f}$ in terms of wavelets requires a precise labeling of each step, by means of shifts and dilations of the basic **unit step** function, denoted by $\varphi_{[0,1[}$ and exhibited in Figure 1.2(a). The unit step function $\varphi_{[0,1[}$ has the values (with the symbol $:=$ defining the left-hand side in terms of the right-hand side)

$$\varphi_{[0,1[}(r) := \begin{cases} 1 & \text{if } 0 \le r < 1, \\ 0 & \text{otherwise.} \end{cases}$$

For a step at the same unit height 1 but with a narrower width w, Figure 1.2(b) shows the step function $\varphi_{[0,w[}$, defined by

$$\varphi_{[0,w[}(r) := \begin{cases} 1 & \text{if } 0 \le r < w, \\ 0 & \text{otherwise.} \end{cases}$$

Similarly, for a step at the same unit height 1, but starting at a different location $r = u$ instead of 0, Figure 1.2(c) shows the step function $\varphi_{[u,w[}$, defined by

$$\varphi_{[u,w[}(r) := \begin{cases} 1 & \text{if } u \le r < w, \\ 0 & \text{otherwise.} \end{cases}$$

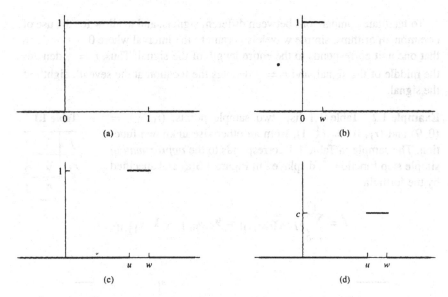

Figure 1.2   (a) $\varphi_{[0,1[}$; (b) $\varphi_{[0,w[}$; (c) $\varphi_{[u,w[}$; (d) $c \cdot \varphi_{[u,w[}$.

Finally, to construct a step function at a different height c, starting at the location u and ending at w, Figure 1.2(d) shows $c \cdot \varphi_{[u,w[}$, a scalar multiple by c of the function $\varphi_{[u,w[}$, so that

$$c \cdot \varphi_{[u,w[}(r) = \begin{cases} c & \text{if } u \leq r < w, \\ 0 & \text{otherwise.} \end{cases}$$

Thus, if a sample point (r_j, s_j) includes a value $s_j = f(r_j)$ at height s_j and at abscissa (time or location) r_j, then that sample point corresponds to the step function

$$s_j \cdot \varphi_{[r_j, r_{j+1}[},$$

which approximates f at height s_j on the interval $[r_j, r_{j+1}[$ from r_j (included) to r_{j+1} (not included). Adding all the step functions corresponding to all the points in the sample yields a formula approximating the simple step function shown in Figure 1.1(c):

$$\tilde{f} = s_0 \cdot \varphi_{[r_0, r_1[} + s_1 \cdot \varphi_{[r_1, r_2[} + \cdots + s_{n-1} \cdot \varphi_{[r_{n-1}, r_n[}$$

$$= \sum_{j=0}^{n-1} s_j \cdot \varphi_{[r_j, r_{j+1}[}.$$

(The notation $\sum_{j=0}^{n-1} s_j \cdot \varphi_{[r_j, r_{j+1}[}$ represents the sum of all the terms $s_j \cdot \varphi_{[r_j, r_{j+1}[}$ from $s_0 \cdot \varphi_{[r_0, r_1[}$ through $s_{n-1} \cdot \varphi_{[r_{n-1}, r_n[}$.)

To facilitate comparisons between different signals, and to allow for the use of common algorithms, simple wavelets pertain to the interval where $0 \leq r < 1$, so that one unit corresponds to the entire length of the signal. Thus, $r = \frac{1}{2}$ denotes the middle of the signal, and $r = \frac{7}{8}$ denotes the location at the seventh eighth of the signal.

Example 1.2   Table 1.1 lists two sample points, $(r_0, s_0) = (0, 9)$ and $(r_1, s_1) = (\frac{1}{2}, 1)$, from an otherwise unknown function. The sample in Table 1.1 corresponds to the *approximating simple step function* $\tilde{f}$, displayed in Figure 1.3(a) and specified by the formula

Table 1.1

j	0	1
r_j	0	$\frac{1}{2}$
s_j	9	1

$$\tilde{f} = \sum_{j=0}^{1} s_j \cdot \varphi_{[r_j, r_{j+1}[} = 9 \cdot \varphi_{[0, \frac{1}{2}[} + 1 \cdot \varphi_{[\frac{1}{2}, 1[}.$$

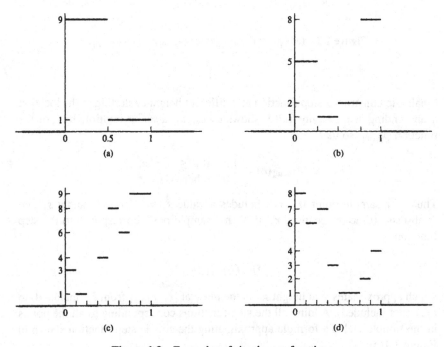

Figure 1.3   Examples of simple step functions.

The first step, $9 \cdot \varphi_{[0, \frac{1}{2}[}$, has height 9 over the interval $[0, \frac{1}{2}[$ starting at 0 (included) and ending at $\frac{1}{2}$ (not included). The second step, $1 \cdot \varphi_{[\frac{1}{2}, 1[}$, has height 1 over the interval $[\frac{1}{2}, 1[$ starting at $\frac{1}{2}$ (included) and ending at 1 (not included). The notation $[0, \frac{1}{2}[$ and $[\frac{1}{2}, 1[$ shows that the value of $\tilde{f}$ at $\frac{1}{2}$ arises from $1 \cdot \varphi_{[\frac{1}{2}, 1[}$, which *includes* $\frac{1}{2}$, but not from $9 \cdot \varphi_{[0, \frac{1}{2}[}$, which *excludes* $\frac{1}{2}$.   □

Example 1.3 The sample in Table 1.2 corresponds to the approximating simple step function $\tilde{g}$, displayed in Figure 1.3(b) and specified by the formula

$$5 \cdot \varphi_{[0,\frac{1}{4}[} + 1 \cdot \varphi_{[\frac{1}{4},\frac{1}{2}[} + 2 \cdot \varphi_{[\frac{1}{2},\frac{3}{4}[} + 8 \cdot \varphi_{[\frac{3}{4},1[}.$$

Table 1.2

j	0	1	2	3
r_j	0	$\frac{1}{4}$	$\frac{1}{2}$	$\frac{3}{4}$
s_j	5	1	2	8

$\square$

Example 1.4 The sample in Table 1.3 corresponds to the approximating simple step function h, displayed in Figure 1.3(c) and specified by the formula

Table 1.3

j	0	1	2	3	4	5	6	7
r_j	0	$\frac{1}{8}$	$\frac{1}{4}$	$\frac{3}{8}$	$\frac{1}{2}$	$\frac{5}{8}$	$\frac{3}{4}$	$\frac{7}{8}$
s_j	3	1	0	4	8	6	9	9

$$\tilde{h} = 3 \cdot \varphi_{[0,\frac{1}{8}[} + 1 \cdot \varphi_{[\frac{1}{8},\frac{1}{4}[} + 0 \cdot \varphi_{[\frac{1}{4},\frac{3}{8}[} + 4 \cdot \varphi_{[\frac{3}{8},\frac{1}{2}[}$$
$$+ 8 \cdot \varphi_{[\frac{1}{2},\frac{5}{8}[} + 6 \cdot \varphi_{[\frac{5}{8},\frac{3}{4}[} + 9 \cdot \varphi_{[\frac{3}{4},\frac{7}{8}[} + 9 \cdot \varphi_{[\frac{7}{8},1[}.$$

$\square$

Slight variations exist for approximations by step functions. For instance, instead of steps extending on only one side of each sample point, other methods may use steps centered at the sample points, extending equally far on both sides of each sample point.

EXERCISES

Exercise 1.1. Write a formula for the step function $\tilde{p}$ plotted in Figure 1.3(d) and corresponding to the sample in Table 1.4.

Table 1.4

j	0	1	2	3	4	5	6	7
r_j	0	$\frac{1}{8}$	$\frac{1}{4}$	$\frac{3}{8}$	$\frac{1}{2}$	$\frac{5}{8}$	$\frac{3}{4}$	$\frac{7}{8}$
s_j	8	6	7	3	1	1	2	4

Exercise 1.2. Plot and write a formula for the step function $\tilde{q}$ corresponding to the sample in Table 1.5.

Table 1.5

j	0	1	2	3	4	5	6	7
r_j	0	$\frac{1}{8}$	$\frac{1}{4}$	$\frac{3}{8}$	$\frac{1}{2}$	$\frac{5}{8}$	$\frac{3}{4}$	$\frac{7}{8}$
s_j	3	1	9	7	7	9	5	7

Exercise 1.3. Verify, through algebra, logic, or cases, that for every number r,

$$\varphi_{[0,w[}(r) = \varphi_{[0,1[}(r/w),$$
$$\varphi_{[u,w[}(r) = \varphi_{[0,1[}\left(\frac{r-u}{w-u}\right).$$

Exercise 1.4. Define Haar's "wavelet" function $\psi_{[0,1[}$ by

$$\psi_{[0,1[} := \varphi_{[0,\frac{1}{2}[} - \varphi_{[\frac{1}{2},1[}.$$

Verify, through algebra, logic, or cases, that for every number r,

$$\psi_{[0,1[}(r) = \begin{cases} 1 & \text{if } 0 \le r < \frac{1}{2}, \\ -1 & \text{if } \frac{1}{2} \le r < 1, \\ 0 & \text{otherwise.} \end{cases}$$

1.2   APPROXIMATION WITH SIMPLE WAVELETS

1.2.1   The Basic Haar Wavelet Transform

Haar's basic transformation expresses the approximating function $\tilde{f}$ with wavelets by replacing an adjacent pair of steps by one wider step and one wavelet. The wider step measures the average of the initial pair of steps, while the wavelet, formed by two alternating steps, measures the difference of the initial pair of steps.

For instance, the *sum* of two adjacent steps with width 1/2 produces the basic unit step function $\varphi_{[0,1[}$, as in Figure 1.4: Indeed,

$$\varphi_{[0,1[} = \varphi_{[0,\frac{1}{2}[} + \varphi_{[\frac{1}{2},1[}.$$

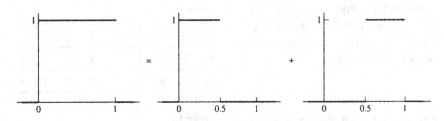

Figure 1.4 $\varphi_{[0,1[} = \varphi_{[0,\frac{1}{2}[} + \varphi_{[\frac{1}{2},1[}.$

Similarly, the *difference* of two such narrower steps gives the corresponding **basic wavelet,** denoted by $\psi_{[0,1[}$ and defined by

$$\psi_{[0,1[} = \varphi_{[0,\frac{1}{2}[} - \varphi_{[\frac{1}{2},1[}.$$

The wavelet $\psi_{[0,1[}$ so defined is a simple step function, with a first step at height 1 followed by a second step at height -1. Thus, from its first step to its second step, the values of the wavelet $\psi_{[0,1[}$ undergo a jump of size -2, as in Figure 1.5.

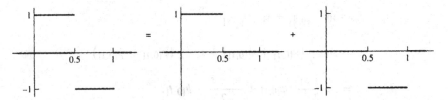

Figure 1.5 $\psi_{[0,1[} = \varphi_{[0,\frac{1}{2}[} - \varphi_{[\frac{1}{2},1[}}$.

Adding and subtracting the two equations just obtained,

$$\begin{cases} \varphi_{[0,1[} = \varphi_{[0,\frac{1}{2}[} + \varphi_{[\frac{1}{2},1[}}, \\ \psi_{[0,1[} = \varphi_{[0,\frac{1}{2}[} - \varphi_{[\frac{1}{2},1[}}, \end{cases}$$

produces the inverse relation, which expresses the narrower steps $\varphi_{[0,\frac{1}{2}[}$ and $\varphi_{[\frac{1}{2},1[}}$ in terms of the basic unit step $\varphi_{[0,1[}$ and wavelet $\psi_{[0,1[}$, as shown in Figure 1.6:

$$\begin{cases} \frac{1}{2}\left(\varphi_{[0,1[} + \psi_{[0,1[}\right) = \varphi_{[0,\frac{1}{2}[}, \\ \frac{1}{2}\left(\varphi_{[0,1[} - \psi_{[0,1[}\right) = \varphi_{[\frac{1}{2},1[}}. \end{cases}$$

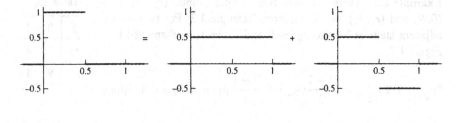

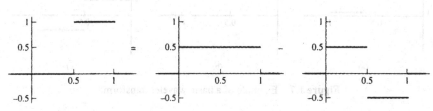

Figure 1.6   *Top:* $\varphi_{[0,\frac{1}{2}[} = \frac{1}{2}\left(\varphi_{[0,1[} + \psi_{[0,1[}\right)$. *Bottom:* $\varphi_{[\frac{1}{2},1[}} = \frac{1}{2}\left(\varphi_{[0,1[} - \psi_{[0,1[}\right)$.

For two adjacent steps at heights s_0 and s_1, the equations just derived yield the following representation with one wider step and one wavelet:

$$\tilde{f} := s_0 \cdot \varphi_{[0,\frac{1}{2}[} + s_1 \cdot \varphi_{[\frac{1}{2},1[}$$

$$= s_0 \cdot \frac{1}{2} \left(\varphi_{[0,1[} + \psi_{[0,1[} \right) + s_1 \cdot \frac{1}{2} \left(\varphi_{[0,1[} - \psi_{[0,1[} \right)$$

$$= \frac{s_0 + s_1}{2} \cdot \varphi_{[0,1[} + \frac{s_0 - s_1}{2} \cdot \psi_{[0,1[}.$$

1.2.2   Significance of the Basic Haar Wavelet Transform

Two sample values s_0 and s_1 measure the value (amplitude, height) of the function $\tilde{f}$ at r_0 and at r_1. In contrast, the results from the basic transform have the following significance.

- The number $(s_0 + s_1)/2$ measures the *average* of the function $\tilde{f}$.

- The number $(s_0 - s_1)/2$ measures the *change* in the function $\tilde{f}$.

The basic transform preserves all the information in the sample, since, while the transform describes the sample differently from the sample values, it also reproduces the sample exactly:

$$s_0 \cdot \varphi_{[0,\frac{1}{2}[} + s_1 \cdot \varphi_{[\frac{1}{2},1[} = \tilde{f} = \frac{s_0 + s_1}{2} \cdot \varphi_{[0,1[} + \frac{s_0 - s_1}{2} \cdot \psi_{[0,1[}.$$

Example 1.5   Table 1.6 lists two sample points, $(r_0, s_0) = (0, 9)$ and $(r_1, s_1) = (\frac{1}{2}, 1)$, from Example 1.3. For two such adjacent steps at heights $s_0 = 9$ and $s_1 = 1$, as displayed in Figure 1.7,

Table 1.6

j	0	1
r_j	0	$\frac{1}{2}$
s_j	9	1

$$9\varphi_{[0,\frac{1}{2}[} + 1\varphi_{[\frac{1}{2},1[} = \frac{9+1}{2}\varphi_{[0,1[} + \frac{9-1}{2}\psi_{[0,1[} = 5\varphi_{[0,1[} + 4\psi_{[0,1[}.$$

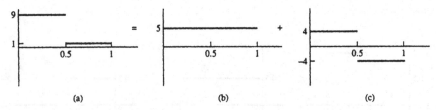

Figure 1.7   Example of a basic wavelet transform.

- The term $5\varphi_{[0,1[}$ means that that the whole sample has an average value (average height) equal to **5**.

- The term $4\psi_{[0,1[}$ means that from its first value to its second value, the sample changes as do **4** basic wavelets: It undergoes a jump of size $4 \cdot (-2) = -8$, effectively from 9 to 1. $\square$

1.2.3 Shifts and Dilations of the Basic Haar Transform

To apply the basic transform starting at a different location u instead of 0, and over an interval extending to w instead of 1, define the shifted and dilated wavelet $\psi_{[u,w[}$ by the midpoint $v := (u+w)/2$:

$$\psi_{[u,w[}(r) := \begin{cases} 1 & \text{if } u \le r < v, \\ -1 & \text{if } v \le r < w. \end{cases}$$

Again, the sum and the difference of two narrower steps give a wider step and a wavelet:

$$\begin{cases} \varphi_{[u,w[} = \varphi_{[u,v[} + \varphi_{[v,w[}, \\ \psi_{[u,w[} = \varphi_{[u,v[} - \varphi_{[v,w[}. \end{cases}$$

Also, adding and subtracting the two equations just obtained yields the inverse relation, expressing the two narrower steps in terms of the wider step and the wavelet:

$$\begin{cases} \frac{1}{2}\left(\varphi_{[u,w[} + \psi_{[u,w[}\right) = \varphi_{[u,v[}, \\ \frac{1}{2}\left(\varphi_{[u,w[} - \psi_{[u,w[}\right) = \varphi_{[v,w[}. \end{cases}$$

The shifted and dilated basic transform just described applies to all the consecutive pairs of values, separated here by semicolons for convenience, in a sample with $2n$ values:

$$s_0, s_1; s_2, s_3; \ldots; s_{2k}, s_{2k+1}; \ldots; s_{2(n-1)}, s_{2n-1}.$$

Example 1.6 Table 1.7 lists four sample points corresponding to the approximating step function from Example 1.3,

Table 1.7

j	0	1	2	3
r_j	0	$\frac{1}{4}$	$\frac{1}{2}$	$\frac{3}{4}$
s_j	5	1	2	8

$$\tilde{f} = 5 \cdot \varphi_{[0,\frac{1}{4}[} + 1 \cdot \varphi_{[\frac{1}{4},\frac{1}{2}[} + 2 \cdot \varphi_{[\frac{1}{2},\frac{3}{4}[} + 8 \cdot \varphi_{[\frac{3}{4},1[}.$$

The basic transform applied to the first pair of steps gives

$$5 \cdot \varphi_{[0,\frac{1}{4}[} + 1 \cdot \varphi_{[\frac{1}{4},\frac{1}{2}[} = \frac{5+1}{2}\varphi_{[0,\frac{1}{2}[} + \frac{5-1}{2}\psi_{[0,\frac{1}{2}[}.$$

Similarly, after a shift by two sample points to the right, the basic transform applied to the second pair gives

$$2 \cdot \varphi_{[\frac{1}{2},\frac{3}{4}[} + 8 \cdot \varphi_{[\frac{3}{4},1[} = \frac{2+8}{2}\varphi_{[\frac{1}{2},1[} + \frac{2-8}{2}\psi_{[\frac{1}{2},1[}.$$

Thus,

$$\tilde{f} = 5 \cdot \varphi_{[0,\frac{1}{4}[} + 1 \cdot \varphi_{[\frac{1}{4},\frac{1}{2}[} + 2 \cdot \varphi_{[\frac{1}{2},\frac{3}{4}[} + 8 \cdot \varphi_{[\frac{3}{4},1[}$$
$$= 3\varphi_{[0,\frac{1}{2}[} + 2\psi_{[0,\frac{1}{2}[} + 5\varphi_{[\frac{1}{2},1[} + (-3)\psi_{[\frac{1}{2},1[}.$$

The coefficients 3, 5, 2, and −3, have the following significance:

- $3\,\varphi_{[0,\frac{1}{2}[}$ indicates that $\tilde{f}$ has an average value 3 over the first half of the interval, from 0 to $\frac{1}{2}$.

- $5\,\varphi_{[\frac{1}{2},1[}$ indicates that $\tilde{f}$ has an average value 5 over the second half of the interval, from $\frac{1}{2}$ to 1.

- $2\psi_{[0,\frac{1}{2}[}$ indicates that $\tilde{f}$ undergoes a jump of size 2 times that of $\psi_{[0,\frac{1}{2}[}$, which jumps down from 1 to −1, for a total of $2 \cdot (-2) = -4$ over the first half of the interval, indeed from 5 to 1.

- $(-3)\psi_{[\frac{1}{2},1[}$ indicates that $\tilde{f}$ undergoes a jump of size −3 times that of $\psi_{[\frac{1}{2},1[}$, which jumps down from 1 to −1, for a total of $(-3) \cdot (-2) = 6$ over the second half of the interval, from 2 to 8.   □

Example 1.7   Table 1.8 reproduces the eight sample points of the function h from Example 1.4.

Table 1.8

j	0	1	2	3	4	5	6	7
r_j	0	$\frac{1}{8}$	$\frac{1}{4}$	$\frac{3}{8}$	$\frac{1}{2}$	$\frac{5}{8}$	$\frac{3}{4}$	$\frac{7}{8}$
s_j	3	1	0	4	8	6	9	9

Applied to consecutive pairs of sample values $(s_0, s_1), (s_2, s_3), \ldots, (s_{2k}, s_{2k+1}),$ $\ldots, (s_6, s_7)$, the basic simple-wavelet transform gives

$$\bar{h} = \frac{3+1}{2}\varphi_{[0,\frac{1}{4}[} + \frac{3-1}{2}\psi_{[0,\frac{1}{4}[} + \frac{0+4}{2}\varphi_{[\frac{1}{4},\frac{1}{2}[} + \frac{0-4}{2}\psi_{[\frac{1}{4},\frac{1}{2}[}$$
$$+ \frac{8+6}{2}\varphi_{[\frac{1}{2},\frac{3}{4}[} + \frac{8-6}{2}\psi_{[\frac{1}{2},\frac{3}{4}[} + \frac{9+9}{2}\varphi_{[\frac{3}{4},1[} + \frac{9-9}{2}\psi_{[\frac{3}{4},1[}. \quad □$$

Remark 1.8   Uppercase letters beginning words in technical phrases will indicate a specific technical meaning. For example, the phrase "Haar Wavelet Transform" designates the specific transform with the specific wavelets described in this chapter.   □

Remark 1.9   Because the Haar Wavelet Transform does not use the abscissa—the first coordinate r_j of the data point (r_j, s_j)—data for the Haar Wavelet Transform can list the ordinates (values) s_j without the abscissae. For example, the data $(s_0, s_1) = (2, 8)$ can replace the entire Table 1.9.   □

Table 1.9

j	0	1
r_j	0	$\frac{1}{2}$
s_j	2	8

EXERCISES

Exercise 1.5. Calculate the Haar Wavelet Transform for the data $(s_0, s_1) = (2, 8)$.

Exercise 1.6. Calculate the Haar Wavelet Transform for the data $(s_0, s_1) = (7, 3)$.

Exercise 1.7. Calculate the basic Haar Wavelet Transform for the first pair and for the last pair in the data $(s_0, s_1, s_2, s_3) = (2, 4, 8, 6)$.

Exercise 1.8. Calculate the basic Haar Wavelet Transform for the first pair and for the last pair in the data $(s_0, s_1, s_2, s_3) = (5, 7, 3, 1)$.

Exercise 1.9. Calculate the basic Haar Wavelet Transform for each pair (s_{2k}, s_{2k+1}) in the array $\vec{s} = (8, 6, 7, 3, 1, 1, 2, 4)$.

Exercise 1.10. Calculate the basic Haar Wavelet Transform for each pair (s_{2k}, s_{2k+1}) in the array $\vec{s} = (3, 1, 9, 7, 7, 9, 5, 7)$.

Exercise 1.11. For each array with four entries $\vec{s} = (s_0, s_1, s_2, s_3)$, consider the averages of the first pair and last pair of entries,

$$(s_0 + s_1)/2,$$

$$(s_2 + s_3)/2.$$

Verify algebraically that the average of both averages produces the average of all the entries in $\vec{s}$:

$$\frac{[(s_0 + s_1)/2] + [(s_2 + s_3)/2]}{2} = \frac{s_0 + s_1 + s_2 + s_3}{4}.$$

Exercise 1.12. For arrays with eight entries $\vec{s} = (s_0, s_1, \ldots, s_6, s_7)$, consider the averages

$$(s_0 + s_1)/2,$$

$$(s_2 + s_3 + s_4 + s_5 + s_6 + s_7)/6.$$

Investigate whether

$$\frac{[(s_0 + s_1)/2] + [(s_2 + s_3 + s_4 + s_5 + s_6 + s_7)/6]}{2}$$

$$\overset{?}{=} \frac{s_0 + s_1 + s_2 + s_3 + s_4 + s_5 + s_6 + s_7}{8}.$$

Either verify such an identity algebraically, or produce an example with specific numbers for which it fails.

Exercise 1.13. For each array $\vec{s} = (s_0, s_1, \ldots, s_6, s_7)$, verify that

$$\frac{\frac{[(s_0+s_1)/2]+[(s_2+s_3)/2]}{2} + \frac{[(s_4+s_5)/2]+[(s_6+s_7)/2]}{2}}{2}$$

$$= \frac{s_0 + s_1 + s_2 + s_3 + s_4 + s_5 + s_6 + s_7}{8}.$$

Thus, the average equals the averages of the averages of the averages.

Exercise 1.14. Generalize and verify, for instance by mathematical induction, a result analogous to that of Exercise 1.13, for arrays with an integral power of two number of entries.

1.3   THE ORDERED FAST HAAR WAVELET TRANSFORM

To analyze a signal or function in terms of wavelets, the Fast Haar Wavelet Transform begins with the initialization of an array with 2^n entries, and then proceeds with n iterations of the basic transform explained in the preceding section.

For each index $\ell \in \{1, \ldots, n\}$, *before* iteration number ℓ, the array will consist of $2^{n-(\ell-1)}$ coefficients of $2^{n-(\ell-1)}$ step functions $\varphi_k^{(n-[\ell-1])}$, defined below. *After* iteration number ℓ, the array will consist of half as many, $2^{n-\ell}$, coefficients of $2^{n-\ell}$ step functions $\varphi_k^{(n-\ell)}$, and $2^{n-\ell}$ coefficients of wavelets $\psi_k^{(n-\ell)}$.

Definition 1.10   For each positive integer n and each index $\ell \in \{0, \ldots, n\}$, define the step functions $\varphi_k^{(n-\ell)}$ and wavelets $\psi_k^{(n-\ell)}$ by

$$\varphi_k^{(n-\ell)}(r) := \varphi_{[0,1[}\left(2^{n-\ell}\left[r - k2^{\ell-n}\right]\right)$$

$$= \begin{cases} 1 & \text{if } k2^{\ell-n} \le r < (k+1)2^{\ell-n}, \\ 0 & \text{otherwise,} \end{cases}$$

$$\psi_k^{(n-\ell)}(r) := \psi_{[0,1[}\left(2^{n-\ell}\left[r - k2^{\ell-n}\right]\right)$$

$$= \begin{cases} 1 & \text{if } k2^{\ell-n} \le r < \left(k + \left[\tfrac{1}{2}\right]\right)2^{\ell-n}, \\ -1 & \text{if } \left(k + \left[\tfrac{1}{2}\right]\right)2^{\ell-n} \le r < (k+1)2^{\ell-n}, \qquad \square \\ 0 & \text{otherwise.} \end{cases}$$

In the foregoing definition, the frequency increases with the index n, as in references [20] and [49]. By contrast, in such references as [7] and [31], the frequency *decreases* as the index *increases*.

1.3.1   Initialization

For Haar's wavelets, the initialization consists only in establishing a one-dimensional *array* $\vec{\mathbf{a}}^{(n)}$, also called a *vector* or a *finite sequence*, of sample values, of the form

$$\vec{\mathbf{a}}^{(n)} = \left(a_0^{(n)}, a_1^{(n)}, \ldots, a_j^{(n)}, \ldots, a_{2^n-2}^{(n)}, a_{2^n-1}^{(n)}\right)$$

$$:= \vec{\mathbf{s}} = \left(s_0, s_1, \ldots, s_j, \ldots, s_{2^n-2}, s_{2^n-1}\right),$$

with a total number of sample values equal to an integral power of two, 2^n, as indicated by the superscript $^{(n)}$. Though indices ranging from 1 through 2^n would also serve the same purpose, indices ranging from 0 through $2^n - 1$ will accommodate a binary encoding with only n binary digits, and will also offer notational simplifications in the exposition. The array corresponds to the sampled step function

$$\tilde{f}^{(n)} = \sum_{j=0}^{2^n-1} a_j^{(n)} \varphi_j^{(n)}.$$

1.3.2 The Ordered Fast Haar Wavelet Transform

The preceding section has demonstrated how a first sweep of the basic transform applies to all the consecutive pairs (s_{2k}, s_{2k+1}) of the initial array of sample values $\vec{a}^{(n)} = \vec{s}$.

In general, the ℓth sweep of the basic transform begins with an array of $2^{n-(\ell-1)}$ values

$$\vec{a}^{(n-[\ell-1])} = \left(a_0^{(n-[\ell-1])}, \dots, a_{2^{n-(\ell-1)}-1}^{(n-[\ell-1])} \right),$$

and applies the basic transform to each pair $(a_{2k}^{(n-[\ell-1])}, a_{2k+1}^{(n-[\ell-1])})$, which gives two new wavelet coefficients

$$a_k^{(n-\ell)} := \frac{a_{2k}^{(n-[\ell-1])} + a_{2k+1}^{(n-[\ell-1])}}{2},$$

$$c_k^{(n-\ell)} := \frac{a_{2k}^{(n-[\ell-1])} - a_{2k+1}^{(n-[\ell-1])}}{2}.$$

These $2^{(n-\ell)}$ pairs of new coefficients represent the *result* of the ℓth sweep, a result that can also be reassembled into two arrays:

$$\vec{a}^{(n-\ell)} := \left(a_0^{(n-\ell)}, a_1^{(n-\ell)}, \dots, a_k^{(n-\ell)}, \dots, a_{2^{n-\ell}-1}^{(n-\ell)} \right),$$

$$\vec{c}^{(n-\ell)} := \left(c_0^{(n-\ell)}, c_1^{(n-\ell)}, \dots, c_k^{(n-\ell)}, \dots, c_{2^{n-\ell}-1}^{(n-\ell)} \right).$$

The arrays related to the ℓth sweep have the following significance.

$\vec{a}^{(n-[\ell-1])}$: The beginning array,

$$\vec{a}^{(n-[\ell-1])} = \left(a_0^{(n-[\ell-1])}, \dots, a_{2^{n-(\ell-1)}-1}^{(n-[\ell-1])} \right),$$

lists the values $a_k^{(n-[\ell-1])}$ of a simple step function $\tilde{f}^{(n-[\ell-1])}$ that approximates the initial function f with $2^{n-(\ell-1)}$ steps of narrower width $2^{(\ell-1)-n}$:

$$\tilde{f}^{(n-[\ell-1])} = \sum_{j=0}^{2^{n-(\ell-1)}-1} a_j^{(n-[\ell-1])} \varphi_j^{(n-[\ell-1])}.$$

$\vec{a}^{(n-\ell)}$: The first array produced by the ℓth sweep,

$$\vec{a}^{(n-\ell)} = \left(a_0^{(n-\ell)}, \ldots, a_{2^{n-\ell}-1}^{(n-\ell)} \right),$$

lists the values $a_k^{(n-\ell)}$ of a simple step function $\tilde{f}^{(n-\ell)}$ that approximates the initial function f with $2^{n-\ell}$ steps of wider width $2^{\ell-n}$,

$$\tilde{f}^{(n-\ell)} = \sum_{j=0}^{2^{n-\ell}-1} a_j^{(n-\ell)} \varphi_j^{(n-\ell)}.$$

$\vec{c}^{(n-\ell)}$: The second array produced by the ℓth sweep,

$$\vec{c}^{(n-\ell)} = \left(c_0^{(n-\ell)}, \ldots, c_{2^{n-\ell}-1}^{(n-\ell)} \right),$$

lists the coefficients $c_k^{(n-\ell)}$ of simple wavelets $\psi_j^{(n-\ell)}$ also of wider width $2^{\ell-n}$,

$$\hat{f}^{(n-\ell)} = \sum_{j=0}^{2^{n-\ell}-1} c_j^{(n-\ell)} \psi_j^{(n-\ell)}.$$

The wavelets given by the second new array, $\vec{c}^{(n-\ell)}$, represent the difference between the finer steps of the initial approximation $\tilde{f}^{(n-[\ell-1])}$ and the coarser steps of $\tilde{f}^{(n-\ell)}$. Thus, each sweep of basic transforms expresses the previous finer approximation as the sum of a new, coarser approximation and a new, lower-frequency, set of wavelets. Nevertheless, because the basic step of Haar's transform does not alter the sampled function but merely expresses it with different wavelets, it follows that the initial approximation $\tilde{f}^{(n-[\ell-1])}$ still equals the sum of the two new approximations, $\tilde{f}^{(n-\ell)}$ and $\hat{f}^{(n-\ell)}$:

$$\tilde{f}^{(n-[\ell-1])} = \tilde{f}^{(n-\ell)} + \hat{f}^{(n-\ell)}.$$

Example 1.11   The array $(s_0, s_1, s_2, s_3) = (5, 1, 2, 8)$ reproduces data from Examples 1.3 and 1.6. To illustrate a common usage [7], the present example will store the final result—the Haar Wavelet Transform of the data—ordered by increasing frequencies: from the lowest frequencies produced last but stored in the first (left) part of the array, through to highest frequencies produced first but stored in the last (right) part of the array.

1.3.2.1   Initialization.   The initial array $\vec{a}^{(2)} = \vec{s} = (5, 1, 2, 8)$ contains the $2^2 = 4$ values of the sample, as in Figure 1.8(a).

1.3.2.2   First Sweep.   Begin with $\vec{a}^{(2)} = (5, 1, 2, 8)$.

$$\vec{a}^{(2-1)} = \left(\frac{5+1}{2}, \frac{2+8}{2}\right) = (3, 5),$$

$$\vec{c}^{(2-1)} = \left(\frac{5-1}{2}, \frac{2-8}{2}\right) = (2, -3),$$

which can be stored in the form

$$\vec{s}^{(2-1)} = \left(\vec{a}^{(2-1)}; \vec{c}^{(2-1)}\right) = (3, 5; 2, -3).$$

The first array, $\vec{a}^{(2-1)} = (3, 5)$, represents a coarse approximation of the initial sample $\vec{a}^{(2)}$, and means that the first half of the sample, $(5, 1)$, has an average value of 3, and that the second half of the sample, $(2, 8)$, has an average value of 5. The second array, $\vec{c}^{(2-1)} = (2, -3)$, means that on the first half of the sample, the values jump downward by 2 times the jump of a wavelet, hence by a total jump of $2 * (-2) = -4$, whereas on the second half of the sample, the values jump by -3 times the jump of a wavelet, hence by a total jump of $(-3) * (-2) = 6$.

1.3.2.3   Second Sweep.   Keep $\vec{c}^{(2-1)}$ and continue with $\vec{a}^{(2-1)} = (3, 5)$.

$$\vec{a}^{(2-2)} = \left(\frac{3+5}{2}\right) = (4),$$

$$\vec{c}^{(2-2)} = \left(\frac{3-5}{2}\right) = (-1),$$

which can be stored in the form

$$\vec{s}^{(2-2)} = \left(\vec{a}^{(2-2)}; \vec{c}^{(2-2)}; \vec{c}^{(2-1)}\right) = (4; -1; 2, -3).$$

The first array, $\vec{a}^{(2-2)} = (4)$, means that the whole sample, $(5, 1, 2, 8)$, has an average value of 4. The second array, $\vec{c}^{(2-2)} = (-1)$, means that at the middle of the sample, the average of the first half, 3, jumps up toward the average of the second half, 5, as does -1 basic wavelet $\psi_{[0,1[}$, in effect a jump of size $(-1) \cdot (-2) = 2$.

1.3.2.4   Results.   The final result from two consecutive sweeps takes the following form:

$$\tilde{f} = 4 \cdot \varphi_{[0,1[} + (-1) \cdot \psi_{[0,1[} + 2 \cdot \psi_{[0,\frac{1}{2}[} + (-3) \cdot \psi_{[\frac{1}{2},1[}$$

$$= 4 \cdot \varphi_0^{(0)} + (-1) \cdot \psi_0^{(0)} + 2 \cdot \psi_0^{(1)} + (-3) \cdot \psi_1^{(1)}.$$

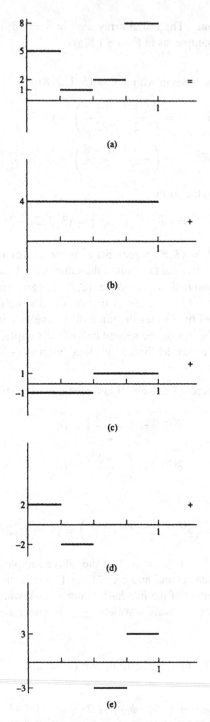

Figure 1.8   Example of a wavelet transform. (a) Data. (b)–(e) Haar Wavelet Transform.

The initial array $\vec{a}^{(2)} = (5, 1, 2, 8)$ contained the values of the sample and of the approximating function $\tilde{f}$. In contrast, the formula just obtained expresses the same $\tilde{f}$ as a sum of higher-frequency wavelets from the first sweep, followed by a lower-frequency wavelet and a global average from the second sweep, as in Figure 1.8(b)–(e).   □

Example 1.12 The array $\vec{s} = (3, 1, 0, 4, 8, 6, 9, 9)$ reproduces the sample from Example 1.4.

1.3.2.5 Initialization. $\vec{a}^{(3)} := \vec{s} = (3, 1, 0, 4, 8, 6, 9, 9)$.

1.3.2.6 First Sweep.

$$\vec{a}^{(3-1)} = \left(\frac{3+1}{2}, \frac{0+4}{2}, \frac{8+6}{2}, \frac{9+9}{2} \right) = (2, 2, 7, 9),$$

$$\vec{c}^{(3-1)} = \left(\frac{3-1}{2}, \frac{0-4}{2}, \frac{8-6}{2}, \frac{9-9}{2} \right) = (1, -2, 1, 0),$$

which can be stored in the form

$$\vec{s}^{(3-1)} = \left(\vec{a}^{(3-1)}; \vec{c}^{(3-1)} \right) = (2, 2, 7, 9; 1, -2, 1, 0).$$

1.3.2.7 Second Sweep.

$$\vec{a}^{(3-1)} = (2, 2, 7, 9),$$

$$\vec{a}^{(3-2)} = \left(\frac{2+2}{2}, \frac{7+9}{2} \right) = (2, 8),$$

$$\vec{c}^{(3-2)} = \left(\frac{2-2}{2}, \frac{7-9}{2} \right) = (0, -1),$$

which can be stored in the form

$$\vec{s}^{(3-2)} = \left(\vec{a}^{(3-2)}; \vec{c}^{(3-2)}; \vec{c}^{(3-1)} \right) = (2, 8; 0, -1; 1, -2, 1, 0).$$

1.3.2.8 Third Sweep.

$$\vec{a}^{(3-2)} = (2, 8),$$

$$\vec{a}^{(3-3)} = \left(\frac{2+8}{2} \right) = (5),$$

$$\vec{c}^{(3-3)} = \left(\frac{2-8}{2} \right) = (-3),$$

which can be stored in the form

$$\vec{s}^{(3-3)} = \left(\vec{a}^{(3-3)}; \vec{c}^{(3-3)}; \vec{c}^{(3-2)}; \vec{c}^{(3-1)}\right) = (5; -3; 0, -1; 1, -2, 1, 0).$$

1.3.2.9 Results. The initial array $\vec{a}^{(3)} = \vec{s}$ represents the approximating function $\tilde{f}$ by its sample values,

$$\tilde{f} = 3 \cdot \varphi_{[0, \frac{1}{8}[} + 1 \cdot \varphi_{[\frac{1}{8}, \frac{1}{4}[} + 0 \cdot \varphi_{[\frac{1}{4}, \frac{3}{8}[} + 4 \cdot \varphi_{[\frac{3}{8}, \frac{1}{2}[}$$
$$+ 8 \cdot \varphi_{[\frac{1}{2}, \frac{5}{8}[} + 6 \cdot \varphi_{[\frac{5}{8}, \frac{3}{4}[} + 9 \cdot \varphi_{[\frac{3}{4}, \frac{7}{8}[} + 9 \cdot \varphi_{[\frac{7}{8}, 1[}.$$

In contrast, the wavelet coefficients $\vec{c}^{(3-\ell)}$ produced by the consecutive sweeps of basic transforms express the same approximating function $\tilde{f}$ in terms of consecutively lower frequencies, ending with a constant step across the entire interval,

$$\tilde{f} = 1 \cdot \psi_{[0, \frac{1}{4}[} + (-2) \cdot \psi_{[\frac{1}{4}, \frac{1}{2}[} + 1 \cdot \psi_{[\frac{1}{2}, \frac{3}{4}[} + 0 \cdot \psi_{[\frac{3}{4}, 1[} \qquad \textit{1st Sweep}$$
$$+ 0 \cdot \psi_{[0, \frac{1}{2}[} + (-1) \cdot \psi_{[\frac{1}{2}, 1[} \qquad \textit{2nd Sweep}$$
$$+ (-3) \cdot \psi_{[0, 1[} + 5 \cdot \varphi_{[0, 1[} \qquad \textit{3rd Sweep}$$

stored in the form $\vec{s}^{(3-3)} = (5; -3; 0, -1; 1, -2, 1, 0)$.

1.3.2.10 Significance. The term produced last, $5 \cdot \varphi_{[0, 1[}$, means that the sample has an average value equal to **5**.

The penultimate term, $-3 \cdot \psi_{[0, 1[}$, indicates that the sample undergoes a jump 3 times the size of, and in the opposite direction from, the wavelet $\psi_{[0, 1[}$ (which jumps downward by 2 at the middle of the interval). Indeed, the sample jumps upward by 6 on average at the middle of the interval: The array $\vec{a}^{(3-2)} = (2, 8)$ shows that the average jumps from 2 on the left-hand half of the interval to 8 on the right-hand half of the interval.

The two terms $0 \cdot \psi_{[0, \frac{1}{2}[} + (-1) \cdot \psi_{[\frac{1}{2}, 1[}$ mean that the sample does not exhibit any average jump at the first quarter of the interval, and exhibits an average jump of $(-1) \cdot (-1) = 1$ at the third quarter.

The four terms $1 \cdot \psi_{[0, \frac{1}{4}[} + (-2) \cdot \psi_{[\frac{1}{4}, \frac{1}{2}[} + 1 \cdot \psi_{[\frac{1}{2}, \frac{3}{4}[} + 0 \cdot \psi_{[\frac{3}{4}, 1[}$ reveal that the sample oscillates, as do the fastest wavelets, with jumps of sizes $-2, 4, -2$, and 0. □

EXERCISES

Exercise 1.15. Calculate the Haar Wavelet Transform for the data $(s_0, s_1) = (3, 9)$.

Exercise 1.16. Calculate the Haar Wavelet Transform for the data $(s_0, s_1) = (1, 7)$.

Exercise 1.17. Calculate the Haar Wavelet Transform for the data $\vec{s} = (2, 4, 8, 6)$.

Exercise 1.18. Calculate the Haar Wavelet Transform for the data $\vec{s} = (5, 7, 3, 1)$.

Exercise 1.19. Calculate the Haar Wavelet Transform for the data $\vec{s} = (8, 6, 7, 3, 1, 1, 2, 4)$.

Exercise 1.20. Calculate the Haar Wavelet Transform for the data $\vec{s} = (3, 1, 9, 7, 7, 9, 5, 7)$.

Exercise 1.21. Assume that the Haar Wavelet Transform of a sample $\vec{s} = (s_0, s_1)$ produces the results $a_0^{(1-1)} := 7$ and $c_0^{(1-1)} := 2$.

(a) Explain how $a_0^{(1-1)} = 7$ relates to the sample (s_0, s_1).

(b) Explain how $c_0^{(1-1)} = 2$ relates to the sample (s_0, s_1).

Exercise 1.22. Assume that the Haar Wavelet Transform of a sample $\vec{s} = (s_0, s_1)$ produces the results $a_0^{(1-1)} := 6$ and $c_0^{(1-1)} := -3$.

(a) Explain how $a_0^{(1-1)} = 6$ relates to the sample (s_0, s_1).

(b) Explain how $c_0^{(1-1)} = -3$ relates to the sample (s_0, s_1).

Exercise 1.23. Assume that the Haar Wavelet Transform of a sample $\vec{s} = (s_0, s_1, s_2, s_3)$ produces the results $\vec{c}^{(2-1)} = (2, 2)$, $\vec{c}^{(2-2)} = (1)$, and $\vec{a}^{(2-2)} = (6)$.

(a) Explain how $a_0^{(2-2)} = 6$ relates to the sample.

(b) Explain how $c_0^{(2-2)} = 1$ relates to the sample.

(c) Explain how $c_0^{(2-1)} = 2$ relates to the sample.

(d) Explain how $c_1^{(2-1)} = 2$ relates to the sample.

Exercise 1.24. Assume that the Haar Wavelet Transform of a sample $\vec{s} = (s_0, s_1, s_2, s_3)$ produces the results $\vec{c}^{(2-1)} = (2, 0)$, $\vec{c}^{(2-2)} = (2)$, and $\vec{a}^{(2-2)} = (4)$.

(a) Explain how $a_0^{(2-2)} = 4$ relates to the sample.

(b) Explain how $c_0^{(2-2)} = 2$ relates to the sample.

(c) Explain how $c_0^{(2-1)} = 2$ relates to the sample.

(d) Explain how $c_1^{(2-1)} = 0$ relates to the sample.

1.4   THE IN-PLACE FAST HAAR WAVELET TRANSFORM

Whereas the presentation in the preceding section conveniently lays out all the steps of the Fast Haar Wavelet Transform, it requires additional arrays at each sweep, and it assumes that the whole sample is known at the start of the algorithm. In contrast, some applications require real-time processing as the signal proceeds, which precludes any knowledge of the whole sample, and some applications involve arrays so large that they do not allow sufficient space for additional arrays at each sweep. The two problems just described, lack of time or space, have a common solution in the In-Place Fast Haar Wavelet Transform presented here, which differs from the preceding algorithm only in its indexing scheme.

1.4.1 In-Place Basic Sweep

For each pair $(a_{2k}^{(n-[\ell-1])}, a_{2k+1}^{(n-[\ell-1])})$, instead of placing its results in two additional arrays, the ℓth sweep of the in-place transform merely *replaces* the pair $(a_{2k}^{(n-[\ell-1])}, a_{2k+1}^{(n-[\ell-1])})$ by the new entries $(a_k^{(n-\ell)}, c_k^{(n-\ell)})$:

1.4.1.1 Initialization. Consider the pair $(a_{2k}^{(n-[\ell-1])}, a_{2k+1}^{(n-[\ell-1])})$.

1.4.1.2 Calculation. Perform the basic transform

$$a_k^{(n-\ell)} := \frac{a_{2k}^{(n-[\ell-1])} + a_{2k+1}^{(n-[\ell-1])}}{2},$$

$$c_k^{(n-\ell)} := \frac{a_{2k}^{(n-[\ell-1])} - a_{2k+1}^{(n-[\ell-1])}}{2}.$$

1.4.1.3 Replacement. Replace the initial pair $(a_{2k}^{(n-[\ell-1])}, a_{2k+1}^{(n-[\ell-1])})$ by the transformed pair $(a_k^{(n-\ell)}, c_k^{(n-\ell)})$.

Example 1.13 For the initial array $\vec{s}^{(1)} := \vec{s} := (9, 1)$, the In-Place Haar Wavelet Transform gives

$$\vec{s}^{(1-1)} = \left(\frac{9+1}{2}, \frac{9-1}{2}\right) = (5, 4). \qquad \square$$

Example 1.14 For the initial array $\vec{s}^{(2)} := \vec{s} := (5, 1, 2, 8)$, the first In-Place basic sweep gives

$$\vec{s}^{(2-1)} = \left(\frac{5+1}{2}, \frac{5-1}{2}, \frac{2+8}{2}, \frac{2-8}{2}\right) = (3, 2, 5, -3). \qquad \square$$

Example 1.15 For the initial array $\vec{s}^{(3)} = \vec{s} = (3, 1, 0, 4, 8, 6, 9, 9)$, the first In-Place basic sweep yields

$$\vec{s}^{(3-1)} = \left(\frac{3+1}{2}, \frac{3-1}{2}, \frac{0+4}{2}, \frac{0-4}{2}, \frac{8+6}{2}, \frac{8-6}{2}, \frac{9+9}{2}, \frac{9-9}{2}\right)$$

$$= (2, 1, 2, -2, 7, 1, \mathbf{9}, \mathbf{0}).$$

For convenience, the entries in **boldface** show the starting array $\vec{a}^{(3-1)}$ for the next sweep, described in the next subsection. $\qquad \square$

1.4.2   The In-Place Fast Haar Wavelet Transform

The in-place basic sweep explained in the preceding subsection extends to a complete algorithm through mere record-keeping. The first few sweeps proceed as follows.

1.4.2.1   Initialization.

$$\vec{s}^{(n)} := \vec{s} = (s_0, s_1, s_2, s_3, \ldots, s_{2k}, s_{2k+1}, \ldots, s_{2^n-2}, s_{2^n-1}).$$

1.4.2.2   First Sweep.

$$\vec{s}^{(n-1)} = \left(\frac{s_0 + s_1}{2}, \frac{s_0 - s_1}{2}, \frac{s_2 + s_3}{2}, \frac{s_2 - s_3}{2}, \ldots, \frac{s_{2k} + s_{2k+1}}{2}, \frac{s_{2k} - s_{2k+1}}{2}, \right.$$

$$\left. \ldots, \frac{s_{2^n-2} + s_{2^n-1}}{2}, \frac{s_{2^n-2} - s_{2^n-1}}{2} \right)$$

$$= \left(a_0^{(n-1)}, c_0^{(n-1)}, a_1^{(n-1)}, c_1^{(n-1)}, a_2^{(n-1)}, c_2^{(n-1)}, a_3^{(n-1)}, c_3^{(n-1)}, \right.$$

$$\left. \ldots, a_k^{(n-1)}, c_k^{(n-1)}, \ldots, a_{2^{n-1}-1}^{(n-1)}, c_{2^{n-1}-1}^{(n-1)} \right).$$

1.4.2.3   Second Sweep.
In the new array $\vec{s}^{(n-1)}$, keep but skip over the wavelet coefficients $c_k^{(n-1)}$, and perform a basic sweep on the array $\vec{a}^{(n-1)}$ at its new location, now occupying every other entry in $\vec{s}^{(n-1)}$:

$$\vec{s}^{(n-2)} = \left(\frac{a_0^{(n-1)} + a_1^{(n-1)}}{2}, c_0^{(n-1)}, \frac{a_0^{(n-1)} - a_1^{(n-1)}}{2}, c_1^{(n-1)}, \right.$$

$$\ldots, \frac{a_2^{(n-1)} + a_3^{(n-1)}}{2}, c_2^{(n-1)}, \frac{a_2^{(n-1)} - a_3^{(n-1)}}{2}, c_3^{(n-1)}, \ldots$$

$$\left. \ldots, \frac{a_{2^{n-1}-2}^{(n-1)} + a_{2^{n-1}-1}^{(n-1)}}{2}, c_{2^{n-1}-2}^{(n-1)}, \frac{a_{2^{n-1}-2}^{(n-1)} - a_{2^{n-1}-1}^{(n-1)}}{2}, c_{2^{n-1}-1}^{(n-1)} \right)$$

$$= \left(a_0^{(n-2)}, c_0^{(n-1)}, c_0^{(n-2)}, c_1^{(n-1)}, a_1^{(n-2)}, c_2^{(n-1)}, c_1^{(n-2)}, c_3^{(n-1)}, \right.$$

$$\left. a_2^{(n-2)}, c_4^{(n-1)}, c_2^{(n-2)}, c_5^{(n-1)}, \ldots, c_{2^{n-2}-1}^{(n-2)}, c_{2^{n-1}-1}^{(n-1)} \right).$$

In general, the In-Place ℓth sweep begins with an array

$$\vec{s}^{(n-[\ell-1])} = \left(a_0^{(n-[\ell-1])}, c_0^{(n-1)}, c_0^{(n-2)}, c_1^{(n-1)}, \right.$$

$$\left. c_0^{(n-3)}, c_2^{(n-1)}, c_1^{(n-2)}, c_3^{(n-1)}, \ldots, c_{2^{n-2}-1}^{(n-2)}, c_{2^{n-1}-1}^{(n-1)} \right),$$

which contains the array

$$\vec{\mathbf{a}}^{(n-[\ell-1])} = \left(a_0^{(n-[\ell-1])}, a_1^{(n-[\ell-1])}, \ldots, a_{2^{n-(\ell-1)}-1}^{(n-[\ell-1])} \right)$$

at the locations $a_k^{(n-[\ell-1])} = s_{2^{\ell-1}k}^{(n-[\ell-1])}$, in other words, at multiples of $2^{\ell-1}$ apart in $\vec{\mathbf{s}}^{(n-[\ell-1])}$, and which the ℓth sweep replaces by

$$a_j^{(n-\ell)} := \frac{a_{2j}^{(n-[\ell-1])} + a_{2j+1}^{(n-[\ell-1])}}{2} = \frac{s_{2^{\ell-1}2j}^{(n-[\ell-1])} + s_{2^{\ell-1}(2j+1)}^{(n-[\ell-1])}}{2},$$

$$c_j^{(n-\ell)} := \frac{a_{2j}^{(n-[\ell-1])} - a_{2j+1}^{(n-[\ell-1])}}{2} = \frac{s_{2^{\ell-1}2j}^{(n-[\ell-1])} - s_{2^{\ell-1}(2j+1)}^{(n-[\ell-1])}}{2},$$

$$s_{2^{\ell-1}2j}^{(n-\ell)} := a_j^{(n-\ell)},$$

$$s_{2^{\ell-1}(2j+1)}^{(n-\ell)} := c_j^{(n-\ell)},$$

so that the new array $\vec{\mathbf{a}}^{(n-\ell)}$ occupies entries at multiples of 2^{ℓ} apart in $\vec{\mathbf{s}}^{(n-\ell)}$, because $a_j^{(n-\ell)} = s_{2^{\ell-1}2j}^{(n-\ell)} = s_{2^{\ell}j}^{(n-\ell)}$. Hence, the foregoing considerations lead to the following algorithm.

Algorithm 1.16   In-Place Fast Haar Wavelet Transform.

DATA:

n (nonnegative integer)

$\vec{\mathbf{s}}$ (array of 2^n numbers)

START.

$I := 1$ (index increment)

$J := 2$ (increment between pairs)

$M := 2^n$ (number of sample values)

FOR $L := 1, \ldots, n$ DO   (loop of basic sweeps)

   $M := M/2$ (halve M)

   FOR $K := 0, \ldots, M - 1$ DO   (loop of values)

      $a_k^{(n-\ell)} := (s_{J \cdot K} + s_{J \cdot K + I})/2$

      $c_k^{(n-\ell)} := (s_{J \cdot K} - s_{J \cdot K + I})/2$

$$s_{J \cdot K} := a_k^{(n-\ell)}$$
$$s_{J \cdot K+1} := c_k^{(n-\ell)}$$

END   (end of the loop of values)
$I := J$ (double I)
$J := 2 * J$ (double J)
END   (end of basic sweeps)
STOP.
RESULT:

$$\vec{s} = \left(a_0^{(n-n)}, c_0^{(n-1)}, c_0^{(n-2)}, c_1^{(n-1)}, \ldots\right),$$
$$c_j^{(n-\ell)} = s_{2^{\ell-1}+2^\ell j} = s_{2^{\ell-1}(2j+1)} \text{ for } j \in \{0, \ldots, 2^{n-\ell} - 1\}. \qquad \square$$

Example 1.17 For the initial array $\vec{s}^{(2)} := \vec{s} := (5, 1, 2, 8)$, the In-Place Fast Haar Wavelet Transform proceeds as follows.

1.4.2.4   Initialization. $\vec{s}^{(2)} := \vec{s} := (5, 1, 2, 8) = \left(a_0^{(2)}, a_1^{(2)}, a_2^{(2)}, a_3^{(2)}\right)$.

1.4.2.5   First Sweep.   The first sweep operates on all the entries of $\vec{s}$:

$$\vec{s}^{(2-1)} = \left(a_0^{(2-1)}, c_0^{(2-1)}, a_1^{(2-1)}, c_1^{(2-1)}\right)$$
$$= \left(\frac{5+1}{2}, \frac{5-1}{2}, \frac{2+8}{2}, \frac{2-8}{2}\right) = (3, 2, 5, -3).$$

1.4.2.6   Second Sweep.   The second sweep operates on the even-indexed entries:

$$\vec{s}^{(2-1)} = (3, 2, 5, -3),$$
$$\vec{s}^{(2-2)} = \left(a_0^{(2-2)}, c_0^{(2-1)}, c_0^{(2-2)}, c_1^{(2-1)}\right)$$
$$= \left(\frac{3+5}{2}, 2, \frac{3-5}{2}, -3\right) = (4, 2, -1, -3). \qquad \square$$

Example 1.18 The array $\vec{s} = (3, 1, 0, 4, 8, 6, 9, 9)$ reproduces the data from Example 1.4.

1.4.2.7   Initialization. $\vec{s}^{(3)} = \vec{s} = (3, 1, 0, 4, 8, 6, 9, 9)$.

1.4.2.8   In-Place Fast Haar Wavelet Transform.

$$\vec{s}^{(3-1)} = \left(a_0^{(3-1)}, c_0^{(3-1)}, a_1^{(3-1)}, c_1^{(3-1)}, a_2^{(3-1)}, c_2^{(3-1)}, a_3^{(3-1)}, c_3^{(3-1)}\right)$$
$$= \left(\frac{3+1}{2}, \frac{3-1}{2}, \frac{0+4}{2}, \frac{0-4}{2}, \frac{8+6}{2}, \frac{8-6}{2}, \frac{9+9}{2}, \frac{9-9}{2}\right)$$

$$= (2, 1, 2, -2, 7, 1, 9, 0),$$

$$\vec{s}^{(3-2)} = \left(a_0^{(3-2)}, c_0^{(3-1)}, c_0^{(3-2)}, c_1^{(3-1)}, a_1^{(3-2)}, c_2^{(3-1)}, c_1^{(3-2)}, c_3^{(3-1)}\right)$$

$$= \left(\frac{2+2}{2}, 1, \frac{2-2}{2}, -2, \frac{7+9}{2}, 1, \frac{7-9}{2}, 0\right)$$

$$= (2, 1, 0, -2, 8, 1, -1, 0),$$

$$\vec{s}^{(3-3)} = \left(a_0^{(3-3)}, c_0^{(3-1)}, c_0^{(3-2)}, c_1^{(3-1)}, c_0^{(3-3)}, c_2^{(3-1)}, c_1^{(3-2)}, c_3^{(3-1)}\right)$$

$$= \left(\frac{2+8}{2}, 1, 0, -2, \frac{2-8}{2}, 1, -1, 0\right)$$

$$= (5, 1, 0, -2, -3, 1, -1, 0). \qquad \square$$

EXERCISES

Exercise 1.25. Calculate the In-Place Fast Haar Wavelet Transform for the data $\vec{s} = (2, 4, 8, 6)$.

Exercise 1.26. Calculate the In-Place Fast Haar Wavelet Transform for the data $\vec{s} = (5, 7, 3, 1)$.

Exercise 1.27. Calculate the In-Place Fast Haar Wavelet Transform for the data $\vec{s} = (8, 6, 7, 3, 1, 1, 2, 4)$.

Exercise 1.28. Calculate the In-Place Fast Haar Wavelet Transform for the data $\vec{s} = (3, 1, 9, 7, 7, 9, 5, 7)$.

Exercise 1.29. Assume that for a sample $\vec{s} = (s_0, s_1, s_2, s_3)$, the *In-Place* Fast Haar Wavelet Transform gives $\vec{s}^{(2-2)} = (5, -1, 2, 0)$.

(a) In the result $\vec{s}^{(2-2)} = (5, -1, 2, 0)$, identify the entry that measures the average of the whole sample.

(b) In the result $\vec{s}^{(2-2)} = (5, -1, 2, 0)$, identify the entry that measures the change from the average over the first half of the sample to the average over the second half.

(c) In the result $\vec{s}^{(2-2)} = (5, -1, 2, 0)$, identify the entry that measures the change from s_0 to s_1.

(d) In the result $\vec{s}^{(2-2)} = (5, -1, 2, 0)$, identify the entry that measures the change from s_2 to s_3.

Exercise 1.30. Assume that for a sample $\vec{s} = (s_0, s_1, s_2, s_3)$, the In-Place Fast Haar Wavelet Transform gives $\vec{s}^{(2-2)} = (6, 1, -2, -1)$.

(a) In the result $\vec{s}^{(2-2)} = (6, 1, -2, -1)$, identify the entry that measures the average of the whole sample.

(b) In the result $\vec{s}^{(2-2)} = (6, 1, -2, -1)$, identify the entry that measures the change from the average over the first half of the sample to the average over the second half.

(c) In the result $\vec{s}^{(2-2)} = (6, 1, -2, -1)$, identify the entry that measures the change from s_0 to s_1.

(d) In the result $\vec{s}^{(2-2)} = (6, 1, -2, -1)$, identify the entry that measures the change from s_2 to s_3.

Exercise 1.31. For each sample with four entries $\vec{s} = (s_0, s_1, s_2, s_3)$, express each entry of its In-Place Haar Wavelet Transform

$$\left(a_0^{(2-2)}, c_0^{(2-1)}, c_0^{(2-2)}, c_1^{(2-1)}\right)$$

with algebraic formulae in terms of the sample (s_0, s_1, s_2, s_3). For example, $a_0^{(2-2)} = (s_0 + s_1 + s_2 + s_3)/4$; derive similar formulae for the remaining three entries, $c_0^{(2-1)}, c_0^{(2-2)}, c_1^{(2-1)}$.

Exercise 1.32. For each sample with four entries $\vec{s} = (s_0, s_1, s_2, s_3)$, assume that the In-Place Fast Haar Wavelet Transform produces

$$\left(a_0^{(2-2)}, c_0^{(2-1)}, c_0^{(2-2)}, c_1^{(2-1)}\right).$$

Derive an algebraic formula in terms of the result for the average of the first half of the sample, (s_0, s_1). In other words, explain how to compute the average of s_0 and s_1 in terms of

$$\left(a_0^{(2-2)}, c_0^{(2-1)}, c_0^{(2-2)}, c_1^{(2-1)}\right).$$

Exercise 1.33. Assume that for some sample with eight entries

$$\vec{s} = (s_0, s_1, s_2, s_3, s_4, s_5, s_6, s_7),$$

the In-Place Fast Haar Wavelet Transform produces the final result $\vec{s}^{(3-3)} :=$ $(4, -1, -1, 2, 0, 1, -2, -2)$.

(a) Determine the average of the whole sample $\vec{s}$.

(b) In the array of results, identify the value of $c_1^{(3-2)}$.

(c) In the array of results, identify the indices k and ℓ such that the entry 0 represents $c_k^{(3-\ell)}$. In other words, determine k and ℓ such that $c_k^{(3-\ell)} = 0$.

(d) Determine the average of the second half of the sample, (s_4, s_5, s_6, s_7).

Exercise 1.34. Assume that for some sample with eight entries $\vec{s} = (s_0, s_1, s_2, s_3, s_4, s_5, s_6, s_7)$, the In-Place Fast Haar Wavelet Transform produces the final result $\vec{s}^{(3-3)} := (5, 1, 1, 0, -3, -1, 0, 1)$.

(a) Determine the average of the whole sample $\vec{s}$.

(b) In the array of results, identify the value of $c_0^{(3-3)}$.

(c) In the array of results, identify the indices k and ℓ such that the entry -1 represents $c_k^{(3-\ell)}$. In other words, determine k and ℓ such that $c_k^{(3-\ell)} = -1$.

(d) Determine the average of the second half of the sample, (s_4, s_5, s_6, s_7).

Exercise 1.35. Write and test a computer program to compute the In-Place Fast Haar Wavelet Transform.

Exercise 1.36. Write and test a computer program to compute the Ordered Fast Haar Wavelet Transform.

1.5   THE IN-PLACE FAST INVERSE HAAR WAVELET TRANSFORM

As described in the preceding section, the Fast Haar Wavelet Transform neither alters nor diminishes the information contained in the initial array $\vec{s} = (s_0, \ldots, s_{2^n-1})$, because each basic transform

$$\begin{cases} a_k^{(\ell)} = (\tfrac{1}{2}) \left(a_{2k}^{(\ell-1)} + a_{2k+1}^{(\ell-1)} \right), \\ c_k^{(\ell)} = (\tfrac{1}{2}) \left(a_{2k}^{(\ell-1)} - a_{2k+1}^{(\ell-1)} \right) \end{cases}$$

admits an inverse transform:

$$\begin{cases} a_{2k}^{(\ell-1)} = a_k^{(\ell)} + c_k^{(\ell)}, \\ a_{2k+1}^{(\ell-1)} = a_k^{(\ell)} - c_k^{(\ell)}. \end{cases}$$

Repeated applications of the basic inverse transform just given, beginning with the wavelet coefficients

$$\vec{s}^{(0)} = \left(a_0^{(n)}, c_0^{(1)}, \ldots, c_{2^{(n-1)}-1}^{(1)} \right),$$

reconstruct the initial array $\vec{s}^{(n)} = \vec{s} = (s_0, \ldots, s_{2^n-1})$.

Algorithm 1.19   In-Place Inverse Haar Wavelet Transform.

DATA:

n	(nonnegative integer)
$\vec{s} = \vec{s}^{(0)}$	(array of 2^n numbers)

START.

$I := 2^{(n-1)}$	(low-pass index increment)
$J := 2 * I$	(pair index increment)
$M := 1$	(pairs of lowest frequency)
FOR $L := n, \ldots, 1$ DO	(loop of basic sweeps)
FOR $K := 0, \ldots, M-1$ DO	(loop of coefficients)

$$a_{2k}^{(\ell-1)} := s_{J \cdot K} + s_{J \cdot K + I}$$
$$a_{2k+1}^{(\ell-1)} := s_{J \cdot K} - s_{J \cdot K + I}$$
$$s_{J \cdot K} := a_{2k}^{(\ell-1)}$$
$$s_{J \cdot K + I} := a_{2k+1}^{(\ell-1)}$$

END	(loop of coefficients)
$J := I$	(halve J)
$I := I/2$	(halve I)
$M := 2 * M$	(double M)
END	(loop of basic sweeps)

STOP.

RESULT:

$$\vec{s} = \vec{s}^{(n)} = (s_0, s_1, s_2, s_3, \ldots, s_{2^n - 1}).$$ $\square$

Example 1.20   For the array of coefficients $\vec{s}^{(0)} = (5, 4)$, the Fast Inverse Haar Wavelet Transform gives

$$\vec{s} = \vec{s}^{(0)} = (5, 4),$$

$$I := 1,$$

$$J := 2,$$

$$K := 0,$$

$$a_{2 \cdot 0}^{(1)} = s_{2 \cdot 0} + s_{2 \cdot 0 + 1} = 5 + 4 = 9,$$

$$a_{2 \cdot 0 + 1}^{(1)} = s_{2 \cdot 0} - s_{2 \cdot 0 + 1} = 5 - 4 = 1,$$

$$\vec{s} = \vec{s}^{(1)} = (a_0^{(1)}, a_1^{(1)}) = (9, 1),$$

which correctly reproduces the initial array $\vec{s}^{(1)} = (9, 1)$. $\square$

Example 1.21   For the wavelet coefficients

$$\vec{s}^{(0)} = (4, 2, -1, -3),$$

the In-Place Fast Inverse Haar Wavelet Transform gives

$$\vec{s} = \vec{s}^{(0)} = (4, 2, -1, -3),$$

$$I := 2,$$

$$J := 4,$$

$$K := 0,$$

$$a_{2 \cdot 0}^{(1)} = s_{4 \cdot 0}^{(0)} + s_{4 \cdot 0 + 2}^{(0)} = 4 + (-1) = 3,$$

$$a_{2 \cdot 0 + 1}^{(1)} = s_{4 \cdot 0}^{(0)} - s_{4 \cdot 0 + 2}^{(0)} = 4 - (-1) = 5,$$

$$\vec{s}^{(1)} = (3, 2, 5, -3),$$

$$I := 1,$$

$$J := 2,$$

$$K := 0,$$

$$a_{2 \cdot 0}^{(2)} = s_{2 \cdot 0}^{(1)} + s_{2 \cdot 0 + 1}^{(1)} = 3 + 2 = 5,$$

$$a_{2 \cdot 0 + 1}^{(2)} = s_{2 \cdot 0}^{(1)} - s_{2 \cdot 0 + 1}^{(1)} = 3 - 2 = 1,$$

$$K := 1,$$

$$a_{2 \cdot 1}^{(2)} = s_{2 \cdot 1}^{(1)} + s_{2 \cdot 1 + 1}^{(1)} = 5 + (-3) = 2,$$

$$a_{2 \cdot 1 + 1}^{(2)} = s_{2 \cdot 1}^{(1)} - s_{2 \cdot 1 + 1}^{(1)} = 5 - (-3) = 8,$$

$$\vec{s}^{(2)} = (a_0^{(2)}, a_1^{(2)}, a_2^{(2)}, a_3^{(2)}) = (5, 1, 2, 8),$$

which correctly reproduces the initial array $\vec{s}^{(2)} = (5, 1, 2, 8)$.   □

EXERCISES

Exercise 1.37. Assume that the In-Place Fast Haar Wavelet Transform of a sample $\vec{s} = (s_0, s_1)$ produces the results $(7, 2)$. Apply the inverse transform to reconstruct the values of the sample, s_0 and s_1.

Exercise 1.38. Assume that the In-Place Fast Haar Wavelet Transform of a sample $\vec{s} = (s_0, s_1)$ produces the results $(6, -3)$. Apply the inverse transform to reconstruct the values of the sample, s_0 and s_1.

Exercise 1.39. Assume that the In-Place Fast Haar Wavelet Transform of a sample $\vec{s} = (s_0, s_1, s_2, s_3)$ produces the results $(6, 2, 1, 2)$. Apply the inverse transform to reconstruct the sample $\vec{s}$.

Exercise 1.40. Assume that the In-Place Fast Haar Wavelet Transform of a sample $\vec{s} = (s_0, s_1, s_2, s_3)$ produces the results $(4, 2, 2, 0)$. Apply the inverse transform to reconstruct the sample $\vec{s}$.

Exercise 1.41. Assume that for some sample with eight entries $\vec{s} = (s_0, s_1, s_2, s_3, s_4, s_5, s_6, s_7)$, the In-Place Fast Haar Wavelet Transform produces the result $(4, -1, -1, 2, 0, 1, -2, -2)$. Apply the inverse transform to reconstruct the sample $\vec{s}$.

Exercise 1.42. Assume that for some sample with eight entries $\vec{s} = (s_0, s_1, s_2, s_3, s_4, s_5, s_6, s_7)$, the In-Place Fast Haar Wavelet Transform produces the result $(5, 1, 1, 0, -3, -1, 0, 1)$. Apply the inverse transform to reconstruct the sample $\vec{s}$.

Exercise 1.43. Assume that the In-Place Fast Haar Wavelet Transform stops at the end of the ℓth sweep. Explain how to reconstruct the initial sample from the result of the ℓth sweep.

Exercise 1.44. Assume that for some sample with eight entries $\vec{s} = (s_0, s_1, s_2, s_3, s_4, s_5, s_6, s_7)$, the In-Place Fast Haar Wavelet Transform stops at the end of the second sweep and gives

$$\vec{s}^{(3-2)} := (3, 1, -1, 1, 7, 1, 1, -1).$$

Reconstruct all the values of the sample.

Exercise 1.45. Write a computer program to compute the In-Place Fast Inverse Haar Wavelet Transform. Test the program by computing the In-Place Fast Haar Wavelet Transform and then the In-Place Fast Inverse Haar Wavelet Transform.

Exercise 1.46. Write a computer program to compute the Ordered Fast Inverse Haar Wavelet Transform. Test the program by computing the Ordered Fast Haar Wavelet Transform and then the Ordered Fast Inverse Haar Wavelet Transform.

1.6   EXAMPLES

This section provides a first demonstration of the practical significance of mathematical wavelets with real data. Any other finite sequence of numbers—including random numbers—might serve the same purpose, but the specific contexts demonstrated here may help in providing suggestions for further applications.

1.6.1   Creek Water Temperature Analysis

This example serves mainly to explain the practical significance of wavelet coefficients.

The following sixteen numbers—also plotted in Figure 1.9—represent semiweekly measurements of temperature, in degrees Fahrenheit, for December 1992 and January 1993 at a fixed common location along Hangman Creek, during a study of riverbank erosion by Mr. Jim Fox, in Spokane, Washington.

32.0, 10.0, 20.0, 38.0, 37.0, 28.0, 38.0, 34.0,
18.0, 24.0, 18.0, 9.0, 23.0, 24.0, 28.0, 34.0

The In-Place Fast Haar Wavelet Transform produces the result

25.9375,	11.0,	−4.0,	−9.0;
−4.625,	4.5,	−1.75,	2.0;
3.6875,	−3.0,	3.75,	4.5,
−5.0,	−0.5,	−3.75,	−3.0.

Equivalently, a rearrangement in increasing frequencies yields the Ordered Fast Haar Wavelet Transform:

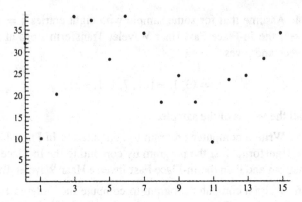

Figure 1.9 Temperature (°F) versus time (half weeks).

25.9375;							
3.6875;							
−4.625	−5;						
−4.0	−1.75	3.75	−3.75;			
11.0	−9.0	4.5	2.0	−3.0	4.5	−0.5	−3.0.

The first coefficient, 25.9375, represents the average temperature for the whole two-month period.

The second coefficient, 3.6875, is the coefficient of the longest wavelet over the whole period, which means that the temperature changed by $3.6875 * (-2) = -7.375$, a *decrease* of 7.375°F, from December to January.

The next two coefficients, −4.625 and −5, represent similar changes of temperature over the first half (first two quarters) and over the second half (last two quarters) of the period. The coefficient −4.625 corresponds to a change of $-4.625 * (-2) = 9.25$, an *increase* of 9.25°F from the first two weeks to the last two weeks in December. The coefficient −5 corresponds to a change of $-5 * (-2) = 10$, an *increase* of 10°F from the first two weeks to the last two weeks in January.

Each of the next four coefficients, −4.0, −1.75, 3.75, and −3.75, represents a change of temperature over two weeks. For instance, the coefficient −4 means that the temperature *increased* by $-4 * (-2) = 8$°F from the first week to the second week of December.

Finally, each of the last eight coefficients,

$$11.0 \quad -9.0 \quad 4.5 \quad 2.0 \quad -3.0 \quad 4.5 \quad -0.5 \quad -3.0$$

represents a change of temperature over one week. For instance, the coefficient 11 means that the temperature changed by $11 * (-2) = -22$°F during the first week of December; indeed, the data show a drop from 32°F down to 10°F.

As a verification, the In-Place Fast Inverse Haar Wavelet Transform reproduced the data exactly.

EXERCISES

Exercise 1.47. Analyze the following measurements of the ground frost depth, in centimeters, at Qualchan on Hangman Creek, for the same period (also by Mr. Jim Fox).

$$22.0, \ 27.0, \ 48.8, \ 47.5, \ 47.0, \ 48.5, \ 48.0, \ 47.0,$$
$$43.0, \ 41.0, \ 41.0, \ 38.0, \ 36.0, \ 47.1, \ 34.0, \ 32.0.$$

Exercise 1.48. Analyze the following measurements of the ground frost depth, in centimeters, at Kracher on Hangman Creek, for the same period (also by Mr. Jim Fox).

$$12.0, \ 16.0, \ 27.0, \ 32.8, \ 33.5, \ 33.5, \ 39.0, \ 39.0,$$
$$40.0, \ 41.3, \ 41.3, \ 42.0, \ 43.0, \ 45.0, \ 35.5, \ 49.0.$$

Exercise 1.49. Analyze the following measurements of river flow, in cubic feet per second, at the US Geological Survey Data Station 1242400 on Hangman Creek, for the same period.

$$10.0, \ 12.0, \ 12.0, \ 7.0, \ 8.0, \ 9.1, \ 8.2, \ 9.4,$$
$$16.0, \ 15.0, \ 13.0, \ 11.0, \ 6.4, \ 9.0, \ 19., \ 118.0.$$

Exercise 1.50. Obtain data of any kind and analyze them with the Haar Wavelet Transform.

1.6.2 Financial Stock Index Event Detection

This example demonstrates the automated use of wavelet transforms—here the automated search for coefficients with large magnitudes—to detect events in large data sets.

The top panel in Figure 1.10 displays on the vertical axis the New York Stock Exchange (NYSE) Composite Index, and on the horizontal axis the date, from 2 January 1981 (business day 0) through 7 February 1988 (business day 2047).

The middle panel in Figure 1.10 shows the coefficients of the In-Place Haar Wavelet Transform. The first coefficient, 111.15, in position 0 (superimposed on the vertical axis), represents the average of the index for the entire period: the coefficient of the slowest function, $\varphi_{[0,1[}$, extending over the whole period. Because in this example the value 111.15 has an order of magnitude larger than the values of the other coefficients, the scale of the graph does not reveal the other coefficients as well as does the bottom panel.

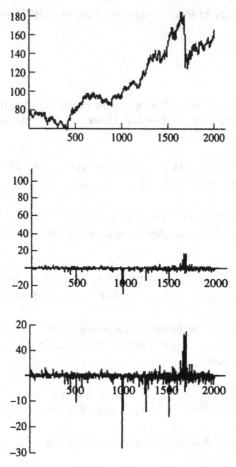

Figure 1.10   New York Stock Exchange Composite Index for 1981–1987. *Top.* Data: index (vertical axis) vs. day (horizontal axis). *Middle.* In-Place Haar Wavelet Transform, average first. *Bottom.* Same transform without average, for details.

The bottom panel in Figure 1.10 shows the coefficients of the In-Place Fast Haar Wavelet Transform in positions from 1 through 2047, both included, but not the average in position 0.

The coefficient with the largest magnitude, -28.96, in position 1024, at the middle of the array of coefficients, corresponds to the slowest wavelet, $\psi_{[0,1[}$, extending over the whole period. Thus, the value -28.96 reflects a *rise* by $(-28.96) * (-2) \approx 58$ points from the first half of the period to the second half of the period (with each half about 3.5-year long).

The next-largest magnitude, 15.255, in position 1717, reflects the *drop* by $15.255 * (-2) \approx 30.5$ points between business days 1716 and 1717: Friday 16 and Monday 19 October 1987.

EXERCISES

Identify the significance of the following wavelet coefficients.

Exercise 1.51. Identify the significance of -10.59 in position 513.

Exercise 1.52. Identify the significance of -14.41 in position 1281.

Exercise 1.53. Identify the significance of -16.26 in position 1537.

Exercise 1.54. Identify the significance of $+15.30$ in position 1716.

CHAPTER 2

Multidimensional Wavelets and Applications

2.0 INTRODUCTION

This chapter extends Haar's wavelets from one-dimensional arrays to multidimensional grids of data, for instance, encodings of photographs, scatter plots, or geographical measurements. Some of the logical derivations involve matrix algebra. For two-dimensional wavelets, encodings can consist of matrices, indexed by rows from top to bottom, and by columns from left to right. Corresponding to the index of the rows, the *first* coordinate axis runs *from top to bottom;* similarly, corresponding to the index of the columns, the *second* coordinate axis runs *from left to right.* Such an indexing scheme amounts to a rotation of the usual mathematical axes by one quarter of a turn clockwise, as shown in Figure 2.1. The same orientation is also common in geophysics, where the first axis indicates depth, and points downward, and the second axis shows such functions of depth as the

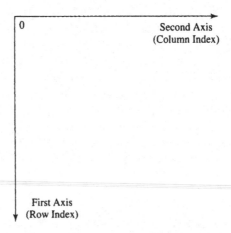

Figure 2.1 Axes matching indices of rows and columns.

36

velocity of seismic waves or the hydrostatic pressure at each depth [10, pp. 57, 61, 71, 72].

2.1 TWO-DIMENSIONAL HAAR WAVELETS

2.1.1 Two-Dimensional Approximation with Step Functions

Just as simple step functions approximate functions in one dimension, simple square-step functions approximate functions in two dimensions. Also, as basic one-dimensional step functions have the value 1 in an interval and 0 everywhere else, each basic square-step function has the value 1 in a selected square, and 0 everywhere else. (For graphing purposes, their graph lies over the two-dimensional square in a three-dimensional space, as in Figure 2.2.)

Example 2.1 The step function $\Phi_{0,0}^{(0)}$ in Figure 2.2 is defined by

$$\Phi_{0,0}^{(0)}(x, y) := \begin{cases} 1 & \text{if } 0 \le x < 1 \text{ and } 0 \le y < 1, \\ 0 & \text{otherwise.} \end{cases} \qquad \Box$$

Example 2.2 Figure 2.3 displays four square-step functions, according to the rotation of axes in force here.

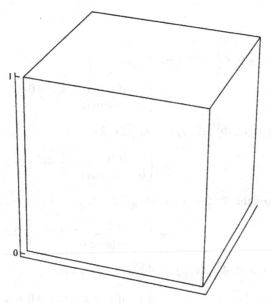

Figure 2.2 The simple square-step function $\Phi_{0,0}^{(0)}$.

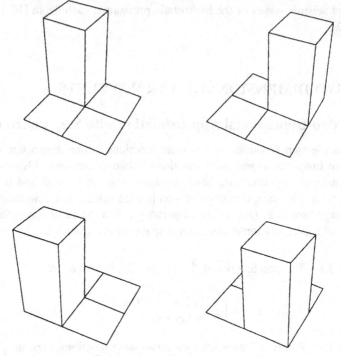

Figure 2.3   *Top left:* $\Phi_{0,0}^{(1)}$. *Top right:* $\Phi_{0,1}^{(1)}$. *Bottom left:* $\Phi_{1,0}^{(1)}$. *Bottom right:* $\Phi_{1,1}^{(1)}$.

Top left. $\Phi_{0,0}^{(1)}(x, y) := \Phi_{0,0}^{(0)}(2x, 2y)$

$$= \begin{cases} 1 & \text{if } 0 \leq x < \tfrac{1}{2} \text{ and } 0 \leq y < \tfrac{1}{2}, \\ 0 & \text{otherwise,} \end{cases}$$

Top right. $\Phi_{0,1}^{(1)}(x, y) := \Phi_{0,0}^{(0)}(2x, 2y - 1)$

$$= \begin{cases} 1 & \text{if } 0 \leq x < \tfrac{1}{2} \text{ and } \tfrac{1}{2} \leq y < 1, \\ 0 & \text{otherwise,} \end{cases}$$

Bottom right. $\Phi_{1,1}^{(1)}(x, y) := \Phi_{0,0}^{(0)}(2x - 1, 2y - 1)$

$$= \begin{cases} 1 & \text{if } \tfrac{1}{2} \leq x < 1 \text{ and } \tfrac{1}{2} \leq y < 1, \\ 0 & \text{otherwise,} \end{cases}$$

Bottom left. $\Phi_{1,0}^{(1)}(x, y) := \Phi_{0,0}^{(0)}(2x - 1, 2y)$

$$= \begin{cases} 1 & \text{if } \tfrac{1}{2} \leq x < 1 \text{ and } 0 \leq y < \tfrac{1}{2}, \\ 0 & \text{otherwise.} \end{cases} \qquad \square$$

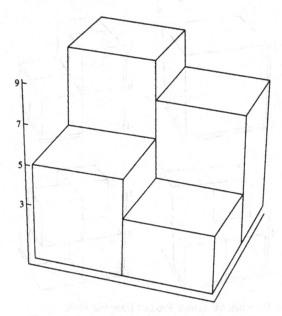

Figure 2.4   Bar graph of a square-step function.

Example 2.3   The function f with sampled values

$$f(0,0) := 9, \quad f(0, \tfrac{1}{2}) = 7,$$
$$f(\tfrac{1}{2}, 0) := 5, \quad f(\tfrac{1}{2}, \tfrac{1}{2}) = 3$$

admits the square-step approximation $\tilde{f}$ shown in Figure 2.4:

$$\tilde{f} := 9\Phi_{0,0}^{(1)} + 7\Phi_{0,1}^{(1)}$$
$$+ 5\Phi_{1,0}^{(1)} + 3\Phi_{1,1}^{(1)}. \qquad \square$$

2.1.2   Tensor Products of Functions

One of the methods to extend one-dimensional wavelets to two-dimensional wavelets employs the products of basic wavelets in the first dimension with basic wavelets in the second dimension.

Definition 2.4   For each pair of functions p and q, the **tensor product** of p and q is the function denoted by $p \otimes q$ and defined by

$$(p \otimes q)(x, y) := p(x) \cdot q(y). \qquad \square$$

Example 2.5   With $p := \varphi_{[0,1[}$ and $q := \varphi_{[0,1[}$ defined by

$$\varphi_{[0,1[}(r) := \begin{cases} 1 & \text{if } 0 \leq r < 1, \\ 0 & \text{otherwise,} \end{cases}$$

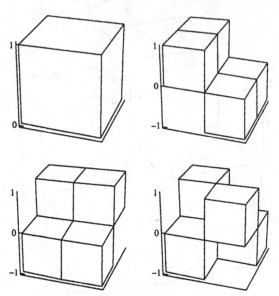

Figure 2.5   Two-Dimensional Tensor-Product Haar Wavelets.
Top left: $\Phi_{0,0}^{(0)} = \varphi_{[0,1[}\otimes\varphi_{[0,1[}$; *Top right:* $\Psi_{0,0}^{h,(0)} = \varphi_{[0,1[} \otimes \psi_{[0,1[}$. *Bottom left:* $\Psi_{0,0}^{v,(0)} = \psi_{[0,1[} \otimes \varphi_{[0,1[}$; *Bottom right:* $\Psi_{0,0}^{d,(0)} = \psi_{[0,1[} \otimes \psi_{[0,1[}$.

the tensor product $\varphi_{[0,1[} \otimes \varphi_{[0,1[}$, displayed in Figure 2.5, becomes

$$\text{(top left) } \Phi_{0,0}^{(0)}(x, y) = (\varphi_{[0,1[} \otimes \varphi_{[0,1[})(x, y)$$

$$= \varphi_{[0,1[}(x) \cdot \varphi_{[0,1[}(y)$$

$$= \begin{cases} 1 & \text{if } 0 \leq x < 1 \text{ and } 0 \leq y < 1, \\ 0 & \text{otherwise.} \end{cases}$$ □

Example 2.6   With $p := \varphi_{[0,1[}$ and $q := \psi_{[0,1[}$ defined by

$$\psi_{[0,1[}(r) := \begin{cases} 1 & \text{if } 0 \leq r < \frac{1}{2}, \\ -1 & \text{if } \frac{1}{2} \leq r < 1, \\ 0 & \text{otherwise,} \end{cases}$$

the tensor product $\Psi_{0,0}^{h,(0)} := p \otimes q = \varphi_{[0,1[} \otimes \psi_{[0,1[}$ in Figure 2.5 is

$$\text{(top right) } \Psi_{0,0}^{h,(0)}(x, y) = (\varphi_{[0,1[} \otimes \psi_{[0,1[})(x, y)$$

$$= \varphi_{[0,1[}(x) \cdot \psi_{[0,1[}(y)$$

$$= \begin{cases} 1 & \text{if } 0 \leq x < 1 \text{ and } 0 \leq y < \frac{1}{2}, \\ -1 & \text{if } 0 \leq x < 1 \text{ and } \frac{1}{2} \leq y < 1, \\ 0 & \text{otherwise.} \end{cases}$$

The superscript ^{h} indicates the correspondence of such wavelets with *horizontal changes* in the data, which cause vertical edges parallel to the first rotated axis of coordinates [7, p. 314].   □

Example 2.7 With $p := \psi_{[0,1[}$ and $q := \varphi_{[0,1[}$, the tensor product $\Psi_{0,0}^{v,(0)} := p \otimes q = \psi_{[0,1[} \otimes \varphi_{[0,1[}$ (see Figure 2.5) becomes

$$\text{(bottom left)} \quad \Psi_{0,0}^{v,(0)}(x, y) = (\psi_{[0,1[} \otimes \varphi_{[0,1[})(x, y)$$

$$= \psi_{[0,1[}(x) \cdot \varphi_{[0,1[}(y)$$

$$= \begin{cases} 1 & \text{if } 0 \le x < \frac{1}{2} \text{ and } 0 \le y < 1, \\ -1 & \text{if } \frac{1}{2} \le x < 1 \text{ and } 0 \le y < 1, \\ 0 & \text{otherwise.} \end{cases}$$

The superscript ^{v} indicates the correspondence of such wavelets with *vertical changes* in the data, which cause horizontal edges parallel to the second rotated axis of coordinates [7, p. 314].   □

Example 2.8 With $p := \psi_{[0,1[}$ and $q := \psi_{[0,1[}$, the tensor product $\Psi_{0,0}^{d,(0)} := p \otimes q = \psi_{[0,1[} \otimes \psi_{[0,1[}$, (see Figure 2.5) becomes

$$\text{(bottom right)} \quad \Psi_{0,0}^{d,(0)}(x, y) = (\psi_{[0,1[} \otimes \psi_{[0,1[})(x, y)$$

$$= \psi_{[0,1[}(x) \cdot \psi_{[0,1[}(y)$$

$$= \begin{cases} 1 & \text{if } 0 \le x < \frac{1}{2} \text{ and } 0 \le y < \frac{1}{2}, \\ 1 & \text{if } \frac{1}{2} \le x < 1 \text{ and } \frac{1}{2} \le y < 1, \\ -1 & \text{if } \frac{1}{2} \le x < 1 \text{ and } 0 \le y < \frac{1}{2}, \\ -1 & \text{if } 0 \le x < \frac{1}{2} \text{ and } \frac{1}{2} \le y < 1, \\ 0 & \text{otherwise.} \end{cases}$$

The superscript ^{d} indicates the correspondence of such wavelets with *diagonal edges* in the data [7, p. 314].   □

Similar two-dimensional wavelets exist for every location and frequency.

Definition 2.9 For all nonnegative integers j (denoting the frequency), and k and ℓ (denoting the location),

$$\Phi_{k,\ell}^{(j)} := \varphi_k^{(j)} \otimes \varphi_\ell^{(j)},$$

$$\Psi_{k,\ell}^{h,(j)} := \varphi_k^{(j)} \otimes \psi_\ell^{(j)},$$

$$\Psi_{k,\ell}^{v,(j)} := \psi_k^{(j)} \otimes \varphi_\ell^{(j)},$$

$$\Psi_{k,\ell}^{d,(j)} := \psi_k^{(j)} \otimes \psi_\ell^{(j)}$$

are the **Two-Dimensional Tensor-Product Haar Wavelets.**   □

Example 2.10 The wavelet $\Psi_{1,3}^{d,(5)} = \psi_1^{(5)} \otimes \psi_3^{(5)}$ represents a diagonal edge in the square $[1 * \frac{1}{32}, 2 * \frac{1}{32}[\times [3 * \frac{1}{32}, 4 * \frac{1}{32}[$, with side $2^{-5} = \frac{1}{32}$ and located at $[1 * \frac{1}{32}, 2 * \frac{1}{32}[$ along the first axis, and $[3 * \frac{1}{32}, 4 * \frac{1}{32}[$ along the second axis. $\square$

2.1.3 The Basic Two-Dimensional Haar Wavelet Transform

Consider a function f approximated by a square-step function $\tilde{f}$ and encoded by a matrix:

$$\tilde{f} = \begin{pmatrix} f(0,0) & f(0,\frac{1}{2}) \\ f(\frac{1}{2},0) & f(\frac{1}{2},\frac{1}{2}) \end{pmatrix} = \begin{pmatrix} s_{0,0} & s_{0,1} \\ s_{1,0} & s_{1,1} \end{pmatrix}.$$

The matrix just displayed encodes the *values* of f, but for some applications, *patterns* of change prove more useful than the values. One method to calculate such patterns expresses the same function in terms of two-dimensional wavelets. For instance, one algorithm begins with a one-dimensional wavelet transform of each row,

$$\begin{pmatrix} s_{0,0} & s_{0,1} \\ s_{1,0} & s_{1,1} \end{pmatrix} \begin{matrix} \to \\ \to \end{matrix} \begin{pmatrix} \frac{s_{0,0}+s_{0,1}}{2} & \frac{s_{0,0}-s_{0,1}}{2} \\ \frac{s_{1,0}+s_{1,1}}{2} & \frac{s_{1,0}-s_{1,1}}{2} \end{pmatrix},$$

and then a one-dimensional wavelet transform on each *new* column,

$$\begin{pmatrix} \frac{s_{0,0}+s_{0,1}}{2} & \frac{s_{0,0}-s_{0,1}}{2} \\ \frac{s_{1,0}+s_{1,1}}{2} & \frac{s_{1,0}-s_{1,1}}{2} \end{pmatrix}$$
$$\downarrow \qquad \qquad \downarrow$$
$$\begin{pmatrix} \frac{(s_{0,0}+s_{0,1})+(s_{1,0}+s_{1,1})}{4} & \frac{(s_{0,0}-s_{0,1})+(s_{1,0}-s_{1,1})}{4} \\ \frac{(s_{0,0}+s_{0,1})-(s_{1,0}+s_{1,1})}{4} & \frac{(s_{0,0}-s_{0,1})-(s_{1,0}-s_{1,1})}{4} \end{pmatrix}.$$

Example 2.11 Consider again the data from Example 2.3. The one-dimensional Haar Wavelet Transform on each row yields

$$\begin{pmatrix} 9 & 7 \\ 5 & 3 \end{pmatrix} \begin{matrix} \to \\ \to \end{matrix} \begin{pmatrix} \frac{9+7}{2} & \frac{9-7}{2} \\ \frac{5+3}{2} & \frac{5-3}{2} \end{pmatrix} = \begin{pmatrix} 8 & 1 \\ 4 & 1 \end{pmatrix},$$

and the one-dimensional wavelet transform on each *new* column gives

$$\begin{pmatrix} 8 & 1 \\ 4 & 1 \end{pmatrix}$$

$$\downarrow \qquad \downarrow$$

$$\begin{pmatrix} \frac{8+4}{2} & \frac{1+1}{2} \\ \frac{8-4}{2} & \frac{1-1}{2} \end{pmatrix}$$

$$= \begin{pmatrix} 6 & 1 \\ 2 & 0 \end{pmatrix}.$$

The coefficients just obtained express the initial data in terms of the tensor-product wavelets, in the sense that

$$\tilde{f} = s_{0,0}\Phi_{0,0}^{(1)} + s_{0,1}\Phi_{0,1}^{(1)}$$
$$+ s_{1,0}\Phi_{1,0}^{(1)} + s_{1,1}\Phi_{1,1}^{(1)}$$
$$= \frac{(s_{0,0} + s_{0,1}) + (s_{1,0} + s_{1,1})}{4} \varphi_{[0,1[} \otimes \varphi_{[0,1[}$$
$$+ \frac{(s_{0,0} - s_{0,1}) + (s_{1,0} - s_{1,1})}{4} \varphi_{[0,1[} \otimes \psi_{[0,1[}$$
$$+ \frac{(s_{0,0} + s_{0,1}) - (s_{1,0} + s_{1,1})}{4} \psi_{[0,1[} \otimes \varphi_{[0,1[}$$
$$+ \frac{(s_{0,0} - s_{0,1}) - (s_{1,0} - s_{1,1})}{4} \psi_{[0,1[} \otimes \psi_{[0,1[}.$$

For example,

$$\tilde{f} = \begin{pmatrix} 9 & 7 \\ 5 & 3 \end{pmatrix}$$

$$= 6 \cdot \varphi_{[0,1[} \otimes \varphi_{[0,1[} + 1 \cdot \varphi_{[0,1[} \otimes \psi_{[0,1[}$$
$$+ 2 \cdot \psi_{[0,1[} \otimes \varphi_{[0,1[} + 0 \cdot \psi_{[0,1[} \otimes \psi_{[0,1[}.$$

The expression just obtained with wavelets provides the following interpretation of the data:

( ) The entry in the upper left-hand corner, 6, measures the average of the four entries in the data: $6 = (9 + 7 + 5 + 3)/4$.

(h) The entry in the upper right-hand corner, 1, corresponds to the "horizontal" change in the data, a drop from left to right measured by 1 wavelet $\varphi_{[0,1[} \otimes \psi_{[0,1[}$, which has a drop of size $1 * 2 = 2$.

(v) The entry in the lower left-hand corner, 2, corresponds to the "vertical" change in the data, a drop from top to bottom, measured by 2 times the wavelet $\psi_{[0,1[} \otimes \varphi_{[0,1[}$. The drop has size $2 \times 2 = 4$ because the wavelet $\psi_{[0,1[} \otimes \varphi_{[0,1[}$ has a drop of size 2.

(*d*) The entry in the lower right-hand corner, 0, corresponds to the absence of any "diagonal" edge in the data, from the upper left to the lower right, measured by 0 times the wavelet $\psi_{[0,1[} \otimes \psi_{[0,1[}$.

Thus, the data decompose as a sum of 6 times the basic square step over the entire square, a "horizontal" drop from left to right of size 2, and a "vertical" drop from top to bottom of size 4.   □

The following example illustrates an application of wavelets to the "compression" of such two-dimensional data as photographs.

Example 2.12   Consider again the data $\tilde{f} := \begin{pmatrix} 9 & 7 \\ 5 & 3 \end{pmatrix}$, shown on the left in Figure 2.6, with Haar Wavelet Transform $\begin{pmatrix} 6 & 1 \\ 2 & 0 \end{pmatrix}$. One method to "compress" such data retains only the wavelet coefficients with the largest magnitude, here 6 and 2, with a list of the corresponding wavelets, here $6 \cdot \varphi_{[0,1[} \otimes \varphi_{[0,1[}$ and $2 \cdot \psi_{[0,1[} \otimes \varphi_{[0,1[}$. Hence, reconstructing the compressed data amounts to computing the values of the resulting combination of wavelets:

$$6 \cdot \varphi_{[0,1[} \otimes \varphi_{[0,1[} + 2 \cdot \psi_{[0,1[} \otimes \varphi_{[0,1[}$$

$$= 6 \cdot \begin{pmatrix} 1 & 1 \\ 1 & 1 \end{pmatrix} + 2 \cdot \begin{pmatrix} 1 & 1 \\ -1 & -1 \end{pmatrix} = \begin{pmatrix} 8 & 8 \\ 4 & 4 \end{pmatrix},$$

as displayed on the right in Figure 2.6. The threshold between discarded and retained coefficients depends upon the accuracy required by the particular application.   □

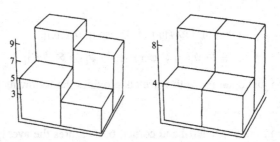

Figure 2.6   *Left.* Bar graph of the data. *Right.* Reconstruction from only the two wavelets with the largest coefficients.

There remains verifying that indeed consecutive one-dimensional wavelet transforms on the rows and then on the new columns yield the coefficients of the tensor-product wavelets. To verify the assertion just made, first express the wavelets in terms of square steps:

$$\Phi_{0,0}^{(0)} = \varphi_{[0,1[} \otimes \varphi_{[0,1[}$$

$$= \Phi_{0,0}^{(1)} + \Phi_{0,1}^{(1)} + \Phi_{1,0}^{(1)} + \Phi_{1,1}^{(1)},$$

$$\Psi_{0,0}^{h,(0)} = \varphi_{[0,1[} \otimes \psi_{[0,1[}$$

$$= \Phi_{0,0}^{(1)} - \Phi_{0,1}^{(1)} + \Phi_{1,0}^{(1)} - \Phi_{1,1}^{(1)},$$

$$\Psi_{0,0}^{v,(0)} = \psi_{[0,1[} \otimes \varphi_{[0,1[}$$

$$= \Phi_{0,0}^{(1)} + \Phi_{0,1}^{(1)} - \Phi_{1,0}^{(1)} - \Phi_{1,1}^{(1)},$$

$$\Psi_{0,0}^{d,(0)} = \psi_{[0,1[} \otimes \psi_{[0,1[}$$

$$= \Phi_{0,0}^{(1)} - \Phi_{0,1}^{(1)} - \Phi_{1,0}^{(1)} + \Phi_{1,1}^{(1)},$$

and then reverse the process, to obtain expressions of the square steps in terms of wavelets. To this end linear algebra provides a standard method [14, § 2.5], which records the coefficients of each of the equations just obtained in the *columns* of a matrix, called a matrix of change of basis, and denoted here by Ω (Greek capital "omega"):

$$\Omega = \begin{array}{c} \\ \Phi_{0,0}^{(1)} \\ \Phi_{0,1}^{(1)} \\ \Phi_{1,0}^{(1)} \\ \Phi_{1,1}^{(1)} \end{array} \begin{pmatrix} \Phi_{0,0}^{(0)} & \Psi_{0,0}^{h,(0)} & \Psi_{0,0}^{v,(0)} & \Psi_{0,0}^{d,(0)} \\ 1 & 1 & 1 & 1 \\ 1 & -1 & 1 & -1 \\ 1 & 1 & -1 & -1 \\ 1 & -1 & -1 & 1 \end{pmatrix}.$$

For example, the second *column* of the matrix Ω means that

$$\Psi_{0,0}^{h,(0)} = \Phi_{0,0}^{(1)} - \Phi_{0,1}^{(1)} + \Phi_{1,0}^{(1)} - \Phi_{1,1}^{(1)}.$$

Linear algebra then shows that the "inverse" matrix Ω^{-1}, computed, for instance, through Gaussian elimination, yields the coefficients of the reverse equations:

$$\Omega^{-1} = \begin{array}{c} \\ \Phi_{0,0}^{(0)} \\ \Psi_{0,0}^{h,(0)} \\ \Psi_{0,0}^{v,(0)} \\ \Psi_{0,0}^{d,(0)} \end{array} \begin{pmatrix} \Phi_{0,0}^{(1)} & \Phi_{0,1}^{(1)} & \Phi_{1,0}^{(1)} & \Phi_{1,1}^{(1)} \\ \frac{1}{4} & \frac{1}{4} & \frac{1}{4} & \frac{1}{4} \\ \frac{1}{4} & -\frac{1}{4} & \frac{1}{4} & -\frac{1}{4} \\ \frac{1}{4} & \frac{1}{4} & -\frac{1}{4} & -\frac{1}{4} \\ \frac{1}{4} & -\frac{1}{4} & -\frac{1}{4} & \frac{1}{4} \end{pmatrix}.$$

For example, the first column of the inverse matrix Ω^{-1} means that

$$\Phi_{0,0}^{(1)} = \frac{1}{4}\Phi_{0,0}^{(0)} + \frac{1}{4}\Psi_{0,0}^{h,(0)} + \frac{1}{4}\Psi_{0,0}^{v,(0)} + \frac{1}{4}\Psi_{0,0}^{d,(0)}.$$

Similarly, from the remaining columns of the inverse matrix Ω^{-1},

$$\Phi_{0,1}^{(1)} = \frac{1}{4}\Phi_{0,0}^{(0)} - \frac{1}{4}\Psi_{0,0}^{h,(0)} + \frac{1}{4}\Psi_{0,0}^{v,(0)} - \frac{1}{4}\Psi_{0,0}^{d,(0)},$$

$$\Phi_{1,0}^{(1)} = \frac{1}{4}\Phi_{0,0}^{(0)} + \frac{1}{4}\Psi_{0,0}^{h,(0)} - \frac{1}{4}\Psi_{0,0}^{v,(0)} - \frac{1}{4}\Psi_{0,0}^{d,(0)},$$

$$\Phi_{1,1}^{(1)} = \frac{1}{4}\Phi_{0,0}^{(0)} - \frac{1}{4}\Psi_{0,0}^{h,(0)} - \frac{1}{4}\Psi_{0,0}^{v,(0)} + \frac{1}{4}\Psi_{0,0}^{d,(0)}.$$

Substituting the results into the approximation $\tilde{f}$ of f yields

$$\begin{aligned}
\tilde{f} &= s_{0,0}\Phi_{0,0}^{(1)} + s_{0,1}\Phi_{0,1}^{(1)} + s_{1,0}\Phi_{1,0}^{(1)} + s_{1,1}\Phi_{1,1}^{(1)} \\
&= \frac{(s_{0,0} + s_{0,1}) + (s_{1,0} + s_{1,1})}{4}\varphi_{[0,1[} \otimes \varphi_{[0,1[} \\
&\quad + \frac{(s_{0,0} - s_{0,1}) + (s_{1,0} - s_{1,1})}{4}\varphi_{[0,1[} \otimes \psi_{[0,1[} \\
&\quad + \frac{(s_{0,0} + s_{0,1}) - (s_{1,0} + s_{1,1})}{4}\psi_{[0,1[} \otimes \varphi_{[0,1[} \\
&\quad + \frac{(s_{0,0} - s_{0,1}) - (s_{1,0} - s_{1,1})}{4}\psi_{[0,1[} \otimes \psi_{[0,1[},
\end{aligned}$$

which is indeed the basic two-dimensional Haar Wavelet Transform.

2.1.4   Two-Dimensional Fast Haar Wavelet Transform

The one-dimensional fast wavelet transform extends to a two-dimensional fast wavelet transform with tensor products, through alternating applications of the one-dimensional transform to each row and then to each new column.

Example 2.13   Consider a function f sampled at $4 \times 4 = 16$ values on a square grid, and approximated by the corresponding square-step function shown in Figure 2.7:

$$\tilde{f} := \begin{pmatrix} 9 & 7 & 6 & 2 \\ 5 & 3 & 4 & 4 \\ 8 & 2 & 4 & 0 \\ 6 & 0 & 2 & 2 \end{pmatrix}.$$

The first step of the two-dimensional transform consists of a first one-dimensional step for each row,

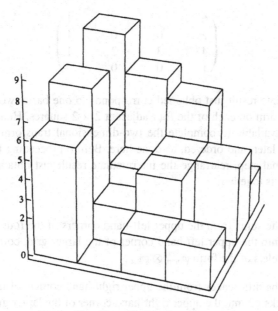

Figure 2.7   Data displayed as a bar graph.

$$
\begin{pmatrix} 9 & 7 & 6 & 2 \\ 5 & 3 & 4 & 4 \\ 8 & 2 & 4 & 0 \\ 6 & 0 & 2 & 2 \end{pmatrix}
\begin{matrix} \rightarrow \\ \rightarrow \\ \rightarrow \\ \rightarrow \end{matrix}
\begin{pmatrix} \frac{9+7}{2} & \frac{9-7}{2} & \frac{6+2}{2} & \frac{6-2}{2} \\ \frac{5+3}{2} & \frac{5-3}{2} & \frac{4+4}{2} & \frac{4-4}{2} \\ \frac{8+2}{2} & \frac{8-2}{2} & \frac{4+0}{2} & \frac{4-0}{2} \\ \frac{6+0}{2} & \frac{6-0}{2} & \frac{2+2}{2} & \frac{2-2}{2} \end{pmatrix}
=
\begin{pmatrix} 8 & 1 & 4 & 2 \\ 4 & 1 & 4 & 0 \\ 5 & 3 & 2 & 2 \\ 3 & 3 & 2 & 0 \end{pmatrix},
$$

followed by a first one-dimensional step for each new column,

$$
\begin{pmatrix} 8 & 1 & 4 & 2 \\ 4 & 1 & 4 & 0 \\ 5 & 3 & 2 & 2 \\ 3 & 3 & 2 & 0 \end{pmatrix}
$$

$$
\downarrow \qquad \downarrow \qquad \downarrow \qquad \downarrow
$$

$$
\begin{pmatrix} \frac{8+4}{2} & \frac{1+1}{2} & \frac{4+4}{2} & \frac{2+0}{2} \\[2mm] \frac{8-4}{2} & \frac{1-1}{2} & \frac{4-4}{2} & \frac{2-0}{2} \\[2mm] \frac{5+3}{2} & \frac{3+3}{2} & \frac{2+2}{2} & \frac{2+0}{2} \\[2mm] \frac{5-3}{2} & \frac{3-3}{2} & \frac{2-2}{2} & \frac{2-0}{2} \end{pmatrix}
$$

$$= \begin{pmatrix} 6 & 1 & 4 & 1 \\ 2 & 0 & 0 & 1 \\ 4 & 3 & 2 & 1 \\ 1 & 0 & 0 & 0 \end{pmatrix}.$$

The intermediate result just obtained corresponds to one basic two-dimensional wavelet transform on each of the four adjacent 2×2 squares. Hence two methods become available to complete the two-dimensional transform: in-place, as demonstrated later, and ordered, as done here. Before proceeding to the second two-dimensional step, rearrange the intermediate result just obtained to collect similar wavelets together:

$\begin{pmatrix} \bullet \end{pmatrix}$ all the sums from the upper left-hand corners of the four 2×2 blocks go into the upper left-hand corner of the larger grid, corresponding to wavelets of the form $\varphi_{m,n} \otimes \varphi_{k,\ell}$;

$\begin{pmatrix} & h \end{pmatrix}$ all the differences from the upper right-hand corners of the four 2×2 blocks go into the upper right-hand corner of the larger grid, for wavelets of the form $\varphi_{m,n} \otimes \psi_{k,\ell}$;

$\begin{pmatrix} & \\ & d \end{pmatrix}$ all the differences from the lower right-hand corners of the four 2×2 blocks go into the lower right-hand corner of the larger grid, for wavelets of the form $\psi_{m,n} \otimes \psi_{k,\ell}$;

$\begin{pmatrix} & \\ v & \end{pmatrix}$ all the differences from the lower left-hand corners of the four 2×2 blocks go into the lower left-hand corner of the larger grid, for wavelets of the form $\psi_{m,n} \otimes \varphi_{k,\ell}$;

$$\begin{pmatrix} 6 & 1 & 4 & 1 \\ 2 & 0 & 0 & 1 \\ & & & \\ 4 & 3 & 2 & 1 \\ 1 & 0 & 0 & 0 \end{pmatrix} \rightarrow \begin{pmatrix} 6 & 4 & 1 & 1 \\ 4 & 2 & 3 & 1 \\ & & & \\ 2 & 0 & 0 & 1 \\ 1 & 0 & 0 & 0 \end{pmatrix}.$$

Finally, perform a two-dimensional wavelet transform *only* on the four entries in the upper left-hand corner:

$$\begin{pmatrix} 6 & 4 \\ 4 & 2 \end{pmatrix} \begin{matrix} \rightarrow \\ \rightarrow \end{matrix} \begin{pmatrix} \frac{6+4}{2} & \frac{6-4}{2} \\ \frac{4+2}{2} & \frac{4-2}{2} \end{pmatrix} = \begin{pmatrix} 5 & 1 \\ 3 & 1 \end{pmatrix}$$

$$\downarrow \qquad \downarrow$$

$$\begin{pmatrix} \frac{5+3}{2} & \frac{1+1}{2} \\ \frac{5-3}{2} & \frac{1-1}{2} \end{pmatrix} = \begin{pmatrix} 4 & 1 \\ 1 & 0 \end{pmatrix}.$$

Thus emerges the completed two-dimensional wavelet transform:

$$\begin{pmatrix} 4 & 1 & 1 & 1 \\ 1 & 0 & 3 & 1 \\ & & & \\ 2 & 0 & 0 & 1 \\ 1 & 0 & 0 & 0 \end{pmatrix}.$$

□

EXERCISES

Exercise 2.1. Design an algorithm to compute an Ordered and In-Place Two-Dimensional Fast Haar Wavelet Transform.

Exercise 2.2. Design an algorithm to compute an Ordered and In-Place Two-Dimensional Fast *Inverse* Haar Wavelet Transform.

Exercise 2.3. Write and test a computer program for the Ordered and In-Place Two-Dimensional Fast Haar Wavelet Transform.

Exercise 2.4. Write and test a computer program for the Ordered and In-Place Two-Dimensional Fast *Inverse* Haar Wavelet Transform.

2.2 APPLICATIONS OF WAVELETS

This section outlines applications of one-dimensional and multidimensional wavelets through general ideas and small numerical examples. A subsequent section will present real data.

2.2.1 Noise Reduction

The distinction between signal and noise depends upon how the measured signal models reality, in other words, upon the assumed relations between the signal and the phenomenon represented by the signal. Thus, depending upon the assumed nature of the noise, several methods exist to eliminate or attenuate noise.

2.2.1.1 Random Noise If the noise has a random nature, then removing wavelets with random coefficients can help in restoring a noiseless signal.

Example 2.14 From the sample $\vec{s} := (3, 1, 0, 4, 8, 6, 9, 9)$, shown in Figure 2.8(a), the In-Place Haar Wavelet Transform produces

$$\vec{s}^{(0)} = \left(a_0^{(0)}, \quad c_0^{(2)}, \quad c_0^{(1)}, \quad c_1^{(2)}, \quad c_0^{(0)}, \quad c_2^{(2)}, \quad c_1^{(1)}, \quad c_3^{(2)} \right)$$
$$= (5, \quad 1, \quad 0, \quad -2, \quad -3, \quad 1, \quad -1, \quad 0).$$

The coefficients $\vec{c}^{(2)}$ corresponding to the highest frequency appear to have random values scattered about an average equal to zero: $\vec{c}^{(2)} = (1, -2, 1, 0)$. If the

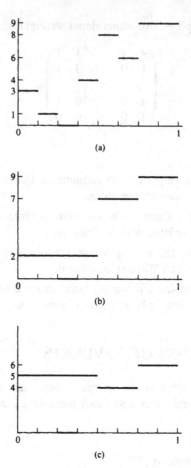

Figure 2.8 Example of noise reduction through wavelets. (a) Initial signal. (b) After removal of the highest frequencies, simulating the attenuation of random noise. (c) After removal of the highest and lowest frequencies, simulating the attenuation of random noise and low rumble.

corresponding wavelets do not belong to the original signal but come from noise, then removing those wavelets restores the original signal. To this effect, setting the corresponding coefficients to zero yields $\vec{c}^{(2)} = (0, 0, 0, 0)$, which gives a corrected Haar Wavelet Transform

$$\vec{s}^{(0)\prime} = (5, 0, 0, 0, -3, 0, -1, 0).$$

Hence, applying the In-Place Inverse Haar Wavelet Transform to $\vec{s}^{(0)\prime}$ produces the corrected signal shown in Figure 2.8(b),

$$\vec{s}^{\prime} = (2, 2, 2, 2, 7, 7, 9, 9),$$

which represents the initial measured sample $\bar{s} := (3, 1, 0, 4, 8, 6, 9, 9)$ minus the highest-frequency wavelets. The measured signal $\bar{s}$ and corrected signal $\bar{s}'$ have identical averages over each consecutive pair of values, (s_0, s_1), (s_2, s_3), ..., (s_6, s_7), but the corrected signal $\bar{s}'$ no longer exhibits any oscillation within such pairs, as shown in Figures 2.8(a) and 2.8(b).   □

2.2.1.2 Band-Specific Noise   If a signal putatively represents a band-limited phenomenon, which may contain frequencies only within a certain range (band), then all frequencies outside that band cannot arise from the phenomenon under consideration but arise from noise. Removing the wavelets with frequencies outside the band restores the initial signal.

Example 2.15   From the sample $\bar{s} := (3, 1, 0, 4, 8, 6, 9, 9)$, shown in Figure 2.8(a), the In-Place Haar Wavelet Transform produces

$$\bar{s}^{(0)} = \left(a_0^{(0)}, \quad c_0^{(2)}, \quad c_0^{(1)}, \quad c_1^{(2)}, \quad c_0^{(0)}, \quad c_2^{(2)}, \quad c_1^{(1)}, \quad c_3^{(2)} \right)$$
$$= (5, \quad 1, \quad 0, \quad -2, \quad -3, \quad 1, \quad -1, \quad 0).$$

For illustration purposes, assume that the phenomenon under consideration involves only frequencies from 2 to 4 cycles over the length of the signal. Thus, such a lower frequency as 1 cycle and such a higher frequency as 8 cycles come from noise. The lower frequency corresponds to the term $c_0^{(0)}$, while the higher frequency comes from the terms $c_j^{(2)}$. Setting all such coefficients to zero gives

$$\bar{s}^{(0)''} = (5, 0, 0, 0, 0, 0, -1, 0).$$

Hence, the In-Place Inverse Haar Wavelet Transform restores the initial signal in the form shown in Figure 2.8(c),

$$\bar{s}'' = (5, 5, 5, 5, 4, 4, 6, 6).$$

The difference with the preceding example lies in the additional removal of the lowest frequency, which has suppressed the average jump from the first half to the second half of the signal.   □

EXERCISES

Exercise 2.5. Assume that the In-Place Haar Wavelet Transform of a sample $\bar{s} = (s_0, s_1)$ produces the results $(7, 2)$. Remove the highest-frequency simple wavelet, and apply the inverse transform to calculate the corrected signal.

Exercise 2.6. Assume that the In-Place Haar Wavelet Transform of a sample $\bar{s} = (s_0, s_1)$ produces the results $(6, -3)$. Remove the highest-frequency simple wavelet, and apply the inverse transform to calculate the corrected signal.

Exercise 2.7. Assume that the In-Place Haar Wavelet Transform of a sample $\bar{s} = (s_0, s_1, s_2, s_3)$ produces the results $(6, 2, 1, 2)$. Remove the two highest-frequency simple wavelets, and apply the inverse transform to calculate the corrected signal.

Exercise 2.8. Assume that the In-Place Haar Wavelet Transform of a sample $\vec{s} = (s_0, s_1, s_2, s_3)$ produces the results $(4, 2, 2, 0)$. Remove the two highest-frequency simple wavelets, and apply the inverse transform to calculate the corrected signal.

Exercise 2.9. Assume that for some sample with eight entries $\vec{s} = (s_0, s_1, s_2, s_3, s_4, s_5, s_6, s_7)$, the In-Place Haar Wavelet Transform produces the result $(4, -1, -1, 2, 0, 1, -2, -2)$. Remove the four highest-frequency simple wavelets, and apply the inverse transform to calculate the corrected signal.

Exercise 2.10. Assume that for some sample with eight entries $\vec{s} = (s_0, s_1, s_2, s_3, s_4, s_5, s_6, s_7)$, the In-Place Haar Wavelet Transform produces the result $(5, 1, 1, 0, -3, -1, 0, 1)$. Remove the four highest-frequency simple wavelets, and apply the inverse transform to calculate the corrected signal.

Exercise 2.11. Assume that the In-Place Haar Wavelet Transform of a sample $\vec{s} = (s_0, s_1)$ produces the results $a_0^{(0)}$ and $c_0^{(0)}$. Starting from the formulae defining $a_0^{(0)}$ and $c_0^{(0)}$, in terms of s_0 and s_1, verify algebraically that removing the highest-frequency simple wavelet amounts to replacing the measured signal by its average.

Exercise 2.12. Assume that the In-Place Haar Wavelet Transform stops at the end of the ℓth sweep. Explain how to remove all the highest-frequency simple wavelets and then how to calculate the corrected signal.

Exercise 2.13. Assume that for some sample with eight entries $\vec{s} = (s_0, s_1, s_2, s_3, s_4, s_5, s_6, s_7)$, the In-Place Haar Wavelet Transform stops at the end of the second sweep and gives

$$\vec{s}^{(2)} = (3, 1, -1, 1, 7, 1, 1, -1).$$

Remove the four highest-frequency simple wavelets, and apply the inverse transform to calculate the corrected signal.

2.2.2   Data Compression

Some applications with large arrays of data may benefit from methods of data compression, to reduce the space necessary to store such data, perhaps at the cost of some degradation of the information in the data. For instance, the same methods used for noise reduction, followed by an appropriate recoding, provide a small but simple compression of the data.

Example 2.16 From the sample $\vec{s} := (3, 1, 0, 4, 8, 6, 9, 9)$, shown in Figure 2.9(a), the In-Place Haar Wavelet Transform produces

$$\vec{s}^{(0)} = \left(a_0^{(0)}, \quad c_0^{(2)}, \quad c_0^{(1)}, \quad c_1^{(2)}, \quad c_0^{(0)}, \quad c_2^{(2)}, \quad c_1^{(1)}, \quad c_3^{(2)} \right)$$
$$= (5, \qquad 1, \qquad 0, \quad -2, \quad -3, \qquad 1, \quad -1, \qquad 0).$$

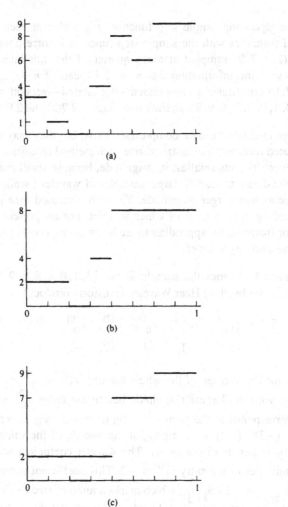

Figure 2.9 Examples of data-compression through wavelets. (a) Initial signal. (b) After retaining only the coefficients with the largest magnitudes. (c) After removal of the highest frequencies, removing the details at the smallest scale.

Hence, setting the coefficients of the highest-frequency simple wavelets to zero produces $\vec{c}^{(2)} = (0, 0, 0, 0)$, which gives a corrected Haar Wavelet Transform

$$\vec{s}^{(0)'} = (5, 0, 0, 0, -3, 0, -1, 0).$$

Finally, applying the In-Place Inverse Haar Wavelet Transform to $\vec{s}^{(0)'}$ produces the corrected signal shown in Figure 2.9(c),

$$\vec{s}' = (2, 2, 2, 2, 7, 7, 9, 9).$$

The corresponding simple step function $\tilde{f}$ sampled at each eighth of the unit interval coincides with the simple step function $\tilde{g}$ corresponding to the sample $\tilde{s} := (2, 2, 7, 9)$ sampled at every quarter of the unit interval. Thus, accompanied with the information that $n = 2$ instead of $n = 3$, the sample $\tilde{s} := (2, 2, 7, 9)$ constitutes a compressed—but altered—version of the initial signal $\tilde{s} := (3, 1, 0, 4, 8, 6, 9, 9)$, as shown in Figures 2.9(a) and 2.9(c).   □

Other methods for data compression exist, but they may require a more sophisticated recoding. For instance, one such method consists in setting to zero the wavelet coefficients smallest in magnitude, because small multiples of wavelets affect the data less than do large multiples of wavelets, while retaining only the coefficients with larger magnitude. Yet such a method then requires a substantial recoding to keep track of which wavelets remain present in the compressed data, for instance, by appending to each remaining coefficient a tag identifying the corresponding wavelet.

Example 2.17 From the sample $\tilde{s} := (3, 1, 0, 4, 8, 6, 9, 9)$, shown in Figure 2.9(a), the In-Place Haar Wavelet Transform produces

$$\tilde{s}^{(0)} = \left(a_0^{(0)}, \quad c_0^{(2)}, \quad c_0^{(1)}, \quad c_1^{(2)}, \quad c_0^{(0)}, \quad c_2^{(2)}, \quad c_1^{(1)}, \quad c_3^{(2)} \right)$$
$$= (5, \quad\quad 1, \quad\quad 0, \quad -2, \quad -3, \quad\quad 1, \quad -1, \quad\quad 0).$$

Except for the average of the whole sample, $a_0^{(0)} = s_0^{(0)} = 5$, the wavelet coefficient with the largest magnitude lies in the entry $s_4^{(0)} = -3$. This coefficient corresponds to the term $s_4^{(0)} \cdot \psi_{[0,1[} = -3 \cdot \psi_{[0,1[}$, which marks a jump of size $(-3) \cdot (-2) = 6$, rising, at the middle of the sample, where the wavelet $\psi_{[0,1[}$ has its discontinuity. The wavelet coefficient with the next-largest magnitude lies in the entry $s_3^{(0)} = -2$. This coefficient corresponds to the term $s_3^{(0)} \cdot \psi_{[\frac{1}{4},\frac{1}{2}[} = -2 \cdot \psi_{[\frac{1}{4},\frac{1}{2}[}$, which marks a jump of size $(-2) \cdot (-2) = 4$, rising, at the third eighth of the sample, where the wavelet $\psi_{[\frac{1}{4},\frac{1}{2}[}$ has its discontinuity. Thus, retaining only the average and the wavelet coefficients with the two largest magnitudes gives a new wavelet transform—compressed but altered—with only three nonzero coefficients,

$$\tilde{s}^{(0)'''} = (5, 0, 0, -2, -3, 0, 0, 0).$$

The In-Place Inverse Haar Wavelet Transform then produces the result displayed in Figure 2.9(b),

$$\tilde{s}''' = (2, 2, 0, 4, 8, 8, 8, 8),$$

which exhibits a rising edge of size 4 at the third eighth, and a rising edge of size 6 from the average over the first half to the average of the second half, but without the smaller oscillations of the initial sample $\tilde{s} = (3, 1, 0, 4, 8, 6, 9, 9)$, as shown in Figures 2.9(a) and 2.9(b).   □

Example 2.18   This example illustrates two methods with two-dimensional tensor-product wavelets to compress two-dimensional data, for instance, photographs. Consider again the data at the top of Figure 2.10:

$$\tilde{f} := \begin{pmatrix} 9 & 7 & 6 & 2 \\ 5 & 3 & 4 & 4 \\ 8 & 2 & 4 & 0 \\ 6 & 0 & 2 & 2 \end{pmatrix},$$

with ordered wavelet transform

$$\begin{pmatrix} 4 & 1 & 1 & 1 \\ 1 & 0 & 3 & 1 \\ & & & \\ 2 & 0 & 0 & 1 \\ 1 & 0 & 0 & 0 \end{pmatrix}.$$

A first method to compress data selects the wavelets with the largest coefficients, for instance

$$\begin{pmatrix} 4 & 1 & 1 & 1 \\ 1 & 0 & 3 & 1 \\ & & & \\ 2 & 0 & 0 & 1 \\ 1 & 0 & 0 & 0 \end{pmatrix}.$$

The coefficients just selected have the following significance:

( ) **4** times a square step on the entire square, added to

(h) **3** times a horizontal drop of size 2, for a total drop by 6, from left to right, in the lower left-hand corner, added to

(v) **2** times a vertical drop of size 2, for a total drop by 4, from top to bottom, in the upper left-hand corner.

Hence, reconstructing the compressed data gives the middle graph in Figure 2.10:

$$4 \cdot \begin{pmatrix} 1 & 1 & 1 & 1 \\ 1 & 1 & 1 & 1 \\ 1 & 1 & 1 & 1 \\ 1 & 1 & 1 & 1 \end{pmatrix} + 3 \cdot \begin{pmatrix} 0 & 0 & 0 & 0 \\ 0 & 0 & 0 & 0 \\ 1 & -1 & 0 & 0 \\ 1 & -1 & 0 & 0 \end{pmatrix} + 2 \cdot \begin{pmatrix} 1 & 1 & 0 & 0 \\ -1 & -1 & 0 & 0 \\ 0 & 0 & 0 & 0 \\ 0 & 0 & 0 & 0 \end{pmatrix}$$

$$= \begin{pmatrix} 4+2 & 4+2 & 4 & 4 \\ 4-2 & 4-2 & 4 & 4 \\ 4+3 & 4-3 & 4 & 4 \\ 4+3 & 4-3 & 4 & 4 \end{pmatrix} = \begin{pmatrix} 6 & 6 & 4 & 4 \\ 2 & 2 & 4 & 4 \\ 7 & 1 & 4 & 4 \\ 7 & 1 & 4 & 4 \end{pmatrix}.$$

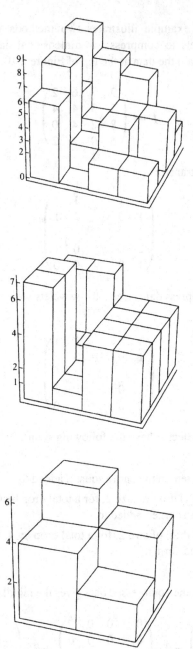

Figure 2.10 Two methods to compress data. *Top.* Bar graph of the data. *Middle.* Compressed data with wavelets with largest coefficients. *Bottom.* Compressed data with wavelets from averages.

Alternatively, a second method to compress data retains only the averages over each of the four quarters of the initial grid, thus blurring the picture. Such averages appear in the upper left-hand corner after the first sweep of the ordered wavelet transform, and they lead to the reconstruction shown at the bottom of Figure 2.10:

$$
\begin{pmatrix} 4 & 1 & 1 & 1 \\ 1 & 0 & 3 & 1 \\ 2 & 0 & 0 & 1 \\ 1 & 0 & 0 & 0 \end{pmatrix} \rightarrow 4 \cdot \begin{pmatrix} 1 & 1 & 1 & 1 \\ 1 & 1 & 1 & 1 \\ 1 & 1 & 1 & 1 \\ 1 & 1 & 1 & 1 \end{pmatrix} + 1 \cdot \begin{pmatrix} 1 & 1 & -1 & -1 \\ 1 & 1 & -1 & -1 \\ 1 & 1 & -1 & -1 \\ 1 & 1 & -1 & -1 \end{pmatrix}
$$

$$
+ 1 \cdot \begin{pmatrix} 1 & 1 & 1 & 1 \\ 1 & 1 & 1 & 1 \\ -1 & -1 & -1 & -1 \\ -1 & -1 & -1 & -1 \end{pmatrix} + 0 \cdot \begin{pmatrix} 1 & 1 & -1 & -1 \\ 1 & 1 & -1 & -1 \\ -1 & -1 & 1 & 1 \\ -1 & -1 & 1 & 1 \end{pmatrix}
$$

$$
= \begin{pmatrix} 4+1+1 & 4+1+1 & 4-1+1 & 4-1+1 \\ 4+1+1 & 4+1+1 & 4-1+1 & 4-1+1 \\ 4+1-1 & 4+1-1 & 4-1-1 & 4-1-1 \\ 4+1-1 & 4+1-1 & 4-1-1 & 4-1-1 \end{pmatrix}
$$

$$
= \begin{pmatrix} 6 & 6 & 4 & 4 \\ 6 & 6 & 4 & 4 \\ 4 & 4 & 2 & 2 \\ 4 & 4 & 2 & 2 \end{pmatrix} \rightarrow \begin{pmatrix} 6 & 4 \\ 4 & 2 \end{pmatrix}.
$$

EXERCISES

Exercise 2.14. Compress the signal $\vec{s} := (2, 4, 8, 6)$.

Exercise 2.15. Compress the signal $\vec{s} := (5, 7, 3, 1)$.

Exercise 2.16. Compress the signal $\vec{s} := (8, 6, 7, 3, 1, 1, 2, 4)$.

Exercise 2.17. Compress the signal $\vec{s} := (3, 1, 9, 7, 7, 9, 5, 7)$.

Exercise 2.18. Compress the signal given by the In-Place Haar Wavelet Transform $\vec{s}^{(0)} := (6, 2, 1, 2)$.

Exercise 2.19. Compress the signal given by the In-Place Haar Wavelet Transform $\vec{s}^{(0)} := (4, 2, 2, 0)$.

Exercise 2.20. Compress the signal given by the In-Place Haar Wavelet Transform $\vec{s}^{(0)} := (4, -1, -1, 2, 0, 1, -2, -2)$.

Exercise 2.21. Compress the signal given by the In-Place Haar Wavelet Transform $\vec{s}^{(0)} := (5, 1, 1, 0, -3, -1, 0, 1)$.

Exercise 2.22. Compress the signal given by the second sweep of the In-Place Haar Wavelet Transform $\vec{s}^{(1)} := (3, 1, -1, 1, 7, 1, 1, -1)$.

Exercise 2.23. Compress the signal given by the first sweep of the In-Place Haar Wavelet Transform $\vec{s}^{(2)} := (2, 1, 2, 1, 8, -1, 8, -1)$.

2.2.3 Edge Detection

A simple method to detect edges, also called jumps or discontinuities, consists in locating the wavelet coefficients with the largest magnitudes, except for the average of the whole sample, because such coefficients mark the location of the largest jumps in the signal.

Example 2.19   As in Example 2.17 (page 54), from the sample

$$\vec{s} := (3, 1, 0, 4, 8, 6, 9, 9)$$

in Figure 2.11(a), the In-Place Haar Wavelet Transform produces

$$
\begin{aligned}
\vec{s}^{(0)} &= \left(a_0^{(0)}, \quad c_0^{(2)}, \quad c_0^{(1)}, \quad c_1^{(2)}, \quad c_0^{(0)}, \quad c_2^{(2)}, \quad c_1^{(1)}, \quad c_3^{(2)} \right) \\
&= (5, \qquad 1, \qquad 0, \qquad -2, \qquad -3, \qquad 1, \qquad -1, \qquad 0).
\end{aligned}
$$

Except for the average of the whole sample, $a_0^{(0)} = s_0^{(0)} = 5$, the wavelet coefficient with the largest magnitude lies in the entry $s_4^{(0)} = -3$. This coefficient corresponds to the term $s_4^{(0)} \cdot \psi_{[0,1[} = -3 \cdot \psi_{[0,1[}$, which marks a jump of size $(-3) \cdot (-2) = 6$, rising, at the middle of the sample, where the wavelet $\psi_{[0,1[}$ has its discontinuity. The wavelet coefficient with the next-largest magnitude lies in the entry $s_3^{(0)} = -2$. This coefficient corresponds to the term $s_3^{(0)} \cdot \psi_{[\frac{1}{4},\frac{1}{2}[} = -2 \cdot \psi_{[\frac{1}{4},\frac{1}{2}[}$, which marks a jump of size $(-2) \cdot (-2) = 4$, rising, at the third eighth of the sample, where the wavelet $\psi_{[\frac{1}{4},\frac{1}{2}[}$ has its discontinuity. Thus, retaining only the average and the wavelet coefficients with the two largest magnitudes gives a new wavelet transform,

$$\vec{s}^{(0)'''} = (5, 0, 0, -2, -3, 0, 0, 0).$$

The In-Place Inverse Haar Wavelet Transform then produces

$$\vec{s}''' = (2, 2, 0, 4, 8, 8, 8, 8),$$

which exhibits a rising edge of size 4 at the third eighth, and a rising edge of size 6 from the average over the first half to the average of the second half, but without the smaller oscillations of the initial sample $\vec{s} = (3, 1, 0, 4, 8, 6, 9, 9)$, as shown in Figures 2.11(a) and 2.11(b).   □

An alternative method of edge detection retains only the highest frequencies, perhaps with the sample average, and discards all the lower frequencies.

Example 2.20   From the sample $\vec{s} := (3, 1, 0, 4, 8, 6, 9, 9)$ in Figure 2.11(a), the In-Place Haar Wavelet Transform produces

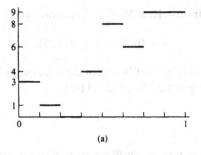

(a)

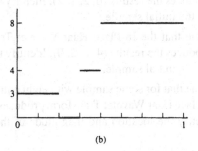

(b)

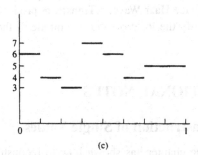

(c)

Figure 2.11   Example of edge-detection through wavelets. (a) Initial signal. (b) After retaining only the coefficients with the largest magnitudes, which retains the sharpest edges. (c) After retaining only the coefficients from the highest frequencies, which retains only the details at the smallest scale.

$$\vec{s}^{(0)} = \left(a_0^{(0)}, \quad c_0^{(2)}, \quad c_0^{(1)}, \quad c_1^{(2)}, \quad c_0^{(0)}, \quad c_2^{(2)}, \quad c_1^{(1)}, \quad c_3^{(2)} \right)$$
$$= (5, \qquad 1, \qquad 0, \qquad -2, \qquad -3, \qquad 1, \qquad -1, \qquad 0).$$

Retaining only the sample average, 5, and the coefficients of the highest-frequency wavelets, with odd indices, and setting to zero all the other coefficients, the method just described gives

$$\vec{s}^{(0)} = (5, 1, 0, -2, 0, 1, 0, 0),$$

whence the In-Place Inverse Haar Wavelet Transform yields

$$\vec{s} = (6, 4, 3, 7, 6, 4, 5, 5),$$

which reproduces the smaller oscillations of the initial sample, here with the same average value, as in Figures 2.11(a) and 2.11(c).   □

EXERCISES

Exercise 2.24. Assume that the In-Place Haar Wavelet Transform of a sample $\vec{s} = (s_0, s_1, s_2, s_3)$ produces the results $(6, 2, 1, 2)$. Identify the location and magnitude of the edges in the initial sample.

Exercise 2.25. Assume that the In-Place Haar Wavelet Transform of a sample $\vec{s} = (s_0, s_1, s_2, s_3)$ produces the results $(4, 2, 2, 0)$. Identify the location and magnitude of the edges in the initial sample.

Exercise 2.26. Assume that for some sample with eight entries $\vec{s} = (s_0, s_1, s_2, s_3, s_4, s_5, s_6, s_7)$, the In-Place Haar Wavelet Transform produces the result $(4, -1, -1, 2, 0, 1, -2, -2)$. Identify the location and magnitude of the edges in the initial sample.

Exercise 2.27. Assume that for some sample with eight entries $\vec{s} = (s_0, s_1, s_2, s_3, s_4, s_5, s_6, s_7)$, the In-Place Haar Wavelet Transform produces the result $(5, 1, 1, 0, -3, -1, 0, 1)$. Identify the location and magnitude of the edges in the initial sample.

2.3   COMPUTATIONAL NOTES

2.3.1   Fast Reconstruction of Single Values

Whereas the preceding chapter has shown how to reconstruct the entire initial sample $\vec{s} = (s_0, \ldots, s_{2^n-1})$ from the In-Place Haar Wavelet Transform $\vec{s}^{(0)} = (a_0^{(0)}, c_0^{(n-1)}, \ldots, c_{2^{(n-1)}-1}^{(n-1)})$, some applications may demand only the reconstruction of one value s_k or a few selected values of the sample.

Example 2.21   Assume that from a sample with four entries $\vec{s} = (s_0, s_1, s_2, s_3)$ of a signal g the In-Place Haar Wavelet Transform produces the result $(4, 2, -1, -3)$, which means that

$$\tilde{g} = 4 \cdot \varphi_{[0,1[} + (-1) \cdot \psi_{[0,1[} + 2 \cdot \psi_{[0,\frac{1}{2}[} + (-3) \cdot \psi_{[\frac{1}{2},1[}.$$

To detect a pattern in the separate reconstruction of the values of the sample, express the values of r_k in binary notation:

$$s_0 = \tilde{g}(r_0) = \tilde{g}(0) = \tilde{g}(0.00_{two})$$

$$= 4 \cdot \varphi_{[0,1[}(0) + (-1) \cdot \psi_{[0,1[}(0) + 2 \cdot \psi_{[0,\frac{1}{2}[}(0) + (-3) \cdot \psi_{[\frac{1}{2},1[}(0)$$

$$= 4 \cdot 1 + (-1) \cdot 1 + 2 \cdot 1 + (-3) \cdot 0 = 5;$$

$$s_1 = \tilde{g}(r_1) = \tilde{g}(\tfrac{1}{4}) = \tilde{g}(0.01_{\text{two}})$$

$$= 4 \cdot \varphi_{[0,1[}(\tfrac{1}{4}) + (-1) \cdot \psi_{[0,1[}(\tfrac{1}{4})$$

$$\quad + 2 \cdot \psi_{[0,\frac{1}{2}[}(\tfrac{1}{4}) + (-3) \cdot \psi_{[\frac{1}{2},1[}(\tfrac{1}{4})$$

$$= 4 \cdot 1 + (-1) \cdot 1 + 2 \cdot (-1) + (-3) \cdot 0 = 1;$$

$$s_2 = \tilde{g}(r_2) = \tilde{g}(\tfrac{1}{2}) = \tilde{g}(0.10_{\text{two}})$$

$$= 4 \cdot \varphi_{[0,1[}(\tfrac{1}{2}) + (-1) \cdot \psi_{[0,1[}(\tfrac{1}{2})$$

$$\quad + 2 \cdot \psi_{[0,\frac{1}{2}[}(\tfrac{1}{2}) + (-3) \cdot \psi_{[\frac{1}{2},1[}(\tfrac{1}{2})$$

$$= 4 \cdot 1 + (-1) \cdot (-1) + 2 \cdot 0 + (-3) \cdot 1 = 2;$$

$$s_3 = \tilde{g}(r_3) = \tilde{g}(\tfrac{3}{4}) = \tilde{g}(0.11_{\text{two}})$$

$$= 4 \cdot \varphi_{[0,1[}(\tfrac{3}{4}) + (-1) \cdot \psi_{[0,1[}(\tfrac{3}{4})$$

$$\quad + 2 \cdot \psi_{[0,\frac{1}{2}[}(\tfrac{3}{4}) + (-3) \cdot \psi_{[\frac{1}{2},1[}(\tfrac{3}{4})$$

$$= 4 \cdot 1 + (-1) \cdot (-1) + 2 \cdot 0 + (-3) \cdot (-1) = 8;$$

Thus, the binary expansion of r_k indicates both which wavelets have a nonzero value at r_k and whether that value equals 1 or -1:

Step 0. Start with the coefficient of $\varphi_{[0,1[}$, here 4.

Express $r_k = 0.b_1, b_2$ with binary digits b_1 and b_2.

Step 1. If $b_1 = 0$, then $0 \le r_k < \tfrac{1}{2}$, where $\psi_{[0,1[}(r_k) = 1$, therefore add the coefficient of $\psi_{[0,1[}$, here -1; hence, focus on the first half, $(4, 2)$, of the array of wavelet coefficients $(4, 2, -1, -3)$. In contrast, if $b_1 = 1$, then $\tfrac{1}{2} \le r_k < 1$, where $\psi_{[0,1[}(r_k) = -1$, therefore subtract the coefficient of $\psi_{[0,1[}$, here -1; hence, focus on the second half, $(-1, -3)$, of the array of wavelet coefficients $(4, 2, -1, -3)$.

Step 2. Refer to the half array of coefficients determined by the preceding step. If $b_2 = 0$, add the last coefficient in that half. In contrast, if $b_2 = 1$, subtract the last coefficient in that half. □

The pattern just observed leads to the following algorithm to reconstruct only one value s_k, or a few selected values of the sample.

Algorithm 2.22 Single Values From In-Place Haar Wavelet Transforms.

DATA:

n	(nonnegative integer)
$\tilde{s}^{(0)}$	(array of 2^n numbers)
$r_k = (0.b_1 b_2 \ldots b_n)_{\text{two}}$	(binary representation of r_k)

START.

$s_k := s_0^{(0)}$ (initialization: coefficient of $\varphi_{[0,1[}$)

$I := 2^{(n-2)}$ (initialization: first half of array)

$J := 2^{(n-1)}$ (initialization: half array length)

FOR $L := 1, \ldots, n$ DO   (loops of binary places)

  IF $b_L = 0$ THEN   (r_k lies in the left-hand half)

    $s_k := s_k + s_J^{(0)}$ (where the wavelet has value $+1$)

    $J := J - I$ (choose the left-hand 2^{1-L}th of the array)

  ELSE $(b_L = 1)$ (r_k lies in the right-hand half)

    $s_k := s_k - s_J^{(0)}$ (where the wavelet has value -1)

    $J := J + I$ (choose the right-hand 2^{1-L}th of the array)

  END IF

  IF $L < n$ THEN

    $I := I/2$ (halve I)

  END IF

END   (loop of binary digits)

STOP.

RESULT: $s_k = s_k^{(n)}$ □

Example 2.23   Assume that for a sample with eight entries $\vec{s} = (s_0, s_1, s_2, s_3, s_4, s_5, s_6, s_7)$, the In-Place Haar Wavelet Transform is $(5, 1, 0, -2, -3, 1, -1, 0)$. Consider the reconstruction of s_5:

DATA:

  $n := 3,$

  $\vec{s}^{(0)} = (5, 1, 0, -2, -3, 1, -1, 0),$

  $r_5 = 5/2^n = \frac{5}{8} = 0.101_{\text{two}}.$

START.

  $s_5 := s_0^{(0)} = 5,$

  $I := 2^{(n-1)} = 2^2 = 4,$

  $J := I = 2^2 = 4,$

  $L := 1:$

    $I := I/2 = 4/2 = 2,$

    $r_5 = 0.101_{\text{two}}, b_1 = 1,$

    $s_5 = s_5 - s_J^{(0)} = s_5 - s_4^{(0)} = 5 - (-3) = 8,$

    $J := J + I = 4 + 2 = 6,$

  $L := 2:$

    $I := I/2 = 2/2 = 1,$

    $r_5 = 0.101_{\text{two}}, b_2 = 0,$

    $s_5 = s_5 + s_J^{(0)} = s_5 + s_6^{(0)} = 8 + (-1) = 7,$

    $J := J - I = 4 + 1 = 5,$

  $L := 3:$

    $r_5 = 0.101_{\text{two}}, b_3 = 1,$

    $s_5 = s_5 - s_J^{(0)} = s_5 - s_5^{(0)} = 7 - 1 = 6.$

RESULT: $s_5 = 6$, which is exactly the data in Example 2.20.   □

2.3.2 Operation Count

Practical uses of the Fast Haar Wavelet Transform may require a knowledge in advance of the total number of arithmetic operations involved, for instance, to estimate in advance the time necessary for the transform. To calculate the total number of arithmetic operations, consider each sweep separately. The ℓth sweep of the Fast Haar Wavelet Transform begins with an array $\vec{s}^{(\ell-1)}$ with 2^n entries, but operates only on the subarray $\vec{a}^{(\ell-1)}$, which contains exactly $2^{n-(\ell-1)}$ entries grouped in $2^{n-\ell}$ consecutive pairs. The ℓth sweep adds and subtracts the two entries in each of the $2^{n-\ell}$ pairs and divides the sum and the difference by 2, which involves the following number of operations:

Number of Operations During the ℓth Sweep.

(+) $2^{n-\ell}$ additions,

(−) $2^{n-\ell}$ subtractions,

(÷) $2^{n-(\ell-1)}$ divisions by 2.

Adding the preliminary totals just obtained for all the sweeps gives the total number of operations for the complete transform.

Total Number of Operations For the Haar Wavelet Transform.

(+) $\sum_{\ell=1}^{n} 2^{n-\ell} = (2^n - 1)/(2 - 1) = 2^n - 1$ additions,

(−) $\sum_{\ell=1}^{n} 2^{n-\ell} = (2^n - 1)/(2 - 1) = 2^n - 1$ subtractions,

(÷) $\sum_{\ell=1}^{n} 2^{n-(\ell-1)} = 2 \cdot (2^n - 1)/(2 - 1) = 2^{n+1} - 2$ divisions by 2.

With binary arithmetic, divisions by 2 amount to a shift by one place in the exponent and thus do not require as many manipulations of digits as additions and subtractions. Consequently, with binary arithmetic, the complete Fast Haar Wavelet Transform of 2^n numbers involves mostly $2^n - 1$ additions and $2^n - 1$ subtractions.

Depending on the method employed to store and retrieve entries in the array $\vec{s}^{(\ell)}$, however, the calculations of the indices $J \cdot K + I$ and access to the entries $s_{J \cdot K + I}$ may also demand substantial resources.

EXERCISES

Exercise 2.28. Prove that setting to zero any *wavelet* coefficient (but not the average) in the Haar Wavelet Transform of any data, and then performing the Inverse Haar Wavelet Transform, produces a modified data that still has the same average as the initial data.

Exercise 2.29. Count the number of arithmetic operations for the Two-Dimensional and Three-Dimensional Fast Haar Wavelet Transforms.

Exercise 2.30. **(a)** Calculate the total number of arithmetic operations for one fast reconstruction of one single value.

(b) Calculate the total number of arithmetic operations to reconstruct all the sample values by reconstructing each value separately. Compare the result with that for the fast wavelet transform.

Exercise 2.31. Prove that the In-Place Haar Wavelet Transform and the Ordered Haar Wavelet Transform involve the same number of arithmetic operations.

Exercise 2.32. Prove that the Fast Inverse Haar Wavelet Transform involves the same number of additions and subtractions as the Fast Haar Wavelet Transform.

Exercise 2.33. Consider the Haar Wavelet Transform $(a_0^{(0)}, c_0^{(0)})$ of a pair of data (s_0, s_1):

$$a_0^{(0)} = \frac{s_0 + s_1}{2}, \qquad c_0^{(0)} = \frac{s_0 - s_1}{2}.$$

(a) Verify, algebraically, logically, or otherwise, that the following alternative algorithm produces the same transform:

$$c_0^{(0)} := \frac{s_0 - s_1}{2}, \qquad a_0^{(0)} := c_0^{(0)} + s_1.$$

(b) Verify, algebraically, logically, or otherwise, that the following alternative algorithm produces the same transform:

$$a_0^{(0)} := \frac{s_0 + s_1}{2}, \qquad c_0^{(0)} := a_0^{(0)} - s_1.$$

(c) (This item requires a working knowledge of floating-point arithmetic.) With the approximate arithmetic of digital computers with finitely many digits, for instance, m digits, verify that if the digital expansions of s_0 and s_1 differ from each other only in their last digit, so that

$$s_0 = 0.d_1 d_2 \ldots d_{m-1} p_m,$$

$$s_1 = 0.d_1 d_2 \ldots d_{m-1} q_m,$$

then the preceding alternative algorithm can produce erroneous results, with incorrect sign or incorrect magnitude. For example, with two $(m = 2)$ binary digits, $s_0 := 0.10_{\text{two}}$ and $s_1 := 0.11_{\text{two}}$, so that

$$\texttt{float}(a_0^{(0)}) = [\texttt{float}(s_0 + s_1)]/2 = [\texttt{float}(1.01_{\text{two}})]/[10_{\text{two}}]$$

rounds to the nearest even number with only two significant digits:

$$[\texttt{float}(1.0_{\text{two}})]/[10_{\text{two}}] = 0.11_{\text{two}},$$

whence the second alternative algorithm yields

$$\texttt{float}(c_0^{(0)}) = \texttt{float}[\texttt{float}(a_0^{(0)}) - s_1]$$

$$= \texttt{float}[0.10_{\text{two}} - 0.11_{\text{two}}] = -0.01_{\text{two}}.$$

In contrast, the basic Haar transform gives the exact value:

$$\texttt{float}(c_0^{(0)}) = \texttt{float}\left(\frac{s_0 - s_1}{2}\right)$$

$$= \texttt{float}\left(\frac{0.10_{two} - 0.11_{two}}{2}\right)$$

$$= \texttt{float}\left(\frac{-0.01_{two}}{2}\right)$$

$$= -0.01_{two} * 2^{-1} = -0.001_{two}.$$

Show that the same considerations apply to the first alternative algorithm, with $-s_1$ instead of s_1.

2.4 EXAMPLES

This section shows applications of mathematical wavelets with real data. Any other finite sequence of—possibly random—numbers might serve the same purpose, but the specific contexts demonstrated here may help in providing suggestions for further applications.

2.4.1 Creek Water Temperature Compression

This example serves mainly to demonstrate the use of wavelets to compress—with some loss of information—data sets.

The following sixteen numbers represent semiweekly measurements of temperature, in degrees Fahrenheit, for December 1992 and January 1993 at a fixed common location along Hangman Creek near Spokane, Washington, during a study of riverbank erosion by Jim Fox, of Spokane, Washington.

32.0, 10.0, 20.0, 38.0, 37.0, 28.0, 38.0, 34.0,

18.0, 24.0, 18., 9.0, 23.0, 24.0, 28.0, 34.0

The complete In-Place Haar Wavelet Transform produces the result

25.9375	11.0	−4.0	−9.0	−4.625	4.5	−1.75	2.0
3.6875	−3.0	3.75	4.5	−5.0	−0.5	−3.75	−3.0

Setting to zero the coefficients with magnitude equal to at most 4 yields the modified In-Place Haar Wavelet Transform

25.9375	11.0	0.0	−9.0	−4.625	4.5	0.0	0.0
0.0	0.0	0.0	4.5	−5.0	0.0	0.0	0.0

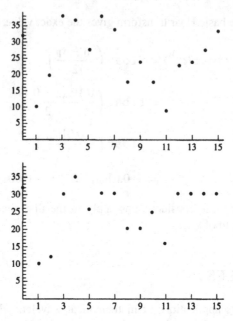

Figure 2.12   Hangman Creek Water Temperature. *Top:* semiweekly records of temperature (°F). *Bottom:* records compressed in the ratio $\frac{1}{2}$, but with alteration of information.

Retaining only the nonzero modified coefficients produces a compressed version of the data, with only one-half as many coefficients as the data had. Applying the In-Place Inverse Haar Wavelet Transform to the compressed (modified) transform produces an approximation to the initial data:

$$32.3125, \quad 10.3125, \quad 12.3125, \quad 30.3125$$
$$35.0625, \quad 26.0625, \quad 30.5625, \quad 30.5625$$
$$20.9375, \quad 20.9375, \quad 25.4375, \quad 16.4375$$
$$30.9375, \quad 30.9375, \quad 30.9375, \quad 30.9375.$$

Figure 2.12 displays the result for comparison.

EXERCISES

Exercise 2.34. Compress the following measurements of the ground frost depth, in centimeters, at Qualchan on Hangman Creek, for the same period (also by Jim Fox).

$$22.0, \ 27.0, \ 48.8, \ 47.5, \ 47.0, \ 48.5, \ 48.0, \ 47.0,$$
$$43.0, \ 41.0, \ 41.0, \ 38.0, \ 36.0, \ 47.1, \ 34.0, \ 32.0$$

Exercise 2.35. Compress the following measurements of the ground frost depth, in centimeters, at Kracher on Hangman Creek, for the same period (also by Jim Fox).

12.0, 16.0, 27.0, 32.8, 33.5, 33.5, 39.0, 39.0,

40.0, 41.3, 41.3, 42.0, 43.0, 45.0, 35.5, 49.0

Exercise 2.36. Compress the following measurements of river flow, in cubic feet per second, at the US Geological Survey Data Station 1242400 on Hangman Creek, for the same period.

10.0, 12.0, 12.0, 7.0, 8.0, 9.1, 8.2, 9.4,

16.0, 15.0, 13.0, 11.0, 6.4, 9.0, 19.0, 118.0

Exercise 2.37. Obtain data of any kind and compress them with the Haar Wavelet Transform.

2.4.2 Financial Stock Index Image Compression

This example demonstrates the use of Haar's Wavelet Transforms to compress large data sets by annihilating small coefficients.

The top left panel in Figure 2.13 displays on the vertical axis the New York Stock Exchange (NYSE) Composite Index, and on the horizontal axis the date, from 2 January 1981 (business day 0) through 7 February 1988 (business day 2047). The top right panel shows the In-Place Haar Wavelet Transform, without the average to rescale details.

The middle right panel in Figure 2.13 shows the coefficients of the In-Place Haar Wavelet Transform after annihilation of all 905 coefficients with relative magnitude less than 1% of the largest coefficient (average excepted). The remaining 1043 coefficients occupy 44% less storage space than the initial 2048 data points. The middle left panel displays the reconstruction after application of the inverse transform to the compressed transform. The compression has altered the information in the data, but for the purpose of displaying graphics the compression shows little difference from the initial data.

The bottom right panel in Figure 2.13 shows the coefficients of the In-Place Haar Wavelet Transform after annihilation of all 2004 coefficients with relative magnitude less than 10% of the largest coefficient (average excepted). The remaining 44 coefficients occupy 98% less storage space than the initial 2048 data points. The bottom left panel displays the reconstruction, after application of the inverse transform to the compressed transform. The compression has further altered the information in the data, but may remain adequate for some graphics.

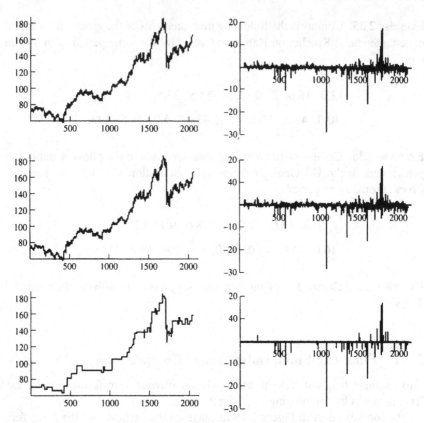

Figure 2.13   New York Stock Exchange Composite Index for 1981–1987. *Top left.* Data: index vs. business day. *Top right.* In-place Haar Wavelet Transform. *Middle left.* Compression ratio $905/2048 \approx 44\%$. *Middle right.* Annihilation of all coefficients with magnitude less than 1% of the wavelet coefficient with the largest magnitude. *Bottom left.* Compression ratio $501/512 \approx 98\%$. *Bottom right.* Annihilation of all coefficients with magnitude less than 10% of the wavelet coefficient with the largest magnitude.

2.4.3   Two-Dimensional Diffusion Analysis

The following data—collected by Dr. Mark F. Dubach [9]—show the diffusion of dopamine one hour after injection into brain tissue, measured on a grid centered near the injection site and with squares of sides 1 mm long. The units correspond to scintillation counts, with one scintillation count representing $4.753 * 10^{-13}$ mole of dopamine. See Table 2.1.

Table 2.1   Densities Measured on a Two-Dimensional Grid

480.0	7022.0	14411.0	51580.0
20910.0	230270.0	28353.0	13138.0
789.0	21260.0	20921.0	11731.0
213.0	1303.0	3765.0	1715.0

Table 2.2   Two-Dimensional *In-Place* Haar Wavelet Transform

9711.0625	−6869.5000	**−2687.9375**	6117.0000
−4404.0000	3598.5000	−5480.5000	−1490.5000
1998.9375	−5390.2500	**−867.0625**	2810.0000
5133.2500	−4845.2500	6793.0000	1785.0000

The Two-Dimensional *In-Place* Haar Wavelet Transform produced the result listed in Table 2.2 and displayed in Figure 2.14.

The coefficient in the upper left-hand corner, **9711.0625**, represents the average of all the measurements. The coefficient −**2687.9375** indicates an increase by −2687.9375 ∗ (−2) = 5375.875 from the left-hand half to the right-hand half of the sample. The coefficient **1998.9375** indicates a change by 1998.9375 ∗ (−2) = −3997.875, a drop, from the top half to the bottom half of the sample. The coefficient −**867.0625** indicates smaller values, −867.0625 ∗ (−2) = 1734.125, along the diagonal ridge from the top left-hand corner to the bottom right-hand half of the sample than along the opposite diagonal. The other coefficients have a similar significance within each 2 × 2 corner block.

A plot of the data, as in Figure 2.14, corroborates the results.

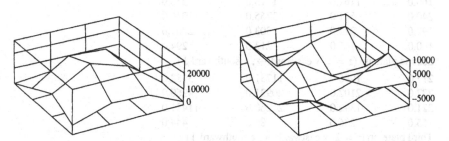

Figure 2.14   *Left.* Plot of dopamine versus location. *Right.* In-Place Two-Dimensional Haar Wavelet Transform.

2.4.4   Three-Dimensional Diffusion Analysis

Listed in Table 2.3 and graphed in Figure 2.15, the following data—from the study by Dr. Mark F. Dubach [9]—show the diffusion of dopamine one hour

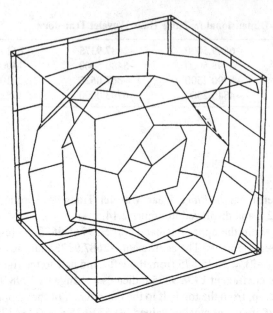

Figure 2.15 Equidensity surfaces of dopamine versus location. The innermost shell approximates the surface where the density has the value 20,000. The intermediate shell approximates the surface where the density has the value 10,000. The outermost shell approximates the surface where the density has the value 1,000.

Table 2.3   Densities Measured on a Three-Dimensional Grid

Bottom plate ($z = 0$, y eastward $\rightarrow$, x southward $\downarrow$):

166.0	1161.0	1715.0	258.0
240.0	581.0	3765.0	1036.0
192.0	224.0	1303.0	2061.0
160.0	166.0	213.0	294.0

Second plate up ($z = 1$, y eastward $\rightarrow$, x southward $\downarrow$):

217.0	1986.0	11731.0	3182.0
287.0	2102.0	20921.0	14960.0
221.0	1392.0	21260.0	11431.0
155.0	236.0	789.0	444.0

Third plate up ($z = 2$, y eastward $\rightarrow$, x southward $\downarrow$):

231.0	1099.0	13138.0	4785.0
358.0	4435.0	28353.0	19742.0
183.0	935.0	23027.0	15531.0
144.0	201.0	2091.0	2137.0

Top plate ($z = 3$, y eastward $\rightarrow$, x southward $\downarrow$):

143.0	406.0	5158.0	1048.0
161.0	532.0	14411.0	6098.0
182.0	218.0	7022.0	4055.0
170.0	169.0	480.0	712.0

Table 2.4   Three-Dimensional *In-Place* Haar Wavelet Transform

Bottom plate ($z = 0$, y eastward →, x southward ↓):

4096.93750	−615.00000	−3501.53125	2337.00000
40.00000	−76.00000	−2974.50000	164.50000
1040.75000	−161.25000	−754.59375	1166.87500
164.00000	−139.50000	4289.37500	1100.87500

Second plate up ($z = 1$, y eastward →, x southward ↓):

−305.50000	281.00000	−5502.50000	−1290.50000
86.50000	−87.50000	2267.50000	−482.50000
−157.75000	151.75000	−3756.62500	−1376.62500
−141.50000	133.00000	−3575.12500	−1270.12500

Third plate up ($z = 2$, y eastward →, x southward ↓):

−820.40625	−697.37500	**817.87500**	3673.37500
−450.87500	414.62500	−5559.37500	−557.62500
−298.03125	−105.50000	261.50000	1273.12500
104.25000	−91.50000	5526.87500	1342.62500

Top plate ($z = 3$, y eastward →, x southward ↓):

610.12500	−538.87500	4912.87500	567.62500
−414.87500	387.62500	−1983.62500	493.12500
90.50000	−96.75000	3814.62500	589.37500
89.00000	−82.25000	3055.62500	542.87500

after injection into brain tissue, measured on a three-dimensional grid of parallelepipeds with sides 1 mm long (horizontal first coordinate, x, and vertical third coordinates, z) and 0.76 mm deep (along the horizontal second coordinate, y). The units correspond to scintillation counts, with one scintillation count representing $4.753 * 10^{-13}$ mole of dopamine.

The Three-Dimensional *In-Place* Haar Wavelet Transform produced the results in Table 2.4.

The first coefficient, **4096.93750**, gives the arithmetic average of all the data. It accompanies the constant function $\varphi_{[0,1[} \otimes \varphi_{[0,1[} \otimes \varphi_{[0,1[}$ over the entire three-dimensional grid of data.

Near the center of Table 2.4, the coefficient **817.87500** corresponds to the wavelet $\psi_{[0,1[} \otimes \psi_{[0,1[} \otimes \psi_{[0,1[}$. Thus, the value **817.87500** contributes to the average of the eight data points closest to the origin,

166.0	1161.0
240.0	581.0
217.0	1986.0
287.0	2102.0,

and the average of the eight data points farthest away from the origin,

$$
\begin{array}{cc}
23027.0 & 15531.0 \\
2091.0 & 2137.0 \\
7022.0 & 4055.0 \\
480.0 & 712.0.
\end{array}
$$

The last coefficient, 542.875, in the lower right-hand corner of the bottom plate in Table 2.4, accompanies the tensor products of wavelets $\psi_{[\frac{1}{2},1[} \otimes \psi_{[\frac{1}{2},1[} \otimes \psi_{[\frac{1}{2},1[}$, which contributes to the drop, diagonally and upward, from the data point 23,027 to 712.

EXERCISES

Exercise 2.38. Design an algorithm to compute an Ordered *or* In-Place Three-Dimensional Fast Haar Wavelet Transform.

Exercise 2.39. Design an algorithm to compute an Ordered *or* In-Place Three-Dimensional Fast *Inverse* Haar Wavelet Transform.

Exercise 2.40. Write and test a computer program for the Ordered *or* In-Place Three-Dimensional Fast Haar Wavelet Transform.

Exercise 2.41. Write and test a computer program for the Ordered *or* In-Place Three-Dimensional Fast *Inverse* Haar Wavelet Transform.

CHAPTER 3

Algorithms for Daubechies Wavelets

3.0 INTRODUCTION

This chapter presents algorithms to calculate transforms with wavelets introduced by Ingrid Daubechies. In contrast to Haar's simple-step wavelets, which exhibit jump discontinuities, Daubechies wavelets are continuous. As a consequence of their continuity, Daubechies wavelets approximate continuous signals more accurately with fewer wavelets than do Haar's wavelets, but at the cost of intricate algorithms based upon a sophisticated theory. Therefore, to ease the transition from Haar wavelets to Daubechies wavelets, the present material postpones to a subsequent chapter the theoretical considerations that led to such wavelets, and focuses first upon a description of algorithms to calculate the Daubechies wavelet transform. Some logical derivations involve matrix algebra.

To clarify and shorten the notation, the concept of the "inner product" of two vectors (one-dimensional arrays) will prove effective.

Definition 3.1 For each pair of one-dimensional arrays of real numbers $\vec{v} = (v_1, v_2, \ldots, v_{n-1}, v_n)$ and $\vec{w} = (w_1, w_2, \ldots, w_{n-1}, w_n)$ with the same length, their **inner product** is the number denoted by $\langle \vec{v}, \vec{w} \rangle$ and defined by the sum of products

$$\langle \vec{v}, \vec{w} \rangle := v_1 w_1 + v_2 w_2 + \cdots + v_{n-1} w_{n-1} + v_n w_n. \qquad \square$$

3.1 CALCULATION OF DAUBECHIES WAVELETS

To design continuous wavelets amenable to a fast transform within an interval with finite length, Ingrid Daubechies introduced a "basic building block," or "scaling" function, denoted here by φ and shown in Figure 3.1. (The phrase "scaling function" will receive an explanation in a subsequent chapter, while the phrase "building block" will be explained in this chapter.) To enable the corresponding

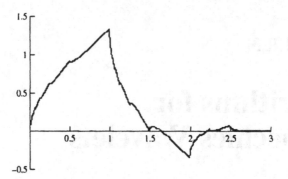

Figure 3.1   Ingrid Daubechies' basic building block φ.

wavelets to analyze or synthesize limited portions of signals, the building block φ equals zero outside the interval from 0 to 3, so that

$$\varphi(r) = 0 \text{ if } r \leq 0 \text{ or } 3 \leq r.$$

Daubechies proved that the building block function φ does not admit any algebraic formula in terms of elementary mathematical functions [6], [7]; indeed, φ belongs to a class of mathematical functions different from the class of functions consisting of polynomials, trigonometric, exponential, elliptic, and special functions from engineering and physics. Yet Daubechies also demonstrated that φ satisfies several algebraic relations that prove as useful as formulae for the purpose of calculations. For instance, starting with the initial values

$$\varphi(0) := 0,$$

$$\varphi(1) := \frac{1 + \sqrt{3}}{2},$$

$$\varphi(2) := \frac{1 - \sqrt{3}}{2}, \tag{3.1}$$

$$\varphi(3) := 0,$$

the building block φ satisfies the recurrence relation

$$\varphi(r) = \frac{1 + \sqrt{3}}{4} \varphi(2r) + \frac{3 + \sqrt{3}}{4} \varphi(2r - 1)$$

$$+ \frac{3 - \sqrt{3}}{4} \varphi(2r - 2) + \frac{1 - \sqrt{3}}{4} \varphi(2r - 3). \tag{3.2}$$

The initial values add up to 1,

$$\varphi(0) + \varphi(1) + \varphi(2) + \varphi(3) = 0 + \frac{1+\sqrt{3}}{2} + \frac{1-\sqrt{3}}{2} + 0 = 1,$$

so that the values of φ may serve as averaging or weighting factors.

Remark 3.2   The approach adopted here—defining φ with initial values and a recursion—resembles a similar approach for other mathematical functions, for example, defining the exponential function exp with initial condition and differential equation

$$\exp(0) = 1,$$

$$\exp' = \exp,$$

then investigating what features such a function must have if it exists, and eventually proving its existence and uniqueness.   □

The following abbreviations will prove convenient:

$$\begin{aligned} h_0 &:= \frac{1+\sqrt{3}}{4}, \\ h_1 &:= \frac{3+\sqrt{3}}{4}, \\ h_2 &:= \frac{3-\sqrt{3}}{4}, \\ h_3 &:= \frac{1-\sqrt{3}}{4}. \end{aligned} \qquad (3.3)$$

With the abbreviations (3.3) just introduced, the recurrence relation (3.2) takes the form of an inner product with (h_0, h_1, h_2, h_3):

$$\varphi(r) = h_0 \cdot \varphi(2r) + h_1 \cdot \varphi(2r-1) + h_2 \cdot \varphi(2r-2) + h_3 \cdot \varphi(2r-3).$$

Example 3.3   This example shows how to calculate other values of φ, from the initial values (3.1), the recursions (3.2) and (3.3):

$$\varphi(\tfrac{1}{2}) = h_3\varphi(1-3) + h_2\varphi(1-2) + h_1\varphi(1-1) + h_0\varphi(1-0)$$

$$= h_3 \cdot 0 + h_2 \cdot 0 + h_1 \cdot 0 + \frac{1+\sqrt{3}}{4} \cdot \frac{1+\sqrt{3}}{2}$$

$$= \frac{2+\sqrt{3}}{4},$$

$$\varphi(\tfrac{3}{2}) = h_3\varphi(3-3) + h_2\varphi(3-2) + h_1\varphi(3-1) + h_0\varphi(3-0)$$

$$= h_3 \cdot 0 + \frac{3-\sqrt{3}}{4} \cdot \frac{1+\sqrt{3}}{2} + \frac{3+\sqrt{3}}{4} \cdot \frac{1-\sqrt{3}}{2} + h_0 \cdot 0$$

$$= 0,$$

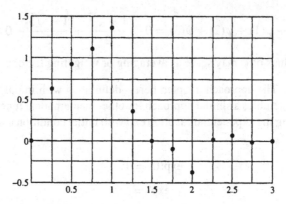

Figure 3.2 Computed by recursion, 12 points on Daubechies' building block φ.

$$\varphi(\tfrac{5}{2}) = h_3\varphi(5-3) + h_2\varphi(5-2) + h_1\varphi(5-1) + h_0\varphi(5-0)$$

$$= \frac{1-\sqrt{3}}{2} \cdot \frac{1-\sqrt{3}}{2} + h_2 \cdot 0 + h_1 \cdot 0 + h_0 \cdot 0$$

$$= \frac{2-\sqrt{3}}{4}.$$

A further recursion gives the values of φ at multiples of $\tfrac{1}{4}$, as illustrated in Figure 3.2:

r	$\tfrac{1}{4}$	$\tfrac{3}{4}$	$\tfrac{5}{4}$	$\tfrac{7}{4}$	$\tfrac{9}{4}$	$\tfrac{11}{4}$
$\varphi(r)$	$\dfrac{5+3\sqrt{3}}{16}$	$\dfrac{9+5\sqrt{3}}{16}$	$\dfrac{1+\sqrt{3}}{8}$	$\dfrac{1-\sqrt{3}}{8}$	$\dfrac{9-5\sqrt{3}}{16}$	$\dfrac{5-3\sqrt{3}}{16}$

Similar recursions yields the 3073 points outlining φ in Figure 3.3. □

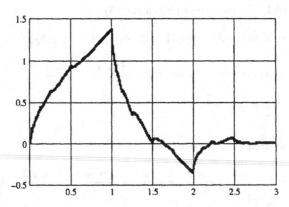

Figure 3.3 Computed by recursion, 3073 points on Daubechies' building block φ.

The function φ serves as the basic building block for its associated wavelet, denoted by ψ, and defined by the following recursion:

$$
\begin{aligned}
\psi(r) &:= -\frac{1+\sqrt{3}}{4}\varphi(2r-1) + \frac{3+\sqrt{3}}{4}\varphi(2r) \\
&\quad -\frac{3-\sqrt{3}}{4}\varphi(2r+1) + \frac{1-\sqrt{3}}{4}\varphi(2r+2) \\
&= -h_0\varphi(2r-1) + h_1\varphi(2r) - h_2\varphi(2r+1) + h_3\varphi(2r+2) \\
&= (-1)^{(1)}h_{1-1}\varphi(2r-1) + (-1)^{(0)}h_{1-0}\varphi(2r-0) \\
&\quad + (-1)^{-1}h_{1-[-1]}\varphi(2r-[-1]) + (-1)^{-2}h_{1-[-2]}\varphi(2r-[-2]).
\end{aligned}
\tag{3.4}
$$

Because $\varphi(r) = 0$ if $r \leq 0$ or $3 \leq r$, it follows that $\psi(r) = 0$ if $2r + 2 \leq 0$ or $3 \leq 2r - 1$, or, equivalently, if $r \leq -1$ or $2 \leq r$. For values r such that $-1 < r < 2$, the recursion (3.4) yields $\psi(r)$.

Example 3.4 This example demonstrates how to calculate other values of the Daubechies wavelet ψ from the initial values of the basic building block φ in equation (3.1) and from the recursion (3.4):

$$
\begin{aligned}
\psi(0) &= -h_0\varphi(2\cdot0-1) + h_1\varphi(2\cdot0) - h_2\varphi(2\cdot0+1) + h_3\varphi(2\cdot0+2) \\
&= -\frac{1+\sqrt{3}}{4}\cdot0 + \frac{3+\sqrt{3}}{4}\cdot0 - \frac{3-\sqrt{3}}{4}\cdot\frac{1+\sqrt{3}}{2} + \frac{1-\sqrt{3}}{4}\cdot\frac{1-\sqrt{3}}{2} \\
&= -0 + 2\frac{\sqrt{3}}{8} + \frac{4-2\sqrt{3}}{8} \\
&= \frac{1-\sqrt{3}}{2},
\end{aligned}
$$

$$
\begin{aligned}
\psi(1) &= -h_0\varphi(2\cdot1-1) + h_1\varphi(2\cdot1) - h_2\varphi(2\cdot1+1) + h_3\varphi(2\cdot1+2) \\
&= -\frac{1+\sqrt{3}}{4}\cdot\frac{1+\sqrt{3}}{2} + \frac{3+\sqrt{3}}{4}\cdot\frac{1-\sqrt{3}}{2} - \frac{3-\sqrt{3}}{4}\cdot0 + \frac{1-\sqrt{3}}{4}\cdot0 \\
&= -\frac{4+2\sqrt{3}}{8} + \frac{0-2\sqrt{3}}{8} \\
&= -\frac{1+\sqrt{3}}{2},
\end{aligned}
$$

$$
\begin{aligned}
\psi(-\tfrac{1}{2}) &= -h_0\varphi(2\cdot[-\tfrac{1}{2}]-1) + h_1\varphi(2\cdot[-\tfrac{1}{2}]) \\
&\quad - h_2\varphi(2\cdot[-\tfrac{1}{2}]+1) + h_3\varphi(2\cdot[-\tfrac{1}{2}]+2) \\
&= -\frac{1+\sqrt{3}}{4}\cdot0 + \frac{3+\sqrt{3}}{4}\cdot0 - \frac{3-\sqrt{3}}{4}\cdot0 + \frac{1-\sqrt{3}}{4}\cdot\frac{1+\sqrt{3}}{2} \\
&= \frac{-1}{4},
\end{aligned}
$$

$$\psi(\tfrac{1}{2}) = -h_0\varphi(2 \cdot [\tfrac{1}{2}] - 1) + h_1\varphi(2 \cdot [\tfrac{1}{2}])$$

$$- h_2\varphi(2 \cdot [\tfrac{1}{2}] + 1) + h_3\varphi(2 \cdot [\tfrac{1}{2}] + 2)$$

$$= -\frac{1+\sqrt{3}}{4} \cdot 0 + \frac{3+\sqrt{3}}{4} \cdot \frac{1+\sqrt{3}}{2} - \frac{3-\sqrt{3}}{4} \cdot \frac{1-\sqrt{3}}{2} + \frac{1-\sqrt{3}}{4} \cdot 0$$

$$= \frac{6+4\sqrt{3}}{8} - \frac{6-4\sqrt{3}}{8}$$

$$= \sqrt{3},$$

$$\psi(\tfrac{3}{2}) = -h_0\varphi(2 \cdot [\tfrac{3}{2}] - 1) + h_1\varphi(2 \cdot [\tfrac{3}{2}])$$

$$- h_2\varphi(2 \cdot [\tfrac{3}{2}] + 1) + h_3\varphi(2 \cdot [\tfrac{3}{2}] + 2)$$

$$= -\frac{1+\sqrt{3}}{4} \cdot \frac{1-\sqrt{3}}{2} + \frac{3+\sqrt{3}}{4} \cdot 0 - \frac{3-\sqrt{3}}{4} \cdot 0 + \frac{1-\sqrt{3}}{4} \cdot 0$$

$$= \tfrac{1}{4},$$

or, in tabular form,

r	$\ldots, -1$	$-\tfrac{1}{2}$	0	$\tfrac{1}{2}$	1	$\tfrac{3}{2}$	$2, \ldots$
$\psi(r)$	$\ldots, 0$	$-\tfrac{1}{4}$	$\dfrac{1-\sqrt{3}}{2}$	$\sqrt{3}$	$-\dfrac{1+\sqrt{3}}{2}$	$-\tfrac{1}{4}$	$0, \ldots$

Similarly, the same recursion yields the 3073 points that outline the function ψ in Figure 3.4.   □

The foregoing definition of the wavelet ψ conforms to Daubechies' initial presentation [7, p. 197]. Alternatively, a shift of ψ by 1 to the right, so that $\psi(r) = 0$ for $r \leq 0$ or $3 \leq r$, as for $\varphi(r)$ [4, pp. 12 and 195], gives the shifted wavelet shown in Figure 3.5.

The recursions just presented yield the values of the building block $\varphi(r)$ and of the wavelet $\psi(r)$ only at integral multiples of positive or negative powers of 2. Such values suffice for equally spaced samples from a signal, because the signal remains unknown between sample values. Because of the importance of such powers of 2, the following notation and terminology will prove convenient.

Definition 3.5   A number r is called a **dyadic number** if, but only if, it is an integral multiple of an integral power of 2. Denote the set of all dyadic numbers by $\mathbb{D}$, and for each negative, zero, or positive integer n, denote by $\mathbb{D}_n$ the set of all integral multiples of 2^{-n}. Equivalently, a dyadic number has a finite binary expansion, and a dyadic number in $\mathbb{D}_n$ has a binary expansion with at most n binary digits past the "binary" point.   □

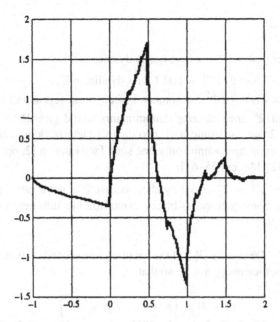

Figure 3.4 Computed by recursion, 3073 points on the Daubechies wavelet ψ.

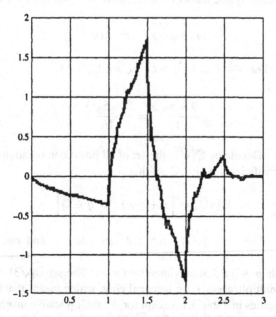

Figure 3.5 The Daubechies wavelet shifted by 1 to the right: $r \mapsto \psi(r-1)$.

Example 3.6

(a) The number $\frac{3}{8} = 3 \cdot 2^{-3} = 0.011_{two}$ is dyadic, in $\mathbb{D}_3$.

(b) The number $4.5 = 9 \cdot 2^{-1} = 100.1_{two}$ is dyadic, in $\mathbb{D}_1$.

(c) The number $\frac{1}{3}$ is *not* a dyadic number. Indeed, if integers m and n existed such that $\frac{1}{3} = m/2^n$, then clearing denominators would give $2^n = 3m$, whence the integer $2^n = 3m$ would have two distinct factorizations with prime integers, but each integer admits only one such factorization [2, pp. 23–24], [23, pp. 135–141], [46, pp. 55–61].

(d) The number $\sqrt{2} = 2^{\frac{1}{2}}$ is *not* dyadic, because it is not a rational number [2, pp. 94–95]; consequently, its binary expansion has infinitely many nonzero digits [2, p. 97]. □

Definition 3.7   Denote by $\mathbb{D}[\sqrt{3}]$ the set of all linear combinations of 1 and $\sqrt{3}$ with dyadic coefficients $p, q \in \mathbb{D}$, so that

$$\mathbb{D}[\sqrt{3}] = \left\{ p + q\sqrt{3} : p, q \in \mathbb{D} \right\}.$$

For each integer n, consider combinations with coefficients in $\mathbb{D}_n$:

$$\mathbb{D}_n[\sqrt{3}] = \left\{ p + q\sqrt{3} : p, q \in \mathbb{D}_n \right\}.$$

Moreover, for each dyadic number $r \in \mathbb{D}$, define the **conjugate** $\bar{r}$ by

$$\overline{p + q\sqrt{3}} := p - q\sqrt{3}. \qquad\qquad □$$

Example 3.8   For $\frac{9+5\sqrt{3}}{16} = (\frac{9}{2}{}^4) + (\frac{5}{2}{}^4)\sqrt{3} \in \mathbb{D}_4[\sqrt{3}]$,

$$\overline{\frac{9+5\sqrt{3}}{16}} = \frac{9-5\sqrt{3}}{16}. \qquad\qquad □$$

Definition 3.9   Denote by $\mathbb{Q}[\sqrt{3}]$ the set of all linear combinations of 1 and $\sqrt{3}$ with rational coefficients $p, q \in \mathbb{Q}$, so that

$$\mathbb{Q}[\sqrt{3}] = \left\{ p + q\sqrt{3} : p, q \in \mathbb{Q} \right\}. \qquad\qquad □$$

Remark 3.10   The set $\mathbb{Q}[\sqrt{3}]$, with ordinary addition and multiplication, is a **number field,** which means that it satisfies all the algebraic properties in Table 3.1, with $p + q\sqrt{3}$ abbreviated by (p, q). The set $\mathbb{D}[\sqrt{3}]$, with ordinary addition and multiplication, is an **integral ring,** which means that it satisfies the algebraic properties in Table 3.1, except for the multiplicative inverse, which still lies in $\mathbb{Q}[\sqrt{3}]$ but may fail to lie in $\mathbb{D}[\sqrt{3}]$. The verification of the properties listed

Table 3.1 Algebraic Properties of $\mathbb{Q}[\sqrt{3}]$ and $\mathbb{D}[\sqrt{3}]$. These properties hold for all (u, v), (x, y), and (p, q) in $\mathbb{Q}[\sqrt{3}]$. All properties but (8) hold for all (u, v), (x, y), (p, q) in $\mathbb{D}[\sqrt{3}]$.

(1) Associativity of $+$	$[(u, v) + (x, y)] + (p, q) = (u, v) + [(x, y) + (p, q)]$
(2) Commutativity of $+$	$(u, v) + (x, y) = (x, y) + (u, v)$
(3) Additive identity	$(x, y) + (0, 0) = (x, y) = (0, 0) + (x, y)$
(4) Additive inverse	$(x, y) + (-x, -y) = (0, 0) = (-x, -y) + (x, y)$
(5) Associativity of $\times$	$[(u, v)(x, y)](p, q) = (u, v)[(x, y)(p, q)]$
(6) Commutativity of $\times$	$(u, v)(x, y) = (x, y)(u, v)$
(7) Multiplicative identity	$(x, y)(1, 0) = (x, y) = (1, 0)(x, y)$
(8) Multiplicative inverse	If $(x, y) \neq 0$,
	then $(1, 0) = (x, y)(x/[x^2 - 3y^2], -y/[x^2 - 3y^2])$
(9) Distributivity	$(u, v)[(x, y) + (p, q)] = [(u, v)(x, y)] + [(u, v)(p, q)]$

in the table proceeds *verbatim* as the same verifications for complex numbers, through straightforward algebraic manipulations. □

Remark 3.11 The algebraic properties of numbers $\mathbb{D}[\sqrt{3}]$ listed in Table 3.1 have the consequence that addition and multiplication with dyadic numbers require only *integer arithmetic* with the dyadic coefficients, or with rational coefficients for division. For addition,

$$(p + q\sqrt{3}) + (r + s\sqrt{3}) = (p + r) + (q + s)\sqrt{3}$$

reduces to

$$(p, q) + (r, s) = (p + r, q + s),$$

whereas for multiplication,

$$(p + q\sqrt{3})(r + s\sqrt{3}) = (pr + 3qs) + (ps + qr)\sqrt{3}$$

reduces to

$$(p, q)(r, s) = (pr + 3qs, ps + qr).$$

For division with rational numbers $p, q \in \mathbb{Q}$,

$$\frac{1}{p + q\sqrt{3}} = \frac{p - q\sqrt{3}}{p^2 - 3q^2}$$

reduces to

$$\frac{1}{(p, q)} = \left(\frac{p}{p^2 - 3q^2}, \frac{-q}{p^2 - 3q^2} \right).$$

In the denominator, $p^2 - 3q^2 \neq 0$ for all rational numbers p and q not both equal to zero. Indeed, if $p^2 - 3q^2 = 0$, then $p^2 = 3q^2$, and clearing denominators on both sides gives for the same number $p^2 = 3q^2$ two prime factorizations, one with an even number of factors 3 on the left-hand side, and one with an odd number of factors 3 on the right-hand side, which cannot occur by the uniqueness of prime factorizations [2, pp. 23–24], [23, pp. 135–141], [46, pp. 55–61]. Thus, if a sequence of arithmetic operations produces a pair (u, v), then only one evaluation of $\sqrt{3}$ at the end yields the result in the form $u + v\sqrt{3}$. This amounts to simplifying all the occurrences of $\sqrt{3}$.   □

EXERCISES

Exercise 3.1. Calculate the values $\varphi(r)$ at odd multiples of $\frac{1}{8}$:

$$r \in \left\{ \frac{1}{8}, \frac{3}{8}, \frac{5}{8}, \frac{7}{8}, \frac{9}{8}, \frac{11}{8}, \frac{13}{8}, \frac{15}{8}, \frac{17}{8}, \frac{19}{8}, \frac{21}{8}, \frac{23}{8} \right\}.$$

Exercise 3.2. Write and program an algorithm to calculate the values $\varphi(r)$ at multiples of 2^{-n} between 0 and 3. Test the algorithm by plotting its results and comparing them with Figure 3.3.

Exercise 3.3. Find the values $\psi(r - 1)$ at odd multiples of $\frac{1}{4}$:

$$r \in \left\{ \frac{1}{4}, \frac{3}{4}, \frac{5}{4}, \frac{7}{4}, \frac{9}{4}, \frac{11}{4} \right\}.$$

Exercise 3.4. Write and program an algorithm to calculate the values $\psi(r)$ at multiples of 2^{-n} between 0 and 3. Test the algorithm by plotting its results and comparing them with Figure 3.4.

Exercise 3.5. Write and test a computer program to calculate the addition, subtraction, and multiplication in $\mathbb{D}_n[\sqrt{3}]$ with only integer arithmetic (without floating-point or any other approximation).

Exercise 3.6. Write and test a computer program to calculate the addition, subtraction, multiplication, and division in $\mathbb{Q}[\sqrt{3}]$ with rational or integer arithmetic (without floating-point or other approximation).

3.2   APPROXIMATION OF SAMPLES WITH DAUBECHIES WAVELETS

By a combination $\tilde{f}$ of shifted building blocks φ and wavelets ψ, Daubechies wavelets can approximate a function f, which may represent any signal. In contrast to Haar's simple wavelets, which interpolate signals at sample points but result in discontinuous step approximations, Daubechies wavelets provide a smoother overall approximation $\tilde{f}$ of a function f known only from a sample

$$\vec{s} = (s_0, s_1, \ldots, s_{2^n-2}, s_{2^n-1}).$$

To simplify the notation, assume that the array samples the values of the signal function f at integer points, so that $s_k = f(k)$. (The change of variable $g(k) := f(k/m)$ allows for any equal spacing, so that $s_k = g(k) = f(k/m)$.) Daubechies wavelets begin by approximating the sample $\vec{s}$ by multiples of shifted basic building blocks φ:

$$\tilde{f}(r) = a_{-2}\varphi(r+2) + a_{-1}\varphi(r+1) + a_0\varphi(r) + a_1\varphi(r-1)$$

$$+ a_2\varphi(r-2) + \cdots + a_{2^n-1}\varphi(r - [2^n - 1]).$$

Shifts $\varphi(r - \ell)$ of $\varphi(r)$ by integers $\ell < -2$ or $\ell > 2^n - 1$ equal zero where $0 \le r \le 2^n$, and, consequently, do not affect the approximation $\tilde{f}$.

3.2.1   Approximate Interpolation

A simple and common choice of the coefficients a_k consists in setting, for each $k \in \{0, \ldots, 2^n - 1\}$,

$$a_k := s_k.$$

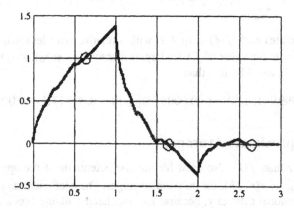

Figure 3.6   Circles mark points where φ takes values near 1, 0, and 0.

The corresponding approximation

$$\tilde{f}(r) = \sum_{k=0}^{2^n-1} s_k \varphi(r - k)$$

nearly interpolates f at the sample points $s_k = f(k)$. Indeed, consider the three points r_0, r_1, and r_2, circled in Figure 3.6, and listed in Table 3.2. The values in

Table 3.2 Locations of r_0, r_1, and r_2, and their values by φ

r	$\varphi(r)$
$650/1024$	$\varphi(650/1024) \approx +0.999\,985$
$1674/1024$	$\varphi(1674/1024) \approx -0.000\,155$
$2698/1024$	$\varphi(2698/1024) \approx +0.000\,170$

Table 3.2 result from automated symbolic and numerical computations based on the recursion (3.2). With

$$r_0 = \frac{650}{1024}, \quad r_1 = \frac{1674}{1024}, \quad r_2 = \frac{2698}{1024}, \tag{3.5}$$

barring any computational errors

$$f(0)\varphi(r_0) \approx f(0),$$
$$f(1)\varphi(r_1) \approx 0,$$
$$f(2)\varphi(r_2) \approx 0,$$

so that for each ℓ, all terms with $k \neq \ell$ nearly cancel, and hence

$$\tilde{f}(r_0 + \ell) = \sum_{k=0}^{2^n-1} f(k)\varphi(r_0 + \ell - k) \approx \cdots + 0 + f(\ell) + 0 + \cdots = f(\ell)$$

nearly interpolates each $f(\ell)$ at $r_0 + \ell$, with a relative error less than $0.000\,4 = 0.04\%$ (relative to the value of f with largest magnitude): $\varphi(r_0)$, $\varphi(r_1)$, and $\varphi(r_2)$ differ from $1, 0$, and 0 by less than

$$|1 - 0.999\,984| + |0 - (-0.000\,156)| + |0 - 0.000\,171| = 0.000\,343.$$

3.2.2 Approximate Averages

Choices other than $f(k)$ also exist for the coefficients a_k of the approximation $\tilde{f}(r) = \sum_{k=0}^{2^n-1} a_k\varphi(r - k)$, for instance, averages. The coefficient a_0 presents the least computational difficulty, because its associated building block φ does not extend beyond the interval under consideration. Thus, a_0 can consist merely in an average of s_0, s_1, s_2, s_3 weighted by the corresponding values of φ:

$$a_0 := s_0 \cdot \varphi(0) + s_1 \cdot \varphi(1) + s_2 \cdot \varphi(2) + s_3 \cdot \varphi(3).$$

Example 3.12 For the sample $\vec{s} = (s_0, s_1, s_2, s_3) := (0, 1, 2, 3)$,

$$a_0 = s_0 \cdot \varphi(0) + s_1 \cdot \varphi(1) + s_2 \cdot \varphi(2) + s_3 \cdot \varphi(3)$$

$$= 0 \cdot 0 + 1 \cdot \frac{1 + \sqrt{3}}{2} + 2 \cdot \frac{1 - \sqrt{3}}{2} + 3 \cdot 0$$

$$= \frac{3 - \sqrt{3}}{2}.$$

The result just obtained means that the approximation $\tilde{f}$ of the sample $\vec{s} = (0, 1, 2, 3)$ contains the term $a_0 \cdot \varphi = \frac{3-\sqrt{3}}{2} \cdot \varphi$. $\qquad \square$

For finite samples, the calculation of the coefficients a_k for $k > 2^n - 4$ requires a preliminary extension of the sample, because the shifted building blocks $\varphi(r-k)$ eventually extend beyond the range of the sample while still overlapping it, so that a weighted average similar to that used for a_0 would call for nonexistent values of the sample. The problem just described—the calculation of the coefficients a_k for shifted building blocks $\varphi(r - k)$ overlapping the edge of the sample—is called the problem of **edge effects.** The following considerations describe several methods to handle such edge effects.

3.3 EXTENSIONS TO ALLEVIATE EDGE EFFECTS

Several methods exist to extend a sample and thereby allow for the computation of the initial coefficients a_k near the edges (start and end) of the sample. Each method has advantages and disadvantages:

METHOD	SPEED	ACCURACY
Zeros	Very fast	Very inaccurate
Periodic	Fast	Accurate
Spline	Slow and requires 2nd array	Very accurate

To demonstrate the need for extensions, the following subsection shows the marked edge effects that result from extensions by zeros, which amounts to ignoring the need for extensions.

The next subsection will then explain the common periodic extension, which requires only a new indexing scheme and already alleviates much of the edge effects.

The last subsection will demonstrate the greater accuracy—with further reduced edge effects—but also the much larger consumption of time and space with extensions by splines.

3.3.1 Zigzag Edge Effects from Extensions by Zeros

Ignoring edge effects by suppressing, in the weighted averages, terms beyond the sample's beginning and end amounts to extending the sample by sequences of zeros before and after the sample. The following example demonstrates how such extensions by zeros produce distorted approximations near the edges of the sample.

Example 3.13 Extending the sample

$$\vec{s} = (s_0, s_1, s_2, s_3) = (f(0), f(1), f(2), f(3)) := (0, 1, 2, 3)$$

by zeros gives an extended sample, here with extensions separated by semicolons for emphasis,

$$(s_{-4}, \quad s_{-3}, \quad s_{-2}, \quad s_{-1}; \quad s_0, \quad s_1, \quad s_2, \quad s_3; \quad s_4, \quad s_5, \quad s_6, \quad s_7)$$
$$(0, \quad\quad 0, \quad\quad 0, \quad\quad 0; \quad\; 0, \quad 1, \quad 2, \quad 3; \quad 0, \quad 0, \quad 0, \quad 0).$$

The calculation of each coefficient a_k consists in a weighted average, also called a "convolution," of the extended sample and the shifted building blocks:

$$a_k = \sum_{r=k+0}^{k+3} \varphi(r - k)s_r.$$

The calculation of a_0 proceeds as in the foregoing section. For $k \neq 0$, aligning the values of r, $\varphi(r - k)$, and s_r may provide a visual aid in understanding such a convolution. Thus, to calculate a_1, form the average of $\vec{s}$ weighted by the values of $\varphi(r - 1)$, or, in other words, calculate the inner product of the extended sample with the building block φ shifted by 1 to the right:

r	-4	-3	-2	-1	0	1	2	3	4	5	6	7
$\varphi(r-1)$	0	0	0	0	0	0	$\dfrac{1+\sqrt{3}}{2}$	$\dfrac{1-\sqrt{3}}{2}$	0	0	0	0
s_r	0	0	0	0	0	1	2	3	0	0	0	0

$$a_1 = s_1 \cdot \varphi(1 - 1) + s_2 \cdot \varphi(2 - 1) + s_3 \cdot \varphi(3 - 1) + s_4 \cdot \varphi(4 - 1)$$

$$= 0 \cdot 0 + 1 \cdot 0 + 2 \cdot \frac{1+\sqrt{3}}{2} + 3 \cdot \frac{1-\sqrt{3}}{2}$$

$$= \frac{5 - \sqrt{3}}{2}.$$

For a_2, form the average of $\vec{s}$ weighted by the values of $\varphi(r - 2)$:

r	-4	-3	-2	-1	0	1	2	3	4	5	6	7
$\varphi(r-2)$	0	0	0	0	0	0	0	$\dfrac{1+\sqrt{3}}{2}$	$\dfrac{1-\sqrt{3}}{2}$	0	0	0
s_r	0	0	0	0	0	1	2	3	0	0	0	0

$$a_2 = s_2 \cdot \varphi(2 - 2) + s_3 \cdot \varphi(3 - 2) + s_4 \cdot \varphi(4 - 2) + s_5 \cdot \varphi(5 - 2)$$

$$= 2 \cdot 0 + 3 \cdot \frac{1+\sqrt{3}}{2} + 0 \cdot \frac{1-\sqrt{3}}{2} + 0 \cdot 0$$

$$= 3 \cdot \frac{1 + \sqrt{3}}{2}.$$

Similarly, the average of $\vec{s}$ weighted by $\varphi(r+1)$ gives a_{-1}:

$$a_{-1} = s_{-1} \cdot \varphi(-1+1) + s_0 \cdot \varphi(0+1) + s_1 \cdot \varphi(1+1) + s_2 \cdot \varphi(2+1)$$

$$= 0 \cdot 0 + 0 \cdot \frac{1+\sqrt{3}}{2} + 1 \cdot \frac{1-\sqrt{3}}{2} + 2 \cdot 0$$

$$= \frac{1-\sqrt{3}}{2}.$$

Finally, the average of $\vec{s}$ weighted by $\varphi(r+2)$ yields a_{-2}:

$$a_{-2} = s_{-2} \cdot \varphi(-2+2) + s_{-1} \cdot \varphi(-1+2) + s_0 \cdot \varphi(0+2) + s_1 \cdot \varphi(1+2)$$

$$= 0 \cdot 0 + 0 \cdot \frac{1+\sqrt{3}}{2} + 0 \cdot \frac{1-\sqrt{3}}{2} + 1 \cdot 0$$

$$= 0.$$

Thus,

$$\tilde{f}(r) = a_{-2}\varphi(r+2) + a_{-1}\varphi(r+1) + a_0\varphi(r)$$

$$+ a_1\varphi(r-1) + a_2\varphi(r-2)$$

$$= 0 \cdot \varphi(r+2) + \frac{1-\sqrt{3}}{2}\varphi(r+1) + \frac{3-\sqrt{3}}{2}\varphi(r)$$

$$+ \frac{5-\sqrt{3}}{2}\varphi(r-1) + 3 \cdot \frac{1+\sqrt{3}}{2}\varphi(r-2).$$

Figure 3.7(b) shows the resulting wavelet approximation $\tilde{f}$, which exhibits distortions near the sample's edges, as indicated by the vertical scale, and corroborated by algebraic verifications:

$$\tilde{f}(3) = 0 \cdot \varphi(3+2) + \frac{1-\sqrt{3}}{2} \cdot \varphi(3+1) + \frac{3-\sqrt{3}}{2} \cdot \varphi(3)$$

$$+ \frac{5-\sqrt{3}}{2} \cdot \varphi(3-1) + 3 \cdot \frac{1+\sqrt{3}}{2} \cdot \varphi(3-2)$$

$$= 0 + 0 + 0 + \frac{5-\sqrt{3}}{2} \cdot \frac{1-\sqrt{3}}{2} + 3 \cdot \frac{1+\sqrt{3}}{2} \cdot \frac{1+\sqrt{3}}{2}$$

$$= 5 \neq 3 = s_3.$$

Similar verifications demonstrate that $\tilde{f}(r) \neq s_r$ near the edges:

r	0	1	2	3
s_r	0	1	2	3
$\tilde{f}(r)$	−0.5	1	2	5

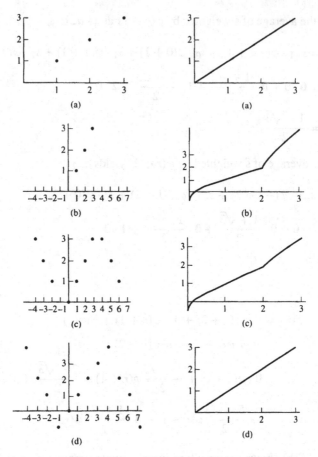

Figure 3.7   Illustration of edge effects. (a) Sample (left) drawn from a signal function (right). (b) Extension of the sample by zeros (left), and resulting wavelet approximation of the signal (right), with distortions of large magnitude near the edges. (c) Mirror reflection of the sample across vertical lines near each end (left), and resulting wavelet approximation of the signal (right), with distortions of moderate magnitude near the edges. (d) Smooth periodic spline extension of the sample at the first and last sample points (left), and result-ing wavelet approximation of the signal (right), with distortions of smaller magnitude near the edges.

The discrepancies between the initial sample and the wavelet approximation near the edges result from the artificial "jump" discontinuities introduced by the exten-sions by zeros at each end.   □

3.3.2   Medium Edge Effects from Mirror Reflections

One method to alleviate edge effects extends the sample not with zeros but with values similar to those in the sample, by reflecting the sample periodically as though through vertical mirrors at both ends.

Example 3.14 Extending the sample

$$\bar{s} = (s_0, s_1, s_2, s_3) = (f(0), f(1), f(2), f(3)) := (0, 1, 2, 3)$$

by mirror reflections across each end of the sample gives an extended sample, here with extensions separated by semicolons for emphasis:

$(s_{-4},$	$s_{-3},$	$s_{-2},$	$s_{-1};$	$s_0,$	$s_1,$	$s_2,$	$s_3;$	$s_4,$	$s_5,$	$s_6,$	$s_7)$
$(3,$	$2,$	$1,$	$0;$	$0,$	$1,$	$2,$	$3;$	$3,$	$2,$	$1,$	$0).$

The calculations of the coefficients a_1, a_0, and a_{-1} remains as in Example 3.13, because the shifted building blocks $\varphi(r-1)$, $\varphi(r)$, and $\varphi(r+1)$ equal zero at the extension points. In contrast, the two coefficients a_2 and a_{-2} acquire new values. To calculate a_2, form the average of $\bar{s}$ weighted by the values of $\varphi(r-2)$:

$$a_2 = s_2 \cdot \varphi(2-2) + s_3 \cdot \varphi(3-2) + s_4 \cdot \varphi(4-2) + s_5 \cdot \varphi(5-2)$$

$$= 2 \cdot 0 + 3 \cdot \frac{1+\sqrt{3}}{2} + 3 \cdot \frac{1-\sqrt{3}}{2} + 2 \cdot 0$$

$$= 3.$$

To calculate a_{-2}, form the average of $\bar{s}$ weighted by $\varphi(r+2)$:

$$a_{-2} = s_{-2} \cdot \varphi(-2+2) + s_{-1} \cdot \varphi(-1+2) + s_0 \cdot \varphi(0+2) + s_1 \cdot \varphi(1+2)$$

$$= 1 \cdot 0 + 0 \cdot \frac{1+\sqrt{3}}{2} + 0 \cdot \frac{1-\sqrt{3}}{2} + 1 \cdot 0$$

$$= 0.$$

Thus,

$$\tilde{f}(r) = a_{-2}\varphi(r+2) + a_{-1}\varphi(r+1) + a_0\varphi(r)$$
$$+ a_1\varphi(r-1) + a_2\varphi(r-2)$$

$$= 0 \cdot \varphi(r+2) + \frac{1-\sqrt{3}}{2} \cdot \varphi(r+1) + \frac{3-\sqrt{3}}{2} \cdot \varphi(r)$$

$$+ \frac{(5-\sqrt{3})}{2} \cdot \varphi(r-1) + 3 \cdot \varphi(r-2).$$

Figure 3.7(c) shows that the resulting wavelet approximation $\tilde{f}$ exhibits smaller edge effects than the one obtained in Example 3.13 through extension by zeros. Algebraic verifications corroborate such an improvement, showing smaller differences near the edges:

r	0	1	2	3
s_r	0	1	2	3
$\tilde{f}(r)$	−0.5	1	2	3.5

Instead of large discrepancies from "jump" discontinuities caused by zeros, the moderate discrepancies just obtained between the sample and the wavelet approximation arise from the "corners" introduced by the mirror reflections.   □

Remark 3.15   The values contained in one period suffice to reconstruct all the other values of a periodic sequence. Therefore, it suffices to retain only the initial data and one mirror reflection of the data at either end, for instance, at the right:

$$\underbrace{s_0, \quad s_1, \quad \ldots, \quad s_{2^n-2}, \quad s_{2^n-1};}_{\text{data}} \underbrace{s_{2^n-1}, \quad s_{2^n-2}, \quad \ldots, \quad s_1, \quad s_0.}_{\text{reflection}}$$

For the computation of the last coefficients, associated to the building blocks that overlap the right-hand edge, yet another reflection—called here a "short" reflection—can prove convenient:

$$\underbrace{s_0, \quad s_1, \quad \ldots, \quad s_{2^n-1};}_{\text{data}} \underbrace{s_{2^n-1}, \quad \ldots, \quad s_1, \quad s_0;}_{\text{reflection}} \underbrace{s_0, \quad s_1.}_{\text{short reflection}}$$

The number of entries in the additional short reflection depends upon the type of wavelets under consideration, and equals the number of recursion coefficients minus two. For the type of Daubechies wavelets considered here, with four recursion coefficients h_0, h_1, h_2, and h_3, just $4 - 2 = 2$ additional values will suffice: s_0 and s_1.

In practice, reflections can follow two alternatives. The first alternative sets up the extended array as just displayed, which involves less programming but requires over twice as much storage as the data does. The second alternative stores only the data but requires additional algorithms to supply individual values from the mirror extensions when needed. The choice depends upon the available equipment and such requirements as speed and reliability from the particular application at hand. In either case, an exercise will show that mirror extensions lead to a symmetry in the coefficients, which then do not require twice as much storage either.   □

3.3.3   Small Edge Effects from Smooth Periodic Extensions

A method more sophisticated than reflections to alleviate edge effects extends the data $s_0, s_1, \ldots, s_{2^n-2}, s_{2^n-1}$ with values *and slopes* similar to those in the data at the edges:

$$\underbrace{s_0, \quad s_1, \quad \ldots, \quad s_{2^n-1};}_{\text{data}} \underbrace{s_{2^n}, \quad \ldots, \quad s_{2^{n+1}-1};}_{\text{extension}} \underbrace{s_0, \quad s_1.}_{\text{short extension}}$$

The short extension merely copies the beginning of the data. The number of entries in the additional short extension depends upon the type of wavelets under consideration, and equals the number of recursion coefficients minus two. For the type of Daubechies wavelets considered here, with four recursion coefficients h_0, h_1, h_2, and h_3, just $4 - 2 = 2$ additional values will suffice: s_0 and s_1.

Example 3.16 Consider the problem of determining the first extension entry s_{2^n} so that the change from the last data entry matches the preceding change between the last two data entries. At the right-hand edge of the data, the data change by

$$s_{2^n-1} - s_{2^n-2},$$

which must match the change of the extension at the same location,

$$s_{2^n} - s_{2^n-1},$$

so that solving for the first extension entry s_{2^n} gives

$$s_{2^n} - s_{2^n-1} = s_{2^n-1} - s_{2^n-2},$$

$$s_{2^n} = 2s_{2^n-1} - s_{2^n-2}. \qquad \square$$

Example 3.17 Consider the problem of determining the last extension entry $s_{2^{n+1}-1}$ so that the change to the first data entry (as in the short extension by periodicity) matches the next change between the first two data entries. At the left-hand edge, the data change by

$$s_1 - s_0,$$

which must match the change of the extension at the same location,

$$s_0 - s_{2^{(n+1)}-1},$$

so that solving for the last extension entry $s_{2^{(n+1)}-1}$ gives

$$s_0 - s_{2^{(n+1)}-1} = s_1 - s_0,$$

$$s_{2^{(n+1)}-1} = 2s_0 - s_1. \qquad \square$$

With both edges of the extension determined, there remains the problem of completing the still missing entries $s_{2^n+1}, \ldots, s_{2^{n+1}-2}$ of the extension in a "smooth" way, which means without introducing extraneous corners or discontinuities. Such a smooth periodic extension can result, for instance, from a "cubic spline," which consists of a cubic polynomial p. The usual polynomial expression with powers of r would fail to take advantage of the situation; in contrast, the following "Newton form" [27, Ch. 6] of the cubic polynomial p takes advantage of the way in which the constraints arise at the edges:

$$p(r) = p_0 + p_1(r - [2^n - 1]) + p_2(r - [2^n - 1])(r - [2^n])$$
$$+ p_3(r - [2^n - 1])(r - [2^n])(r - [2^{(n+1)} - 1]),$$

subject to the following conditions to ensure the periodicity of the extended sample:

$$p(2^n - 1) = s_{2^n - 1},$$

$$p(2^n) = s_{2^n} = 2s_{2^n - 1} - s_{2^n - 2},$$

$$p(2^{(n+1)} - 1) = s_{2^{(n+1)} - 1} = 2s_0 - s_1,$$

$$p(2^{(n+1)}) = s_0.$$

The conditions just stated form a linear system of equations for the coefficients p_0, p_1, p_2, and p_3, obtained by substituting the formula for p on the left-hand side of each constraint and canceling such factors as $([2^n - 1] - [2^n - 1])$, $(2^n - [2^n])$, and $([2^{(n+1)} - 1] - [2^{(n+1)} - 1])$:

$$p_0 = s_{2^n - 1},$$

$$p_0 + p_1(2^n - [2^n - 1]) = s_{2^n} = 2s_{2^n - 1} - s_{2^n - 2},$$

$$p_0 + p_1([2^{(n+1)} - 1] - [2^n - 1]) + p_2$$

$$([2^{(n+1)} - 1] - [2^n - 1])([2^{(n+1)} - 1] - [2^n]) = s_{2^{(n+1)} - 1} = 2s_0 - s_1,$$

$$p_0 + p_1(2^{n+1} - [2^n - 1]) + p_2(2^{n+1} - [2^n - 1])$$

$$(2^{n+1} - [2^n]) + p_3(2^{n+1} - [2^n - 1])$$

$$(2^{n+1} - [2^n])(2^{n+1} - [2^{(n+1)} - 1]) = s_0.$$

Solving the "triangular" linear system just obtained from the first equation toward the last equation, substituting solutions from one equation into the next equation, gives

$$p_0 = s_{2^n - 1},$$

$$p_1 = s_{2^n - 1} - s_{2^n - 2},$$

$$p_2 = \frac{2s_0 - s_1 - s_{2^n - 1} - 2^n p_1}{2^n(2^n - 1)},$$

$$p_3 = \frac{s_0 - s_{2^n - 1} - (2^n + 1)p_1}{(2^n + 1)(2^n)} - p_2.$$

The resulting cubic polynomial p then provides a means to compute the extended values by the definition

$$s_k := p(k)$$

for k ranging from $2^n + 1$ through $2^{(n+1)} - 2$.

Example 3.18   A smooth periodic extension of the sample

$$\bar{s} = (s_0, s_1, s_2, s_3) = (f(0), f(1), f(2), f(3)) := (0, 1, 2, 3)$$

can proceed as follows:

$$s_{2(n+1)-1} = 2s_0 - s_1 = 2 \cdot 0 - 1 = -1,$$

$$s_{2n} = 2s_{2n-1} - s_{2n-2} = 2 \cdot 3 - 2 = 4,$$

which gives the endpoints of the extended sample:

$$\underbrace{0, \quad 1, \quad 2, \quad 3;}_{\text{data}} \quad \underbrace{4, \quad ?, \quad ?, \quad -1;}_{\text{extension}} \quad \underbrace{0, \quad 1,}_{\text{short extension}}$$

where the question marks (?) indicate the position of yet unknown values of the extension. In principle, such values need only fill the gap smoothly, for instance, by means of a cubic polynomial:

$$\underbrace{0, \quad 1, \quad 2, \quad 3;}_{\text{data}} \quad \underbrace{4, \quad 2, \quad 1, \quad -1;}_{\text{extension}} \quad \underbrace{0, \quad 1.}_{\text{short extension}}$$

The calculations of the coefficients a_1, a_0, and a_{-1} remains as in Example 3.13, because the shifted building blocks $\varphi(r - 1)$, $\varphi(r)$, and $\varphi(r + 1)$ equal zero at the extension points. In contrast, the two new coefficients a_2 and a_{-2} acquire new values. To calculate a_2, form the average of $\vec{s}$ weighted by the values of $\varphi(r - 2)$:

$$a_2 = s_2 \cdot \varphi(2 - 2) + s_3 \cdot \varphi(3 - 2) + s_4 \cdot \varphi(4 - 2) + s_5 \cdot \varphi(5 - 2)$$

$$= 2 \cdot 0 + 3 \cdot \frac{1 + \sqrt{3}}{2} + 4 \cdot \frac{1 - \sqrt{3}}{2} + 2 \cdot 0$$

$$= \frac{7 - \sqrt{3}}{2}.$$

To calculate a_{-2}, use periodicity at the left-hand edge, and form the average of $\vec{s}$ weighted by the values of $\varphi(r + 2)$:

$$a_{-2} = s_{-2} \cdot \varphi(-2 + 2) + s_{-1} \cdot \varphi(-1 + 2) + s_0 \cdot \varphi(0 + 2) + s_1 \cdot \varphi(1 + 2)$$

$$= 2 \cdot 0 + -1 \cdot \frac{1 + \sqrt{3}}{2} + 0 \cdot \frac{1 - \sqrt{3}}{2} + 1 \cdot 0$$

$$= -\frac{1 + \sqrt{3}}{2}.$$

Thus,

$$\tilde{f}(r) = a_{-2}\varphi(r + 2) + a_{-1}\varphi(r + 1) + a_0\varphi(r) + a_1\varphi(r - 1) + a_2\varphi(r - 2)$$

$$= -\frac{1 + \sqrt{3}}{2} \cdot \varphi(r + 2) + \frac{1 - \sqrt{3}}{2} \cdot \varphi(r + 1) + \frac{3 - \sqrt{3}}{2} \cdot \varphi(r)$$

$$+ \frac{5 - \sqrt{3}}{2} \cdot \varphi(r - 1) \frac{7 - \sqrt{3}}{2} \cdot \varphi(r - 2).$$

Figure 3.7(d) shows that the resulting wavelet approximation $\tilde{f}$ exhibits smaller edge effects than the one obtained in Example 3.13 through extension by zeros, and Example 3.14 through mirror reflections. Algebraic verifications corroborate such an improvement:

r	0	1	2	3
s_r	**0**	**1**	**2**	**3**
$\tilde{f}(r)$	0	1	2	3

Instead of larger discrepancies caused by "jump" discontinuities or "corners" at the edges of the initial sample, the smaller discrepancies just obtained result from smoother transitions, with equal slopes from the sample to its extensions.   □

Remark 3.19   Example 3.18 demonstrates how Daubechies wavelets can reproduce affine functions (slanting lines) *exactly*, in contrast to the coarse stair steps of Haar's wavelets.   □

EXERCISES

Exercise 3.7. From the conjugate symmetry of the values of φ,

$$\varphi(1) = \frac{1 + \sqrt{3}}{2}, \quad \varphi(2) = \frac{1 - \sqrt{3}}{2}, \quad \varphi(1) = \overline{\varphi(2)},$$

prove that for mirror reflections of rational data a similar conjugate symmetry exists for the coefficients:

$$a_{[2^{n+1}-4]-k} = \overline{a_k}.$$

Exercise 3.8. Write and test a computer program to produce extensions by mirror reflections, with an extended array *or* with an algorithm to produce the reflected entries when needed.

Exercise 3.9. Write and test a computer program to calculate the coefficients p_0, p_1, p_2, and p_3 of the cubic polynomial p for smooth periodic extensions. See also [27, Ch. 6] or [39, Ch. 2].

Exercise 3.10. Write and test a computer program to produce smooth periodic extensions, with an extended array *or* an algorithm to produce the reflected entries when needed. See [27, Ch. 6].

Exercise 3.11. Investigate smooth periodic extensions through three pieces of straight lines joined continuously—instead of a cubic polynomial—with the same slopes as the sample's at both ends.

Exercise 3.12. Investigate smooth periodic extensions through a cubic polynomial with the same slopes (which can require calculus) as the sample's difference quotients at both ends.

3.4 THE FAST DAUBECHIES WAVELET TRANSFORM

This section demonstrates the calculation of a fast wavelet transform of a sample with Daubechies wavelets. For the present version of the fast wavelet transform to succeed, the sample *must* contain a number of values equal to an integral power of two, $N = 2^n$:

$$\vec{s} = (s_0, s_1, \ldots, s_{2^n-2}, s_{2^n-1}).$$

If the number of values does not equal an integral power of two, then it becomes necessary first to shorten or extend the sample smoothly to a length equal to a power of two before proceeding with the fast wavelet transform. Regardless of the methods used to obtain the periodic extension of the sample, and then the coefficients $a_0, \ldots, a_{2^{n+1}-1}$, the basic wavelet transform consists in replacing the $2N = 2^{(n+1)}$ basic building blocks

$$\tilde{f}(r) = \sum_{k=0}^{2^{(n+1)}-1} a_k \cdot \varphi(r - k)$$

by an *equivalent* combination of $N = 2^n$ slower building blocks $\varphi([r/2] - k)$ and $N = 2^n$ slower wavelets $\psi([r/2 - 1] - k)$:

$$\tilde{f}(r) = \sum_{k=0}^{2^n-1} a_k^{(n-1)} \cdot \varphi([r/2] - k) + \sum_{k=0}^{2^n-1} c_k^{(n-1)} \cdot \psi([r/2 - 1] - k).$$

The superscripts $^{(n-1)}$ indicate a frequency lower than for the initial coefficients, which can also be relabeled with a superscript $^{(n)}$:

$$a_k^{(n)} := a_k.$$

Subsequent passes of the wavelet transform will produce yet lower frequencies. As in the definition of the Daubechies wavelet ψ, the shift from $r/2$ to $r/2 - 1$ ensures that φ and ψ equal zero outside the same interval. The calculation of the new coefficients $a_k^{(n-1)}$ and $c_k^{(n-1)}$ relies upon the recurrence relations

$$\varphi(r) = h_0 \cdot \varphi(2r) + h_1 \cdot \varphi(2r - 1) + h_2 \cdot \varphi(2r - 2) + h_3 \cdot \varphi(2r - 3),$$

$$\psi(r) = -h_0 \cdot \varphi(2r - 1) + h_1 \cdot \varphi(2r) - h_2 \cdot \varphi(2r + 1) + h_3 \cdot \varphi(2r + 2).$$

The substitution of $[r/2]$ for r gives

$$\varphi([r/2]) = h_0 \cdot \varphi(r) + h_1 \cdot \varphi(r - 1) + h_2 \cdot \varphi(r - 2) + h_3 \cdot \varphi(r - 3),$$

$$\psi([r/2]) = -h_0 \cdot \varphi(r - 1) + h_1 \cdot \varphi(r) - h_2 \cdot \varphi(r + 1) + h_3 \cdot \varphi(r + 2).$$

Hence, the algorithm for the Daubechies wavelet transform arises from an interpretation of the recursions

$$\varphi([r/2]) = h_0 \cdot \varphi(r) + h_1 \cdot \varphi(r-1) + h_2 \cdot \varphi(r-2) + h_3 \cdot \varphi(r-3),$$

$$\psi([r/2]-1) = h_3 \cdot \varphi(r) - h_2 \cdot \varphi(r-1) + h_1 \cdot \varphi(r-2) - h_0 \cdot \varphi(r-3),$$

$$\vdots$$

$$\varphi([r/2]-k) = h_0 \cdot \varphi(r-2k) + h_1 \cdot \varphi(r-1-2k)$$
$$+ h_2 \cdot \varphi(r-2-2k) + h_3 \cdot \varphi(r-3-2k),$$

$$\psi([r/2]-1-k) = h_3 \cdot \varphi(r-2k) - h_2 \cdot \varphi(r-1-2k)$$
$$+ h_1 \cdot \varphi(r-2-2k) - h_0 \cdot \varphi(r-3-2k),$$

as a change of basis with matrix (where all blank entries equal zero)

$$
{}_D\Omega =
\begin{array}{c}
\\
\varphi(r) \\
\varphi(r-1) \\
\varphi(r-2) \\
\varphi(r-3) \\
\varphi(r-4) \\
\varphi(r-5) \\
\vdots
\end{array}
\begin{pmatrix}
\varphi(\frac{r}{2}) & \psi(\frac{r}{2}-1) & \varphi(\frac{r}{2}-1) & \psi(\frac{r}{2}-1-1) \\
h_0 & h_3 & & & \cdots \\
h_1 & -h_2 & & & \cdots \\
h_2 & h_1 & h_0 & h_3 & \cdots \\
h_3 & -h_0 & h_1 & -h_2 & \cdots \\
& & h_2 & h_1 & \cdots \\
& & h_3 & -h_0 & \cdots \\
\vdots & \vdots & \vdots & \vdots & \ddots
\end{pmatrix},
$$

where each column lists the coefficients expressing the function at the top of the column in terms of functions in the left-hand margin. For instance, the third *column* records that

$$\varphi([r/2]-1) = h_0\varphi(r-2) + h_1\varphi(r-3) + h_2\varphi(r-4) + h_3\varphi(r-5).$$

The inverse matrix ${}_D\Omega^{-1}$ expresses the reverse change of basis, which will express each function in the left-hand margin in terms of the building blocks and wavelets listed at the top. The calculation of the inverse matrix uses the following algebraic relations for the inner products of columns of the matrix ${}_D\Omega$:

$$2 = h_0^2 + h_1^2 + h_2^2 + h_3^2,$$

$$0 = h_0h_3 - h_1h_2 + h_2h_1 - h_3h_0, \tag{3.6}$$

$$0 = h_2h_0 + h_3h_1.$$

Direct substitutions of the values of h_0, h_1, h_2, and h_3 verify the relations just listed. With I denoting the identity matrix, which has only 1's on the diagonal and only 0's off the diagonal,

$$
I =
\begin{pmatrix}
1 & 0 & 0 & \cdots \\
0 & 1 & 0 & \cdots \\
0 & 0 & 1 & \cdots \\
\vdots & \vdots & & \ddots
\end{pmatrix},
$$

the relations (3.6) mean that $_D\Omega^T {}_D\Omega = 2I$, so that $_D\Omega^{-1} = (\frac{1}{2})_D\Omega^T$. In other words, the *rows* of $_D\Omega^{-1}$ are the *columns* of $_D\Omega$ multipied by $\frac{1}{2}$ (with all blank entries equal to zero):

$$2_D\Omega^{-1}$$

$$
\begin{array}{c}
\\
\varphi(\frac{r}{2}) \\
\psi(\frac{r}{2}-1) \\
\varphi(\frac{r}{2}-1) \\
= \psi(\frac{r}{2}-2) \\
\varphi(\frac{r}{2}-2) \\
\psi(\frac{r}{2}-3) \\
\vdots
\end{array}
\begin{array}{cccccc}
\varphi(r) & \varphi(r-1) & \varphi(r-2) & \varphi(r-3) & \varphi(r-4) & \\
\left(\begin{array}{ccccc}
h_0 & h_1 & h_2 & h_3 & \cdots \\
h_3 & -h_2 & h_1 & -h_0 & \cdots \\
 & & h_0 & h_1 & h_2 & \cdots \\
 & & h_3 & -h_2 & h_1 & \cdots \\
 & & & & h_0 & \cdots \\
 & & & & h_3 & \cdots \\
\vdots & \vdots & \vdots & \vdots & \vdots &
\end{array} \right),
\end{array}
$$

Thus, a function

$$\tilde{f}(r) = \sum_k a_k^{(n)} \varphi(r-k)$$

is a linear combination of the columns $\varphi(r-k)$ of the matrix $_D\Omega^{-1}$. Consequently, calculating the wavelet transform amounts to changing from the basis with $\varphi(r-k)$ to the basis with $\varphi([r/2]-k)$ and $\psi([r/2-1]-k)$, which reduces to multiplying the array of coefficients $\bar{\mathbf{a}}$ by the inverse matrix $_D\Omega^{-1} = (\frac{1}{2})_D\Omega^T$, and reading the new coefficients in the rows of the products

$$\left(a_0^{(n-1)}, c_0^{(n-1)}, a_1^{(n-1)}, c_1^{(n-1)}, a_2^{(n-1)}, c_2^{(n-1)}, \ldots \right) = {}_D\Omega^{-1}\bar{\mathbf{a}}^{(n)},$$

$$
\begin{pmatrix}
a_0^{(n-1)} \\
c_0^{(n-1)} \\
a_1^{(n-1)} \\
c_1^{(n-1)} \\
a_2^{(n-1)} \\
c_2^{(n-1)} \\
\vdots
\end{pmatrix}
= \frac{1}{2}
\begin{pmatrix}
h_0 & h_1 & h_2 & h_3 & & & \cdots \\
h_3 & -h_2 & h_1 & -h_0 & & & \cdots \\
 & & h_0 & h_1 & h_2 & h_3 & \cdots \\
 & & h_3 & -h_2 & h_1 & -h_0 & \cdots \\
 & & & & h_0 & h_1 & \cdots \\
 & & & & h_3 & -h_2 & \cdots \\
\vdots & \vdots & \vdots & \vdots & \vdots & \vdots &
\end{pmatrix}
\begin{pmatrix}
a_0^{(n)} \\
a_1^{(n)} \\
a_2^{(n)} \\
a_3^{(n)} \\
a_4^{(n)} \\
a_5^{(n)} \\
\vdots
\end{pmatrix}.
$$

The formula just displayed in terms of matrices provides a concise yet still explicit description of the Daubechies wavelet transform. In contrast, individual formulae for all the coefficients constitute a lengthy list. For instance, the first two rows yield the coefficients $a_0^{(n-1)}$ of $\varphi([r/2])$ and $c_0^{(n-1)}$ of $\psi([r/2]-1)$:

$$a_0^{(n-1)} = (\tfrac{1}{2})(h_0, h_1, h_2, h_3)(a_0^{(n)}, a_1^{(n)}, a_2^{(n)}, a_3^{(n)})^T,$$

$$c_0^{(n-1)} = (\tfrac{1}{2})(h_3, -h_2, h_1, -h_0)(a_0^{(n)}, a_1^{(n)}, a_2^{(n)}, a_3^{(n)})^T.$$

If the initial sample $\vec{s} = (s_0, s_1, \ldots, s_{2^n-1})$ has 2^n entries, then any periodic extension has 2^{n+1} entries, which allows for $2^{n-1} - 1$ shifts by two indices each. After $2^{n-1} - 1$ shifts by two indices, the last block of coefficients matches the last two entries of the periodic extensions, $a_{2^{n+1}-2}^{(n)}, a_{2^{n+1}-1}^{(n)}$, followed by the short extension $a_{2^{n+1}}^{(n)}, a_{2^{n+1}+1}^{(n)}$. Thus, the last two rows yield the coefficients $a_{2^n-1}^{(n-1)}$ of $\varphi([r/2] - 2^{n-1})$ and $c_{2^n-1}^{(n-1)}$ of $\psi([r/2 - 1] - 2^{n-1})$:

$$a_{2^n-1}^{(n-1)} = (\tfrac{1}{2})(h_0, h_1, h_2, h_3)(a_{2^{n+1}-2}^{(n)}, a_{2^{n+1}-1}^{(n)}, a_{2^{n+1}}^{(n)}, a_{2^{n+1}+1}^{(n)})^T,$$

$$c_{2^n-1}^{(n-1)} = (\tfrac{1}{2})(h_3, -h_2, h_1, -h_0)(a_{2^{n+1}-2}^{(n)}, a_{2^{n+1}-1}^{(n)}, a_{2^{n+1}}^{(n)}, a_{2^{n+1}+1}^{(n)})^T.$$

Remark 3.20   If the initial sample has 2^n entries, then its periodic extension has 2^{n+1} entries in the first period and hence continues indefinitely with repeating blocks of lengths 2^{n+1}. Consequently, continuing the wavelet transform past index $2^{n+1} - 1$ would reproduce the initial 2^{n-1} coefficients $a_0^{(N-1)}, \ldots, a_{2^{n-1}-1}^{(n-1)}$. In other words, the new coefficients already have period 2^{n-1}:

$$
\begin{array}{cccc}
a_0 & a_1 & a_2 & a_3 \\
h_0 & h_1 & h_2 & h_3
\end{array}
$$

$$\underbrace{}_{a_0^{(n-1)}}$$

$$
\begin{array}{ccccccccc}
\cdots & a_{2^n-2} & a_{2^n-1} & a_{2^n} & a_{2^n+1} & \cdots & a_{2^{n+1}-4} & a_{2^{n+1}-3} \\
 & & & & & & a_{2^{n+1}-2} & a_{2^{n+1}-1} & a_0 & a_1 \\
 & & & & & & h_0 & h_1 & h_2 & h_3
\end{array}
$$

$$\underbrace{\phantom{a_{2^{n+1}-2} \quad a_{2^{n+1}-1} \quad a_0 \quad a_1}}_{a_{2^{n-1}-1}^{(n-1)}}$$

and the next shift by two indices would reproduce the first coefficient, $a_0^{(n-1)}$. In particular, the periodic sequence of coefficients $(a_k^{(n-1)})_{k=0}^\infty$ does not suffer from edge effects. The periodicity of the coefficients will prove relevant for the subsequent passes of the direct transform, and for the inverse transform, which can then use periodic extensions in the same order, without reflection:

$$a_0^{(n-1)}, \ldots, a_{2^{n-1}-1}^{(n-1)}; \quad a_0^{(n-1)}, \ldots, a_{2^{n-1}-1}^{(n-1)}; \cdots$$

In other words, at the ℓth step, the wavelet transform multiplies the periodic sequence of $2^{n+1-\ell}$ coefficients $a_k^{(n+1-\ell)}$ by the matrix $_D\Omega^{-1}$, and produces a periodic sequence of $2^{n-\ell}$ coefficients $a_k^{(n-\ell)}$. The inverse transform will later multiply such periodic sequences by $_D\Omega$ to reconstruct the initial sample.   □

Example 3.21   With $n := 2$ and $N = 2^n = 2^2 = 4$, consider

$$\vec{s} = (s_0, s_1, s_2, s_3) := (0, 1, 2, 3).$$

Step 1: extend the sample periodically. Here, $N = 2^n = 4$ and $2N = 2^{(n+1)} = 8$, with, among many possible periodic extensions,

$$\underbrace{0,\quad 1,\quad 2,\quad 3;}_{\text{data}}\quad \underbrace{4,\quad 2,\quad 1,\quad -1;}_{\text{extension}}\quad \underbrace{0,\quad 1.}_{\text{short extension}}$$

Step 2: calculate the coefficients of the basic building blocks. The calculations in the preceding section produced the coefficients

$$\tilde{f}(r) = \frac{3 - \sqrt{3}}{2} \cdot \varphi(r) + \frac{5 - \sqrt{3}}{2} \cdot \varphi(r - 1)$$

$$+ \frac{7 - \sqrt{3}}{2} \cdot \varphi(r - 2) + (3 + \sqrt{3}) \cdot \varphi(r - 3)$$

$$+ \frac{3 + \sqrt{3}}{2} \cdot \varphi(r - 4) + \sqrt{3} \cdot \varphi(r - 5)$$

$$- \frac{1 + \sqrt{3}}{2} \cdot \varphi(r - 6) + \frac{1 - \sqrt{3}}{2} \cdot \varphi(r - 7).$$

Hence, calculations give

$$a_1^{(n-1)} = (\tfrac{1}{2})\langle(h_0, h_1, h_2, h_3),\ (a_{2\times 1}, a_{2\times 1+1}, a_{2\times 1+2}, a_{2\times 1+3})\rangle$$

$$= \frac{1 + \sqrt{3}}{8} \cdot \frac{7 - \sqrt{3}}{2} + \frac{3 + \sqrt{3}}{8} \cdot (3 + \sqrt{3})$$

$$+ \frac{3 - \sqrt{3}}{8} \cdot \frac{3 + \sqrt{3}}{2} + \frac{1 - \sqrt{3}}{8} \cdot \sqrt{3}$$

$$= \frac{7 + 5\sqrt{3}}{4},$$

$$c_0^{(n-1)} = (\tfrac{1}{2})\langle(h_3, -h_2, h_1, -h_0),\ (a_{2\times 0}, a_{2\times 0+1}, a_{2\times 0+2}, a_{2\times 0+3})\rangle$$

$$= \frac{1 - \sqrt{3}}{8} \cdot \frac{3 - \sqrt{3}}{2} - \frac{3 - \sqrt{3}}{8} \cdot \frac{5 - \sqrt{3}}{2}$$

$$+ \frac{3 + \sqrt{3}}{8} \cdot \frac{7 - \sqrt{3}}{2} - \frac{1 + \sqrt{3}}{8} \cdot (3 + \sqrt{3})$$

$$= -\tfrac{3}{8}.$$

Similar calculations produce the remaining coefficients

$$\left(a_0^{(n-1)}, a_1^{(n-1)}, a_2^{(n-1)}, a_3^{(n-1)}\right) = \left(\frac{18 - 5\sqrt{3}}{8}, \frac{7 + 5\sqrt{3}}{4}, \frac{8 + 3\sqrt{3}}{8}, 1 - \sqrt{3}\right),$$

$$\left(c_0^{(n-1)}, c_1^{(n-1)}, c_2^{(n-1)}, c_3^{(n-1)}\right) = \left(\frac{-3}{8}, \frac{1 - \sqrt{3}}{4}, \frac{1 - 6\sqrt{3}}{8}, 0\right).$$

Repeating the foregoing calculations with the new periodic sample

$$a_0^{(n-1)}, \qquad a_1^{(n-1)}, \qquad a_2^{(n-1)}, \qquad a_3^{(n-1)}; \qquad a_0^{(n-1)}, \qquad a_1^{(n-1)}, \qquad \ldots$$

$$= \frac{18 - 5\sqrt{3}}{8}, \frac{7 + 5\sqrt{3}}{4}, \frac{8 + 3\sqrt{3}}{8}, 1 - \sqrt{3}; \frac{18 - 5\sqrt{3}}{8}, \frac{7 + 5\sqrt{3}}{4}, \ldots$$

gives

$$\left(a_0^{(n-2)}, a_1^{(n-2)} \right) = \left(\frac{61 + 21\sqrt{3}}{32}, \frac{7(5 - 3\sqrt{3})}{32} \right),$$

$$\left(c_0^{(n-2)}, c_1^{(n-2)} \right) = \left(\frac{35 - 11\sqrt{3}}{32}, \frac{3(-9 + \sqrt{3})}{32} \right).$$

Repeating the foregoing calculations with the new periodic sample

$$a_0^{(n-2)}, \qquad a_1^{(n-2)}; \qquad a_0^{(n-2)}, \qquad a_1^{(n-2)}, \qquad \ldots$$

$$= \frac{61 + 21\sqrt{3}}{32}, \frac{7(5 - 3\sqrt{3})}{32}, \frac{61 + 21\sqrt{3}}{32}, \frac{7(5 - 3\sqrt{3})}{32}, \ldots$$

gives

$$(a_0^{(n-3)}) = \left(\frac{3}{2} \right),$$

$$(c_0^{(n-3)}) = \left(\frac{13 + 21\sqrt{3}}{32} \right).$$

The number of passes, 3, exceeds $n = 2$ by 1, because the periodic extension has 2^{n+1} entries instead of 2^n entries in the data. The coefficients just obtained mean that

$$\tilde{f}(r) = a_0\varphi(r) + a_1\varphi(r - 1) + a_2\varphi(r - 2) + a_3\varphi(r - 3)$$

$$+ a_4\varphi(r - 4) + a_5\varphi(r - 5) + a_6\varphi(r - 6) + a_7\varphi(r - 7)$$

$$= a_0^{(n-1)}\varphi([r/2]) + a_1^{(n-1)}\varphi([r/2] - 2 \cdot 1)$$

$$+ a_2^{(n-1)}\varphi([r/2] - 2 \cdot 2) + a_3^{(n-1)}\varphi([r/2] - 2 \cdot 3)$$

$$+ c_0^{(n-1)}\psi([r/2 - 1]) + c_1^{(n-1)}\psi([r/2 - 1] - 2 \cdot 1)$$

$$+ c_2^{(n-1)}\psi([r/2 - 1] - 2 \cdot 2) + c_3^{(n-1)}\psi([r/2 - 1] - 2 \cdot 3)$$

$$= a_0^{(n-2)}\varphi([r/4]) + a_1^{(n-2)}\varphi([r/4] - 4 \cdot 1)$$

$$+ c_0^{(n-2)}\psi([r/4 - 1]) + c_1^{(n-2)}\psi([r/4 - 1] - 4 \cdot 1)$$

$$+ c_0^{(n-1)} \psi([r/2 - 1]) + c_1^{(n-1)} \psi([r/2 - 1] - 2 \cdot 1)$$

$$+ c_2^{(n-1)} \psi([r/2 - 1] - 2 \cdot 2) + c_3^{(n-1)} \psi([r/2 - 1] - 2 \cdot 3)$$

$$= a_0^{(n-3)} \varphi([r/8])$$

$$+ c_0^{(n-3)} \varphi([r/8 - 1])$$

$$+ c_0^{(n-2)} \psi([r/4 - 1]) + c_1^{(n-2)} \psi([r/4 - 1] - 4 \cdot 1)$$

$$+ c_0^{(n-1)} \psi([r/2 - 1]) + c_1^{(n-1)} \psi([r/2 - 1] - 2 \cdot 1)$$

$$+ c_2^{(n-1)} \psi([r/2 - 1] - 2 \cdot 2) + c_3^{(n-1)} \psi([r/2 - 1] - 2 \cdot 3). \qquad \square$$

EXERCISES

Exercise 3.13. Calculate the Daubechies Wavelet Transform of the sample

$$\vec{s} = (s_0, s_1, s_2, s_3) := (1, 1, 1, 1).$$

Exercise 3.14. Write and test a computer program to compute The Fast Daubechies Wavelet Transform.

3.5 THE FAST INVERSE DAUBECHIES WAVELET TRANSFORM

The Fast Inverse Daubechies Wavelet Transform starts from a wavelet expansion of the following form, here with frequencies sorted line by line:

$$\tilde{f}(r) = a_0^{(-1)} \varphi([r/2^{n+1}]) + c_0^{(-1)} \psi([r/2^{n+1} - 1])$$

$$+ c_0^{(0)} \psi([r/2^n - 1]) + c_1^{(0)} \psi([r/2^n - 1] - 2^n \cdot 1) + \cdots$$

$$+ c_0^{(1)} \psi([r/2^{n-1} - 1]) + c_1^{(1)} \psi([r/2^{n-1} - 1] - 2^{n-1} \cdot 1) + \cdots$$

$$\vdots$$

$$+ c_0^{(n-1)} \psi([r/2 - 1]) + c_1^{(n-1)} \psi([r/2 - 1] - 2 \cdot 1) + \cdots.$$

Hence the Fast Inverse Daubechies Wavelet Transform reconstructs the expansion of the same signal $\tilde{f}$ back with basic building blocks:

$$\tilde{f}(r) = a_0 \varphi(r) + a_1 \varphi(r - 1) + a_2 \varphi(r - 2) + a_3 \varphi(r - 3)$$

$$+ a_4 \varphi(r - 4) + a_5 \varphi(r - 5) + a_6 \varphi(r - 6) + a_7 \varphi(r - 7) + \cdots.$$

A concise yet still explicit way to derive formulae for the inverse transform utilizes the formula for the direct transform with matrices:

$$\left(a_0^{(n-1)}, c_0^{(n-1)}, a_1^{(n-1)}, c_1^{(n-1)}, a_2^{(n-1)}, c_2^{(n-1)}, \ldots \right) = {}_D\Omega^{-1}\vec{a}^{(n)},$$

$$
\begin{pmatrix}
a_0^{(n-1)} \\
c_0^{(n-1)} \\
a_1^{(n-1)} \\
c_1^{(n-1)} \\
a_2^{(n-1)} \\
c_2^{(n-1)} \\
\vdots
\end{pmatrix}
= \frac{1}{2}
\begin{pmatrix}
h_0 & h_1 & h_2 & h_3 & & & \cdots \\
h_3 & -h_2 & h_1 & -h_0 & & & \cdots \\
 & & h_0 & h_1 & h_2 & h_3 & \cdots \\
 & & h_3 & -h_2 & h_1 & -h_0 & \cdots \\
 & & & & h_0 & h_1 & \cdots \\
 & & & & h_3 & -h_2 & \cdots \\
\vdots & \vdots & \vdots & \vdots & \vdots & \vdots &
\end{pmatrix}
\begin{pmatrix}
a_0^{(n)} \\
a_1^{(n)} \\
a_2^{(n)} \\
a_3^{(n)} \\
a_4^{(n)} \\
a_5^{(n)} \\
\vdots
\end{pmatrix}.
$$

Hence, a multiplication of both sides by ${}_D\Omega$ yields a formula for the inverse transform:

$$\vec{a}^{(n)} = {}_D\Omega \left(a_0^{(n-1)}, c_0^{(n-1)}, a_1^{(n-1)}, c_1^{(n-1)}, a_2^{(n-1)}, c_2^{(n-1)}, \ldots \right),$$

$$
\begin{pmatrix}
a_0^{(n)} \\
a_1^{(n)} \\
a_2^{(n)} \\
a_3^{(n)} \\
a_4^{(n)} \\
a_5^{(n)} \\
\vdots
\end{pmatrix}
=
\begin{pmatrix}
h_0 & h_3 & & & \cdots \\
h_1 & -h_2 & & & \cdots \\
h_2 & h_1 & h_0 & h_3 & \cdots \\
h_3 & -h_0 & h_1 & -h_2 & \cdots \\
 & & h_2 & h_1 & \cdots \\
 & & h_3 & -h_0 & \cdots \\
\vdots & \vdots & \vdots & \vdots &
\end{pmatrix}
\begin{pmatrix}
a_0^{(n-1)} \\
c_0^{(n-1)} \\
a_1^{(n-1)} \\
c_1^{(n-1)} \\
a_2^{(n-1)} \\
c_2^{(n-1)} \\
\vdots
\end{pmatrix}.
$$

The formula just displayed also reveals a shift by two indices, in the sense that the first complete inner product gives the second pair of coefficients; indeed, the second and third rows give a_2 and a_3:

$$a_2^{(n)} = h_2 a_0^{(n-1)} + h_1 c_0^{(n-1)} + h_0 a_1^{(n-1)} + h_3 c_1^{(n-1)},$$

$$a_3^{(n)} = h_3 a_0^{(n-1)} - h_0 c_0^{(n-1)} + h_1 a_1^{(n-1)} - h_2 c_1^{(n-1)}.$$

Similarly, the fourth and fifth rows give a_4 and a_5:

$$a_4^{(n)} = h_2 a_1^{(n-1)} + h_1 c_1^{(n-1)} + h_0 a_2^{(n-1)} + h_3 c_2^{(n-1)},$$

$$a_5^{(n)} = h_3 a_1^{(n-1)} - h_0 c_1^{(n-1)} + h_1 a_2^{(n-1)} - h_2 c_2^{(n-1)}.$$

Generally, the rows with indices $2k$ and $2k + 1$ give a_{2k} and a_{2k+1}:

$$a_{2k}^{(n)} = h_2 a_{k-1}^{(n-1)} + h_1 c_{k-1}^{(n-1)} + h_0 a_k^{(n-1)} + h_3 c_k^{(n-1)},$$

$$a_{2k+1}^{(n)} = h_3 a_{k-1}^{(n-1)} - h_0 c_{k-1}^{(n-1)} + h_1 a_k^{(n-1)} - h_2 c_k^{(n-1)}.$$

There remain the first two rows. By periodicity, the first two rows coincide with the rows with indices $2 * (2^{n-1} - 1)$ and $2 * (2^{n-1} - 1) + 1$, which correspond to $k = 2^{n-1} - 1$:

$$a_0^{(n)} = h_2 a_{2^{n-1}-1}^{(n-1)} + h_1 c_{2^{n-1}-1}^{(n-1)} + h_0 a_0^{(n-1)} + h_3 c_0^{(n-1)},$$

$$a_1^{(n)} = h_3 a_{2^{n-1}-1}^{(n-1)} - h_0 c_{2^{n-1}-1}^{(n-1)} + h_1 a_0^{(n-1)} - h_2 c_0^{(n-1)}.$$

Thus, it suffices to place the last two entries $a_{2^{n-1}-1}^{(n-1)}$ and $c_{2^{n-1}-1}^{(n-1)}$ in front of the data $a_0^{(n-1)}, c_0^{(n-1)}, \ldots$ to get the formula

$$\vec{a}^{(n)} = D\Omega(\vec{a}^{(n-1)}, \vec{c}^{(n-1)}),$$

$$
\begin{pmatrix} a_0^{(n)} \\ a_1^{(n)} \\ a_2^{(n)} \\ a_3^{(n)} \\ a_4^{(n)} \\ a_5^{(n)} \\ \vdots \end{pmatrix}
=
\begin{pmatrix} h_2 & h_1 & h_0 & h_3 & \cdots \\ h_3 & -h_0 & h_1 & -h_2 & \cdots \\ & & h_2 & h_1 & \cdots \\ & & h_3 & -h_0 & \cdots \\ \vdots & \vdots & \vdots & \vdots \end{pmatrix}
\begin{pmatrix} a_{2^{n-1}-1}^{(n-1)} \\ c_{2^{n-1}-1}^{(n-1)} \\ a_0^{(n-1)} \\ c_0^{(n-1)} \\ a_1^{(n-1)} \\ c_1^{(n-1)} \\ a_2^{(n-1)} \\ c_2^{(n-1)} \\ \vdots \end{pmatrix}.
$$

Example 3.22 Consider the wavelet expansion from Example 3.21:

$$\tilde{f}(r) = (\tfrac{3}{2})\varphi([r/8]) + \tfrac{13+21\sqrt{3}}{32}\psi([r/8 - 1])$$

$$+ \frac{35 - 11\sqrt{3}}{32}\psi([r/4 - 1]) + \frac{3(-9 + \sqrt{3})}{32}\psi([r/4 - 1] - 4 \cdot 1)$$

$$+ -\frac{3}{8}\psi([r/2 - 1]) + \frac{1 - \sqrt{3}}{4}\psi([r/2 - 1] - 2 \cdot 1)$$

$$+ \frac{1 - 6\sqrt{3}}{8}\psi([r/2 - 1] - 2 \cdot 2) + 0\psi([r/2 - 1] - 2 \cdot 3).$$

The computation of the inverse transform proceeds as follows.

Step 1. Extend $a_0^{(-1)} = (\frac{3}{2})$ and $c_0^{(-1)} = \frac{13+21\sqrt{3}}{32}$ periodically,

$$\left(a_0^{(-1)}, c_0^{(-1)}; a_0^{(-1)}, c_0^{(-1)}; \ldots\right) = \left(\frac{3}{2}, \frac{13+21\sqrt{3}}{32}; \frac{3}{2}, \frac{13+21\sqrt{3}}{32}; \ldots\right).$$

In the first step, the last two entries coincide with all the other pairs. Hence reconstruct $(a_0^{(0)}, a_1^{(0)})$:

$$
\begin{aligned}
a_0^{(0)} &= h_2 a_0^{(-1)} + h_1 c_0^{(-1)} + h_0 a_0^{(-1)} + h_3 c_0^{(-1)} \\
&= \frac{3-\sqrt{3}}{4}(\frac{3}{2}) + \frac{3+\sqrt{3}}{4}\frac{13+21\sqrt{3}}{32} + \frac{1+\sqrt{3}}{4}(\frac{3}{2}) + \frac{1-\sqrt{3}}{4}\frac{13+21\sqrt{3}}{32} \\
&= \frac{61+21\sqrt{3}}{32};
\end{aligned}
$$

$$
\begin{aligned}
a_1^{(0)} &= h_3 a_0^{(-1)} - h_0 c_0^{(-1)} + h_1 a_0^{(-1)} - h_2 c_0^{(-1)} \\
&= \frac{1-\sqrt{3}}{4}(\frac{3}{2}) - \frac{1+\sqrt{3}}{4}\frac{13+21\sqrt{3}}{32} + \frac{3+\sqrt{3}}{4}(\frac{3}{2}) - \frac{3-\sqrt{3}}{4}\frac{13+21\sqrt{3}}{32} \\
&= \frac{35-21\sqrt{3}}{32} = \frac{7(5-3\sqrt{3})}{32}.
\end{aligned}
$$

Step 2. Extend the coefficients $(a_0^{(0)}, a_1^{(0)})$ just found and the coefficients $(c_0^{(0)}, c_1^{(0)})$ from the wavelet transform periodically, beginning with the last pair:

$$
\begin{aligned}
&\left(a_1^{(0)}, c_1^{(0)}; a_0^{(0)}, c_0^{(0)}, a_1^{(0)}, c_1^{(0)}; a_0^{(0)}, c_0^{(0)}, a_1^{(0)}, c_1^{(0)}; \ldots\right) \\
&= \left(\frac{7(5-3\sqrt{3})}{32}, \frac{3(-9+\sqrt{3})}{32}; \frac{61+21\sqrt{3}}{32}, \frac{35-11\sqrt{3}}{32}, \frac{7(5-3\sqrt{3})}{32},\right. \\
&\qquad \left. \frac{3(-9+\sqrt{3})}{32}; \frac{61+21\sqrt{3}}{32}, \frac{35-11\sqrt{3}}{32}, \frac{7(5-3\sqrt{3})}{32}, \frac{3(-9+\sqrt{3})}{32}; \ldots\right),
\end{aligned}
$$

and reconstruct $(a_0^{(1)}, a_1^{(1)}, a_2^{(1)}, a_3^{(1)})$:

$$
\begin{aligned}
a_0^{(1)} &= h_2 a_1^{(0)} + h_1 c_1^{(0)} + h_0 a_0^{(0)} + h_3 c_0^{(0)} \\
&= \frac{3-\sqrt{3}}{4}\frac{7(5-3\sqrt{3})}{32} + \frac{3+\sqrt{3}}{4}\frac{3(-9+\sqrt{3})}{32} \\
&\quad + \frac{1+\sqrt{3}}{4}\frac{61+21\sqrt{3}}{32} + \frac{1-\sqrt{3}}{4}\frac{35-11\sqrt{3}}{32} \\
&= \frac{18-5\sqrt{3}}{8};
\end{aligned}
$$

$$a_1^{(1)} = h_3 a_1^{(0)} - h_0 c_1^{(0)} + h_1 a_0^{(0)} - h_2 c_0^{(0)}$$

$$= \frac{1-\sqrt{3}}{4} \frac{7(5-3\sqrt{3})}{32} - \frac{1+\sqrt{3}}{4} \frac{3(-9+\sqrt{3})}{32}$$

$$+ \frac{3+\sqrt{3}}{4} \frac{61+21\sqrt{3}}{32} - \frac{3-\sqrt{3}}{4} \frac{35-11\sqrt{3}}{32}$$

$$= \frac{7+5\sqrt{3}}{4}.$$

For the next two coefficients, $a_2^{(1)}$ and $a_3^{(1)}$, shift indices by 1 in the periodic extensions:

$$a_2^{(1)} = h_2 a_0^{(0)} + h_1 c_0^{(0)} + h_0 a_1^{(0)} + h_3 c_1^{(0)}$$

$$= \frac{3-\sqrt{3}}{4} \frac{61+21\sqrt{3}}{32} + \frac{3+\sqrt{3}}{4} \frac{35-11\sqrt{3}}{32}$$

$$+ \frac{1+\sqrt{3}}{4} \frac{7(5-3\sqrt{3})}{32} + \frac{1-\sqrt{3}}{4} \frac{3(-9+\sqrt{3})}{32}$$

$$= \frac{8+3\sqrt{3}}{8};$$

$$a_3^{(1)} = h_3 a_0^{(0)} - h_0 c_0^{(0)} + h_1 a_1^{(0)} - h_2 c_1^{(0)}$$

$$= \frac{1-\sqrt{3}}{4} \frac{61+21\sqrt{3}}{32} - \frac{1+\sqrt{3}}{4} \frac{35-11\sqrt{3}}{32}$$

$$+ \frac{3+\sqrt{3}}{4} \frac{7(5-3\sqrt{3})}{32} - \frac{3-\sqrt{3}}{4} \frac{3(-9+\sqrt{3})}{32}$$

$$= 1 - \sqrt{3}.$$

Step 3. Extend $(a_0^{(1)}, a_1^{(1)}, a_2^{(1)}, a_3^{(1)})$ and $(c_0^{(1)}, c_1^{(1)}, c_2^{(1)}, c_3^{(1)})$ periodically, beginning with the last pair $(a_3^{(1)}, c_3^{(1)})$, reconstruct

$$\left(a_0^{(2)}, a_1^{(2)}, a_2^{(2)}, a_3^{(2)}, a_4^{(2)}, a_5^{(2)}, a_6^{(2)}, a_7^{(2)}\right),$$

and verify that they agree with the coefficients from Example 3.21:

$$a_0^{(2)} = h_2 a_3^{(1)} + h_1 c_3^{(1)} + h_0 a_0^{(1)} + h_3 c_0^{(1)}$$

$$= \frac{3-\sqrt{3}}{4}(1-\sqrt{3}) + \frac{3+\sqrt{3}}{4} 0 + \frac{1+\sqrt{3}}{4} \frac{18-\sqrt{3}}{8}$$

$$+ \frac{1-\sqrt{3}}{4} \frac{-3}{8}$$

$$= \frac{3 - \sqrt{3}}{2};$$

$$a_1^{(2)} = h_3 a_3^{(1)} - h_0 c_3^{(1)} + h_1 a_0^{(1)} - h_2 c_0^{(1)}$$

$$= \frac{1 - \sqrt{3}}{4}(1 - \sqrt{3}) - \frac{1 + \sqrt{3}}{4}0 + \frac{3 + \sqrt{3}}{4}\frac{18 - \sqrt{3}}{8}$$

$$- \frac{3 - \sqrt{3}}{4}\frac{-3}{8}$$

$$= \frac{5 - \sqrt{3}}{2}.$$

For the next two coefficients, $a_2^{(2)}$ and $a_3^{(2)}$, shift indices by 1 in the periodic extensions:

$$a_2^{(2)} = h_2 a_0^{(1)} + h_1 c_0^{(1)} + h_0 a_1^{(1)} + h_3 c_1^{(1)}$$

$$= \frac{3 - \sqrt{3}}{4}\frac{18 - 5\sqrt{3}}{8} + \frac{3 + \sqrt{3}}{4}\frac{-3}{8} + \frac{1 + \sqrt{3}}{4}\frac{7 + 5\sqrt{3}}{4}$$

$$+ \frac{1 - \sqrt{3}}{4}\frac{1 - \sqrt{3}}{4}$$

$$= \frac{7 - \sqrt{3}}{2};$$

$$a_3^{(2)} = h_3 a_0^{(1)} - h_0 c_0^{(1)} + h_1 a_1^{(1)} - h_2 c_1^{(1)}$$

$$= \frac{1 - \sqrt{3}}{4}\frac{18 - 5\sqrt{3}}{8} - \frac{1 + \sqrt{3}}{4}\frac{-3}{8} + \frac{3 + \sqrt{3}}{4}\frac{7 + 5\sqrt{3}}{4}$$

$$- \frac{3 - \sqrt{3}}{4}\frac{1 - \sqrt{3}}{4}$$

$$= \frac{3 + \sqrt{3}}{2}.$$

Similarly, shift again by 1 for $a_4^{(0)}$ and $a_5^{(0)}$, and then again by 1 for $a_6^{(0)}$ and $a_7^{(0)}$.
□

Remark 3.23 The inverse wavelet transform reproduces only the *coefficients* of the shifted basic building blocks of the expansion of $\tilde{f}$. If such coefficients were selected to coincide with the values of the sample, then $s_k = a_k$. In contrast, if the coefficients a_k represent averages of the values of the sample, then it suffices to evaluate each shifted building block and then add the sum to approximate the *values* of $\tilde{f}$. For instance, if $a_k = s_{k+1}\varphi(1) + s_{k+2}\varphi(2)$, then $s_k \approx a_{k-1}\varphi(1) + a_{k-2}\varphi(2)$.
□

Example 3.24 With the same numerical example as in Examples 3.21 and 3.22,

$$\tilde{f}(2) = \frac{3 - \sqrt{3}}{2} \cdot \varphi(2) + \frac{5 - \sqrt{3}}{2} \cdot \varphi(2 - 1)$$

$$+ \frac{7 - \sqrt{3}}{2} \cdot \varphi(2 - 2) + \frac{3 + \sqrt{3}}{2} \cdot \varphi(2 - 3)$$

$$+ \frac{3 + \sqrt{3}}{2} \cdot \varphi(2 - 4) + \frac{\sqrt{3}}{2} \cdot \varphi(2 - 5)$$

$$- \frac{1 + \sqrt{3}}{2} \cdot \varphi(2 - 6) + \frac{1 - \sqrt{3}}{2} \cdot \varphi(2 - 7).$$

$$= \frac{3 - \sqrt{3}}{2} \cdot \frac{1 - \sqrt{3}}{2} + \frac{5 - \sqrt{3}}{2} \cdot \frac{1 + \sqrt{3}}{2} + 0 + 0 + 0 + 0 + 0 + 0$$

$$= 2.$$

In this particular example, $\tilde{f}(2) = f(2) = 2$, because the initial sample $(0, 1, 2, 3)$ came from an affine function f, which Daubechies wavelets can reproduce exaclty, as proved in a subsequent chapter. In general, however, $\tilde{f}$ constitutes only an approximation of f, so that $\tilde{f}(r) \approx f(r)$, with the accuracy of the approximation depending upon the rate of the initial sample of f. □

EXERCISES

Exercise 3.15. Calculate the Fast Inverse Daubechies Wavelet Transform from the wavelet coeffcients

$$\left(a_0^2, c_0^{(1)}, c_0^2, c_1^{(1)} \right) := (1, 0, 0, 0).$$

Exercise 3.16. Write and test a computer program to compute the Fast Inverse Daubechies Wavelet Transform. Test the program by computing the Ordered Daubechies Wavelet Transform, then the inverse transform, and finally comparing with the data.

3.6 MULTIDIMENSIONAL DAUBECHIES WAVELET TRANSFORMS

For a square array $S = (s_{i,j})$ with 2^n rows and 2^n columns, exactly as the Haar transform, one of Daubechies' two-dimensional wavelet transforms consists in performing a first pass on each row, from $(a_{i,j}^{(n)})_{j=0}^{2^n-1}$ to $(a_{i,j}^{(n-1)})_{j=0}^{2^{n-1}-1}$ and $(c_{i,j}^{(n-1)})_{j=0}^{2^{n-1}-1}$, and then a first pass on each of the new columns. Also, as the two-dimensional Haar transform, such a two-dimensional Daubechies wavelet transform produces the coefficients of tensor-product wavelets, pictured in Figures 3.8 and 3.9:

$\Phi_{k,\ell}^{(n-1)} = \varphi_k \otimes \varphi_\ell$ in the upper left-hand corner of each 2×2 block,

$\Psi_{k,\ell}^{h,(N-1)} = \varphi_k \otimes \psi_\ell$ in the upper right-hand corner of each 2×2 block,

$\Psi_{k,\ell}^{v,(N-1)} = \psi_k \otimes \varphi_\ell$ in the lower left-hand corner of each 2×2 block,

$\Psi_{k,\ell}^{d,(N-1)} = \psi_k \otimes \psi_\ell$ in the lower right-hand corner of each 2×2 block.

The next pass then repeats the first pass on the smaller array, with 2^{n-1} rows and 2^{n-1} columns, of the coefficients of each $\Phi_{k,\ell}^{(n-1)}$ in the upper left-hand corner of each 2×2 block.

Indeed, if $s_{i,j} = f(i,j)$ and

$$\tilde{f} = \sum_{i=0}^{2^n-1} \sum_{j=0}^{2^n-1} a_{i,j} \varphi_i^{(n)} \otimes \varphi_j^{(n)},$$

and if the faster building blocks $\varphi_m^{(n)}$ relate to slower functions γ_ℓ—which represent a shorthand for both slower building blocks $\varphi_k^{(n-1)}$ and slower wavelets $\psi_\ell^{(n-1)}$—through linear equations

$$\varphi_m^{(n)} = \sum_{\ell=0}^{2^n-1} (D\Omega^{-1})_{m,\ell} \gamma_\ell,$$

then

$$\tilde{f} = \sum_{i=0}^{2^n-1} \sum_{j=0}^{2^n-1} a_{i,j} \varphi_i^{(n)} \otimes \varphi_j^{(n)}$$

$$= \sum_{i=0}^{2^n-1} \sum_{j=0}^{2^n-1} a_{i,j} \sum_{k=0}^{2^n-1} (D\Omega^{-1})_{i,k} \gamma_k \otimes \sum_{\ell=0}^{2^n-1} (D\Omega^{-1})_{j,\ell} \gamma_\ell$$

$$= \sum_{k=0}^{2^n-1} \sum_{\ell=0}^{2^n-1} \sum_{i=0}^{2^n-1} \sum_{j=0}^{2^n-1} (D\Omega^{-1})_{i,k} (D\Omega^{-1})_{j,\ell} a_{i,j} \gamma_k \otimes \gamma_\ell,$$

where the inner sum

$$\sum_{j=0}^{2^n-1} (D\Omega^{-1})_{j,\ell} a_{i,j} = a_{i,\ell}^{(n-1)}$$

represents one pass of the one-dimensional Daubechies Wavelet Transform on each row with index i, and then the next inner sum

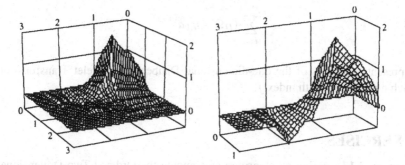

Figure 3.8 Daubechies' two-dimensional tensor-product scaling function $\Phi = \varphi \otimes \varphi$ (left), and an "exploded view" (right).

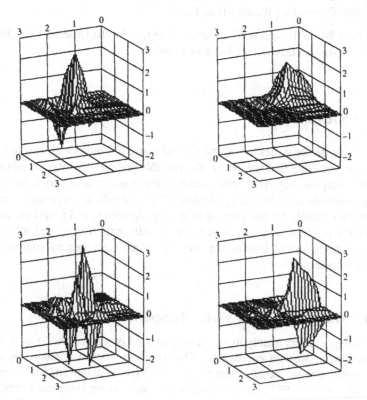

Figure 3.9 Two-Dimensional Tensor-Product Daubechies Wavelets, in the following order.

$$\Psi^v = \psi \otimes \varphi, \quad \Phi = \varphi \otimes \varphi,$$
$$\Psi^d = \psi \otimes \psi, \quad \Psi^h = \varphi \otimes \psi.$$

$$\sum_{i=0}^{2^n-1} (D\Omega^{-1})_{i,k} a_{i,\ell}^{(n-1)}$$

represents one pass of the one-dimensional Daubechies Wavelet Transform on each new column with index j.

EXERCISES

Exercise 3.17. Design an algorithm to compute the Ordered Two-Dimensional Fast Daubechies Wavelet Transform.

Exercise 3.18. Design an algorithm to compute the Ordered Two-Dimensional Fast *Inverse* Daubechies Wavelet Transform.

Exercise 3.19. Write and test a computer program for the Ordered Two-Dimensional Fast Daubechies Wavelet Transform.

Exercise 3.20. Write and test a computer program for the Ordered Two-Dimensional Fast *Inverse* Daubechies Wavelet Transform.

3.7   EXAMPLES

This section demonstrates the practical significance of Daubechies wavelets with real data. Any other finite sequence of—possibly random—numbers might serve the same purpose, but the specific contexts demonstrated here may help in providing suggestions for further applications. The first subsection provides several sets of data, mainly for the purpose of testing algorithms and programs, and to explain the practical significance of wavelet coefficients. The second subsection demonstrate the use of Daubechies wavelets to compress data by removing the highest frequencies.

3.7.1   Hangman Creek Water Temperature Analysis

This example serves mainly to test algorithms and programs, and to explain the significance of wavelet coefficients. The following sixteen numbers represent semiweekly measurements of temperature, in degrees Fahrenheit, for December 1992 and January 1993 at a fixed common location along Hangman Creek near Spokane, Washington, in a study of riverbank erosion by Jim Fox, of Spokane, Washington.

32.0, 10.0, 20.0, 38.0, 37.0, 28.0, 38.0, 34.0,
18.0, 24.0, 18.0, 9.0, 23.0, 24.0, 28.0, 34.0.

A mirror extension and an additional short reflection produce the extended array

$$\underbrace{32.0\ 10.0\ 20.0\ 38.0\ 37.0\ 28.0\ 38.0\ 34.0\ 18.0\ 24.0\ 18.0\ 9.0\ 23.0\ 24.0\ 28.0\ 34.0}_{\text{data}},$$

$$\underbrace{34.0\ 28.0\ 24.0\ 23.0\ 9.0\ 18.0\ 24.0\ 18.0\ 34.0\ 38.0\ 28.0\ 37.0\ 38.0\ 20.0\ 10.0\ 32.0}_{\text{mirror reflection}},$$

$$\underbrace{32.0\ 10.0}_{\text{short reflection}}.$$

From the approximately interpolating coefficients $a_k := s_k$ equal to the data, the complete Ordered Daubechies Wavelet Transform produced the following result, sorted here by frequency, from the lowest frequency on the top line to the highest frequency on the bottom line (split into four quarter lines):

25.937500,   −0.064378;

−2.017716,   2.031207;

−8.502922,   0.902389,   7.039151,   −2.277719;

0.787219,   0.534696,   −1.366025,   3.009855, ...
... − 3.850159,   3.871994,   −4.035136,   −7.702443;

−5.660254,   4.470671,   3.042468,   −6.415064, ...
...2.122595,   2.334936,   −0.957532,   2.598076, ...
... − 1.207532,   −6.665064,   4.372595,   2.084936, ...
... − 5.207532,   7.220671,   −11.660254,   9.526279.

The first coefficient, $a_0^{(-1)} = 25.937500$, equals the arithmetic average of the data, exactly as the first coefficient of Haar's wavelet transform of the same data.

As a verification, the Daubechies inverse wavelet transform applied to the result just obtained reconstructed the initial periodic extension of the data exactly.

EXERCISES

Exercise 3.21. Compute the Daubechies wavelet transform of the following measurements of the ground frost depth, in centimeters, at Qualchan on Hangman Creek, for the same period (also by Jim Fox), and verify the result through the Daubechies inverse wavelet transform.

22.0, 27.0, 48.8, 47.5, 47.0, 48.5, 48.0, 47.0,
43.0, 41.0, 41.0, 38.0, 36.0, 47.1, 34., 32.0

Exercise 3.22. Compute the Daubechies wavelet transform of the following measurements of the ground frost depth, in centimeters, at Kracher on Hangman Creek, for the same period (also by Jim Fox), and verify the result through Daubechies inverse wavelet transform.

$$12.0, \ 16.0, \ 27.0, \ 32.8, \ 33.5, \ 33.5, \ 39.0, \ 39.0,$$
$$40.0, \ 41.3, \ 41.3, \ 42.0, \ 43.0, \ 45.0, \ 35.5, \ 49.0$$

Exercise 3.23. Compute the Daubechies wavelet transform of the following measurements of river flow, in cubic feet per second, at the US Geological Survey Data Station 1242400 on Hangman Creek, for the same period, and verify the result through the Daubechies inverse wavelet transform.

$$10.0, \ 12.0, \ 12.0, \ 7.0, \ 8.0, \ 9.1, \ 8.2, \ 9.4,$$
$$16.0, \ 15.0, \ 13.0, \ 11.0, \ 6.4, \ 9.0, \ 19.0, \ 118.0$$

Exercise 3.24. Obtain data of any kind and analyze them with the Daubechies Wavelet Transform.

3.7.2   Financial Stock Index Image Compression

This example demonstrates the use of the Daubechies Wavelet Transforms to compress large data sets by removing high frequencies.

The top left panel in Figure 3.10 displays on the vertical axis the New York Stock Exchange (NYSE) Composite Index, and on the horizontal axis the date, from 2 January 1981 (business day 0) through 7 February 1988 (business day 2047).

The middle left panel shows the mirror reflection and short extension of the data. The middle right panel shows the Ordered Daubechies Wavelet Transform, without the average, thus beginning with the coefficient of the longest wavelet, and ending with the 1024 coefficients of the fastest wavelets on the right-hand half of the graph.

The bottom right panel in Figure 3.10 shows the coefficients of the Ordered Daubechies Wavelet Transform after annihilation of all the coefficients from the highest frequency—now represented by the horizontal segment at height zero on the right-hand half of the graph—in other words, after omission of all the wavelet coefficients from the first pass. The remaining averages from the first pass thus occupy 50% less storage space than the initial data. The bottom left panel displays the reconstruction, after application of the inverse transform to the compressed transform. The compression has altered the information in the intial data, but for the purpose of displaying graphics the compression shows little difference from the initial data.

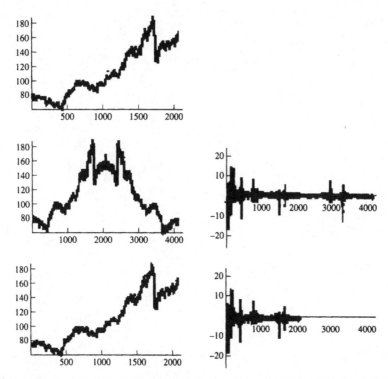

Figure 3.10 New York Stock Exchange Composite Index for 1981–1987. *Top left*. Data: index vs. business day. *Middle left*. Mirror reflection and short extension of the data. *Middle right*. Ordered Daubechies Wavelet Transform. *Bottom left*. Compression ratio $\frac{1}{2} = 50\%$. *Bottom right*. Annihilation of all coefficients from the highest frequency.

EXERCISES

Exercise 3.25. Identify the significance of the first coefficient, $a_0^{(-1)} = 111.149$, of the Daubechies Wavelet Transform of the data in Figure 3.10.

Exercise 3.26. Compute the Daubechies Wavelet Transform of the data on two-dimensional and three-dimensional diffusion given in the preceding chapter.

PART B

Basic Fourier Analysis

CHAPTER 4

Inner Products and Orthogonal Projections

4.0 INTRODUCTION

A wavelet representation of a signal constitutes an orthogonal projection, like a shadow, of the signal on a space of wavelets. Therefore, this chapter reviews the linear algebra underlying the concept, theory, and applications of orthogonal projections. The theory reviewed here also appear in several textbooks, for instance, in linear algebra [14], [15], [24], or in functional analysis [45].

4.1 LINEAR SPACES

4.1.1 Number Fields

Various applications require various types of "numbers." For example, physics and engineering use "real" numbers and "complex" numbers, while cryptography also uses "rational" numbers and systems containing only finitely many "modular" integers. Daubechies wavelets rely on yet another system of numbers, based on combinations of rational numbers and the square root of 3. Nevertheless, all such systems of numbers—called "number fields"—share certain algebraic features, which lend themselves to a common treatment. In other words, the abstract concept of "number field" allows the same theory to apply to various applications, and most importantly, it enables users to recognize that some concepts arising from some applications may also prove useful in other applications. With digital computers, a review of the concept of number field also offers an opportunity to demonstrate that the "machine numbers" used in such computers do *not* form a field, whence arises the need for some analysis of error, studied in numerical analysis [27], [39].

117

Table 4.1   Algebraic Properties of Number Fields

The following properties must hold for all elements $h, k, \ell \in \mathbb{F}$.

(1)	Associativity of $+$	$[h + \ell] + k = h + [\ell + k]$
(2)	Commutativity of $+$	$h + \ell = \ell + h$
(3)	Additive identity	$\ell + 0_{\mathbb{F}} = \ell = 0_{\mathbb{F}} + \ell$
(4)	Additive inverse	$\mathbb{F}$ contains k with $\ell + k = 0_{\mathbb{F}} = k + \ell$
(5)	Associativity of $\times$	$[h\ell]k = h[\ell k]$
(6)	Commutativity of $\times$	$h\ell = \ell h$
(7)	Multiplicative identity	$\ell 1_{\mathbb{F}} = \ell = 1_{\mathbb{F}}\ell$
(8)	Multiplicative inverse	If $\ell \neq 0_{\mathbb{F}}$,
		then $\mathbb{F}$ contains k with $k\ell = 1_{\mathbb{F}}$
(9)	Distributivity	$h[\ell + k] = [h\ell] + [hk]$

Definition 4.1   A **number field** consists of a set $\mathbb{F}$ containing at least two distinct elements $0_{\mathbb{F}}$ and $1_{\mathbb{F}}$, with two binary operations $+$ and $\times$, which are functions

$$+ : \mathbb{F} \times \mathbb{F} \to \mathbb{F}, \quad (r, s) \mapsto r + s,$$

$$\times : \mathbb{F} \times \mathbb{F} \to \mathbb{F}, \quad (r, s) \mapsto r \times s \quad (\text{or} \quad r \cdot s \quad \text{or} \quad rs),$$

such that all the properties in Table 4.1 hold. □

Example 4.2   The set $\mathbb{Q}$ of all rational numbers, with ordinary addition and multiplication, forms a number field. □

Example 4.3   The set $\mathbb{R}$ of all real numbers, with ordinary addition and multiplication, forms a number field. □

Example 4.4   The set $\mathbb{C}$ of all complex numbers, with ordinary addition and multiplication, forms a number field. □

Example 4.5   The set $\mathbb{Q}[\sqrt{3}]$ of all combinations $p + q\sqrt{3}$ with rational numbers p and q, with the addition and multiplication of real numbers, is a number field, where for each nonzero $p + q\sqrt{3}$,

$$\frac{1}{p + q\sqrt{3}} = \frac{p}{[p^2 - 3q^2]} - \frac{q}{[p^2 - 3q^2]}\sqrt{3}. \qquad \square$$

Example 4.6   The set $\mathbb{Z}_2 := \{\bar{0}, \bar{1}\}$ forms a field, called the field of "integers modulo two," with the following operations:

$+$	$\bar{0}$	$\bar{1}$
$\bar{0}$	$\bar{0}$	$\bar{1}$
$\bar{1}$	$\bar{1}$	$\bar{0}$

$\times$	$\bar{0}$	$\bar{1}$
$\bar{0}$	$\bar{0}$	$\bar{0}$
$\bar{1}$	$\bar{0}$	$\bar{1}$

□

Counterexample 4.7 The set $\mathbb{Z}$ of all negative, zero, and positive integers, with ordinary addition and multiplication is *not* a number field, because no integer (except 1 and -1) has an integer multiplicative inverse. For instance, for 2, there does not exist any integer n such that $2n = 1$.   □

The following theorem demonstrates how some of the familiar features of numbers extend to all types of number fields.

Theorem 4.8 *Each number field $\mathbb{F}$ shares the following features.*

(A) The field $\mathbb{F}$ contains only one additive identity.

(B) For each $h \in \mathbb{F}$, the field $\mathbb{F}$ contains only one additive inverse for h.

(C) For each element $h \in \mathbb{F}$, the equalities $0_{\mathbb{F}}h = 0_{\mathbb{F}} = h0_{\mathbb{F}}$ hold.

(D) The field $\mathbb{F}$ contains only one multiplicative identity.

(E) For each nonzero $h \in \mathbb{F}$, the field $\mathbb{F}$ contains only one multiplicative inverse for h.

PROOF:

(A) Assume that $\mathbb{F}$ contains elements 0_1 and 0_2 such that $\ell + 0_1 = \ell = 0_1 + \ell$ and $\ell + 0_2 = \ell = 0_2 + \ell$ for every $\ell \in \mathbb{F}$. Then the following argument confirms that $0_1 = 0_2$. Setting $\ell := 0_2$ in $\ell + 0_1 = \ell = 0_1 + \ell$ gives $0_2 + 0_1 = 0_2 = 0_1 + 0_2$. Similarly, setting $\ell := 0_1$ in $\ell + 0_2 = \ell = 0_2 + \ell$ gives $0_1 + 0_2 = 0_1 = 0_2 + 0_1$. Hence, combining the two results just obtained yields $0_2 = 0_1 + 0_2 = 0_1$.

(B) Assume that for some element $h \in \mathbb{F}$, the number field $\mathbb{F}$ contains elements k_1 and k_2 such that $h + k_1 = 0_{\mathbb{F}} = k_1 + h$ and $h + k_2 = 0_{\mathbb{F}} = k_2 + h$. Hence,

$$k_2 = k_2 + 0_{\mathbb{F}} = k_2 + (h + k_1) = (k_2 + h) + k_1 = 0_{\mathbb{F}} + k_1 = k_1.$$

(C)

$$0_{\mathbb{F}}h = (0_{\mathbb{F}} + 0_{\mathbb{F}})h = (0_{\mathbb{F}}h) + (0_{\mathbb{F}}h).$$

Because $0_{\mathbb{F}}h \in \mathbb{F}$, the number field $\mathbb{F}$ also contains an additive inverse ℓ for $0_{\mathbb{F}}h$. Hence,

$$0_{\mathbb{F}}h = (0_{\mathbb{F}}h) + (0_{\mathbb{F}}h),$$

$$(0_{\mathbb{F}}h) + \ell = [(0_{\mathbb{F}}h) + (0_{\mathbb{F}}h)] + \ell,$$

$$0_{\mathbb{F}} = (0_{\mathbb{F}}h) + \ell = [(0_{\mathbb{F}}h) + (0_{\mathbb{F}}h)] + \ell$$

$$= (0_{\mathbb{F}}h) + [(0_{\mathbb{F}}h) + \ell] = (0_{\mathbb{F}}h) + 0_{\mathbb{F}} = (0_{\mathbb{F}}h),$$

which means that $0_{\mathbb{F}}h = 0_{\mathbb{F}}$. The similar equality $h0_{\mathbb{F}} = 0_{\mathbb{F}}$ then follows from commutativity.

The last two properties form the object of exercises.   □

EXERCISES

Exercise 4.1. Prove that every field $\mathbb{F}$ contains only one multiplicative identity.

Exercise 4.2. Prove that for each nonzero $h \in \mathbb{F}$, the field $\mathbb{F}$ contains only one multiplicative inverse for h.

4.1.2 Linear Spaces

Several algebraic features of vectors in the plane or in space extend to functions in various sets and to yet more abstract mathematical objects, all called "linear spaces" or "vector spaces."

Definition 4.9 A **linear space over a field** $\mathbb{F}$ consists of a set V containing at least one element 0_V, with two binary operations $\boxplus$ and $\boxdot$, which are functions

$$\boxplus: V \times V \to V, \quad (v, w) \mapsto v \boxplus w,$$

$$\boxdot: \mathbb{F} \times V \to V, \quad (r, v) \mapsto r \boxdot v \quad (\text{or} \quad rv),$$

such that all the properties in Table 4.2 hold.

Also, a **vector** is an element of a linear space. $\qquad\square$

Table 4.2   Algebraic Properties of Linear Spaces

The following properties must hold for all $r, s \in \mathbb{F}$ and $u, v, w \in V$.

(1)	Associativity of $\boxplus$	$[u \boxplus v] \boxplus w = u \boxplus [v \boxplus w]$
(2)	Commutativity of $\boxplus$	$u \boxplus v = v \boxplus u$
(3)	Additive identity	$v \boxplus 0_V = v = 0_V \boxplus v$
(4)	Additive inverse	$\mathbb{F}$ contains w with $v \boxplus w = 0_V = w \boxplus v$
(5)	Associativity of $\boxdot$	$[rs] \boxdot v = r \boxdot [s \boxdot v]$
(7)	Multiplicative identity	$1_\mathbb{F} \boxdot v = v$
(8)	Left distributivity	$r \boxdot [u \boxplus v] = [r \boxdot u] \boxplus [r \boxdot v]$
(9)	Right distributivity	$[r +_\mathbb{F} s] \boxdot v = [r \boxdot v] \boxplus [s \boxdot v]$

Remark 4.10 The word "vector" merely designates an element of a linear space. $\qquad\square$

Example 4.11 Each number field $\mathbb{F}$ is a linear space over itself with $\boxplus = +_\mathbb{F}$, $\boxdot = \times_\mathbb{F}$, and $0_V = 0_\mathbb{F}$. $\qquad\square$

Example 4.12 For each positive integer n and each number field $\mathbb{F}$, the set $\mathbb{F}^n := \{(x_1, \ldots, x_n) : x_1, \ldots, x_n \in \mathbb{F}\}$ is a linear space over $\mathbb{F}$, with "coordinatewise" operations:

$$(x_1, \ldots, x_n) \boxplus (z_1, \ldots, z_n) := (x_1 + z_1, \ldots, x_n + z_n),$$

$$r \boxdot (x_1, \ldots, x_n) := (r \cdot x_1, \ldots, r \cdot x_n),$$

$$0_{\mathbb{F}^n} := (0_\mathbb{F}, \ldots, 0_\mathbb{F}).$$

A "vector" in $\mathbb{F}^n$ is a finite *sequence* $(x_1, \ldots, x_n)$ of elements $x_1, \ldots, x_n$ in $\mathbb{F}$, which is also a *function* $x : \{1, \ldots, n\} \to \mathbb{F}$ with $x(i) = x_i$. □

Remark 4.13 In Example 4.12, the symbols ⊞ and ⊡ denote operations in the linear space, whereas $+$ and $\cdot$ denote operations in the number field; thus, the assignment $:=$ defines the new operations in the linear space in terms of the previously established operations in the number field. Common usage often employs $+$ for both $+$ and ⊞, and similarly $\cdot$ for both $\cdot$ and ⊡. The next few examples will maintain the distinction for specificity. □

Example 4.14 For each number field $\mathbb{F}$, for each linear space W with field $\mathbb{F}$ and with operations $\oplus$ and $\odot$ for W, and for each nonempty set S, the set $V := W^S$ of all functions from S into W,

$$W^S := \{f : S \to W\}$$

is a linear space with the same field $\mathbb{F}$, with "pointwise" operations ⊞ and ⊡: For all $f, g \in W^S$ and for each $x \in S$,

$$(f \boxplus g)(x) := (f(x)) \oplus (g(x)),$$
$$(r \boxdot f)(x) := r \odot (f(x)),$$
$$0_V : S \to W, \quad x \mapsto 0_W.$$

Thus, a "vector" in W^S is a function $f : S \to W$. □

Example 4.15 For all real numbers $a < b$, the set $C^0([a, b], \mathbb{R})$ of all real-valued continuous functions defined on the closed interval $[a, b]$ is a linear space over the field $\mathbb{R}$. For instance, the square root function restricted to the unit interval, $\sqrt{\ } : [0, 1] \to \mathbb{R}, x \mapsto \sqrt{x}$, is an element of (a "vector" in) the linear space $C^0([0, 1], \mathbb{R})$. □

Definition 4.16 A **linear subspace** of a linear space V over a field $\mathbb{F}$ consists of a subset $W \subseteq V$ that is also a linear space with the same element 0_V and with the same binary operations ⊞ and ⊡ already existing for V. □

Example 4.17 For all real numbers $a < b$, the set $C^1([a, b], \mathbb{R})$ of all real-valued *differentiable* functions defined on the closed interval $[a, b]$ is a linear subspace of the linear space $C^0([a, b], \mathbb{R})$. □

The following theorem demonstrates some of the features of all linear spaces similar to some of the features of number fields.

Theorem 4.18 *Each linear space V over any number field $\mathbb{F}$ shares the following features.*

(F) The linear space V contains only one additive identity.

(G) For each $v \in V$, the linear space V contains only one additive inverse for v.

(H) For each element $v \in V$, the equality $0_\mathbb{F} v = 0_V$ holds.

PROOF: The proof for the corresponding features of number fields extends *verbatim.*

(F) Assume that V contains elements 0_1 and 0_2 such that $v + 0_1 = v = 0_1 + v$ and $v + 0_2 = v = 0_2 + v$ for every $v \in V$. Then the following argument confirms that $0_1 = 0_2$. Setting $v := 0_2$ in $v + 0_1 = v = 0_1 + v$ gives $0_2 + 0_1 = 0_2 = 0_1 + 0_2$. Similarly, setting $v := 0_1$ in $v + 0_2 = v = 0_2 + v$ gives $0_1 + 0_2 = 0_1 = 0_2 + 0_1$. Hence, combining the two results just obtained yields $0_2 = 0_1 + 0_2 = 0_1$.

(G) Assume that for some element $v \in V$, the linear space V contains elements w_1 and w_2 such that $v + w_1 = 0_V = w_1 + v$ and $v + w_2 = 0_V = w_2 + v$. Hence,

$$w_2 = w_2 + 0_V = w_2 + (v + w_1) = (w_2 + v) + w_1 = 0_V + w_1 = w_1.$$

(H)

$$0_\mathbb{F} v = (0_\mathbb{F} + 0_\mathbb{F})v = (0_\mathbb{F} v) + (0_\mathbb{F} v).$$

Because $0_\mathbb{F} v \in V$, the linear space V also contains an additive inverse u for $0_\mathbb{F} v$. Hence,

$$0_\mathbb{F} v = (0_\mathbb{F} v) + (0_\mathbb{F} v),$$

$$(0_\mathbb{F} v) + u = [(0_\mathbb{F} v) + (0_\mathbb{F} v)] + u,$$

$$0_V = (0_\mathbb{F} v) + u = [(0_\mathbb{F} v) + (0_\mathbb{F} v)] + u$$

$$= (0_\mathbb{F} v) + [(0_\mathbb{F} v) + u] = (0_\mathbb{F} v) + 0_V = (0_\mathbb{F} v),$$

which means that $0_\mathbb{F} v = 0_V$. $\square$

4.1.3   Linear Maps

The context of linear spaces lends itself to operations called "linear functions," which correspond to addition and multiplication by a constant proportionality factor in each direction.

Definition 4.19   A **linear function**, also called **linear map, linear mapping, linear operator,** or **linear transformation,** is a function $L : V \rightarrow W$ from a linear space V to a linear space W, with the same number field $\mathbb{F}$ for V and W, and such that

$$L(r \cdot u + s \cdot v) = r \cdot L(u) + s \cdot L(v)$$

for all $r, s \in \mathbb{F}$ and all $u, v \in V$. (The symbols $+$ and $\cdot$ on the left-hand side denotes addition and multiplication for V; the symbol $+$ on the right-hand side denotes addition and multiplication for W.) $\square$

Example 4.20 The function $\ell : \mathbb{Q} \to \mathbb{Q}$ defined by $\ell(x) = 2x$ is linear. Indeed, here $\mathbb{F} = \mathbb{Q}$ and $V = \mathbb{Q} = W$, and

$$\ell(ru + sv) = 2(ru + sv) = 2(ru) + 2(sv) = r(2u) + s(2v)$$
$$= rL(u) + sL(v)$$

for all $r, s \in \mathbb{F} = \mathbb{Q}$ and all $u, v \in V = \mathbb{Q}$. □

Counterexample 4.21 The function $f : \mathbb{Q} \to \mathbb{Q}$ defined by $f(x) = 2x + 1$ is *not* linear, because

$$f(ru + sv) \neq rf(u) + sf(v)$$

for some $r, s \in \mathbb{F} = \mathbb{Q}$ and some $u, v \in V = \mathbb{Q}$. Indeed,

$$f(ru + sv) = 2(ru + sv) + 1 = 2(ru) + 2(sv) + 1$$
$$= r(2u) + 1 + s(2v) + 1 - 1 = rf(u) + sf(v) + 1 - r - s$$
$$\neq rf(u) + sf(v).$$ □

Linear functions share certain characteristics that facilitate calculations, for instance, the following one.

Proposition 4.22 *For each number field* $\mathbb{F}$, *for all linear spaces V and W with field* $\mathbb{F}$, *and for each linear function $L : V \twoheadrightarrow W$,*

$$L(0_V) = 0_W.$$

PROOF: Apply the relations

$$0_{\mathbb{F}} 0_V = 0_V,$$
$$0_V + 0_V = 0_V,$$
$$0_{\mathbb{F}} w = 0_W,$$
$$0_W + 0_W = 0_W.$$

Thus,

$$L(0_V) = L(0_{\mathbb{F}} 0_V + 0_{\mathbb{F}} 0_V) = 0_{\mathbb{F}} L(0_V) + 0_{\mathbb{F}} L(0_V) = 0_W + 0_W = 0_W.$$ □

Further examples of linear functions include rotations and symmetries in space, and differentiation and integration in calculus.

4.2 PROJECTIONS

A type of linear function relevant to wavelets consists of orthogonal projections based on inner products, which represent or approximate signals through combinations of wavelets.

4.2.1 Inner Products

The dot product of vectors in the plane and in space also extends to other types of linear spaces, for instance, to spaces of functions, where such generalized dot products—called "inner products"—provide a means to define and measure a "distance" or degree of closeness between two functions.

Table 4.3   Algebraic Properties of Inner Products

The following properties must hold for all $r, s \in \mathbb{F}$ and $u, v, w \in V$.

(1)	Nonnegativity of $\langle\,,\,\rangle$	$\langle v, v \rangle \geq 0_{\mathbb{F}}$
(2)	Positivity of $\langle\,,\,\rangle$	If $\langle v, v \rangle = 0_{\mathbb{F}}$, then $v = 0_V$
(3)	Linearity of $\langle\,,\,\rangle$	$\langle r \boxdot u \boxplus s \boxdot v, w \rangle = r\langle u, w \rangle +_{\mathbb{F}} s\langle v, w \rangle$
(4)	Hermitian $\langle\,,\,\rangle$	$\langle v, w \rangle = \overline{\langle w, v \rangle}$

Definition 4.23   An **inner product** (or a "scalar" or "dot" product) defined on a linear space V over a field $\mathbb{F} \subseteq \mathbb{C}$ is a function

$$\langle\,,\,\rangle : V \times V \to \mathbb{F}, \quad (v, w) \mapsto \langle v, w \rangle,$$

which satisfies all the properties in Table 4.3, where $\overline{(p, q)} = (p, -q)$ denotes *complex* conjugation for each $(p, q) \in \mathbb{C}$.   □

Example 4.24   For each positive integer n and each number field $\mathbb{F} \subseteq \mathbb{C}$, the set $\mathbb{F}^n := \{(x_1, \ldots, x_n) : x_1, \ldots, x_n \in \mathbb{F}\}$ is a linear space over $\mathbb{F}$, with dot product

$$\langle (x_1, \ldots, x_n), (z_1, \ldots, z_n) \rangle := x_1\overline{z_1} + \cdots + x_n\overline{z_n}.$$ □

Example 4.25   For each pair of real numbers $a < b$, the linear space

$$C^0([a, b], \mathbb{C})$$

of all continuous functions $f\,[a, b] \to \mathbb{C}$, with $f(t) = u(t) + iv(t)$, has an inner product defined by

$$\langle f, g \rangle := \int_a^b f(t)\overline{g(t)}\, dt.$$ □

Theorem 4.26   (*Cauchy–Schwarz Inequality.*) *For each inner product $\langle\,,\,\rangle$ on any linear space over any field $\mathbb{F} \subseteq \mathbb{C}$, and for all $v, w \in V$,*

$$|\langle v, w \rangle|^2 \leq \langle v, v \rangle \cdot \langle w, w \rangle,$$

with equality if, but only if, v or w is a multiple of the other: $\mathbb{F}$ contains an element r such that $v = rw$ or $w = rv$.

PROOF: If $w = 0_V$, then $|\langle v, w \rangle|^2 = 0_{\mathbb{F}} = \langle v, v \rangle \cdot \langle w, w \rangle$. If $w \neq 0_V$, then

$$0 \leq \left\langle v - \frac{\langle v, w \rangle}{\langle w, w \rangle} \cdot w, v - \frac{\langle v, w \rangle}{\langle w, w \rangle} \cdot w \right\rangle$$

$$= \langle v, v \rangle - \frac{\langle v, w \rangle}{\langle w, w \rangle} \langle w, v \rangle - \frac{\overline{\langle v, w \rangle}}{\langle w, w \rangle} \langle v, w \rangle + \frac{\langle v, w \rangle}{\langle w, w \rangle} \cdot \frac{\overline{\langle v, w \rangle}}{\langle w, w \rangle} \langle w, w \rangle$$

$$= \langle v, v \rangle + (-1 - 1 + 1) \cdot \frac{\langle v, w \rangle \overline{\langle v, w \rangle}}{\langle w, w \rangle}$$

$$= \langle v, v \rangle - \frac{|\langle v, w \rangle|^2}{\langle w, w \rangle}.$$

Multiplying both sides by the positive number $\langle w, w \rangle$ and then adding $|\langle v, w \rangle|^2$ to both sides yields the Cauchy–Schwarz inequality.

Finally, the first inequality in the proof shows that the equality $|\langle v, w \rangle|^2 = \langle v, v \rangle \cdot \langle w, w \rangle$ holds if, but only if, $v - \frac{\langle v, w \rangle}{\langle w, w \rangle} \cdot w = 0$, which means that $v = \frac{\langle v, w \rangle}{\langle w, w \rangle} \cdot w$. □

Definition 4.27 For each linear space V with an inner product $\langle \, , \, \rangle$, the **norm** induced by the inner product is the function $\| \ \| : V \to \mathbb{R}$ defined for every $v \in V$ by

$$\|v\| := \sqrt{\langle v, v \rangle}.$$ □

Theorem 4.28 *(Triangle Inequality.) For each inner product $\langle \, , \, \rangle$ on any linear space over any field $\mathbb{F} \subseteq \mathbb{C}$, and for all elements $v, w \in V$,*

$$\|v + w\| \leq \|v\| + \|w\|,$$

with equality if, but only if, v or w equals a nonnegative multiple of the other: $\mathbb{F}$ contains an element $r \geq 0$ such that $v = rw$ or $w = rv$.

PROOF: Apply the definition of the norm and the Cauchy–Schwarz inequality with $\mathfrak{Re}$ and $\mathfrak{Im}$ denoting the real and imaginary parts of complex numbers:

$$\|v + w\|^2 = \langle v + w, v + w \rangle = \langle v, v \rangle + \langle v, w \rangle + \langle w, v \rangle + \langle w, w \rangle$$

$$= \|v\|^2 + \langle v, w \rangle + \overline{\langle v, w \rangle} + \|w\|^2$$

$$= \|v\|^2 + 2\mathfrak{Re}(\langle v, w \rangle) + \|w\|^2$$

$$\leq \|v\|^2 + 2|\langle v, w \rangle| + \|w\|^2$$

$$\leq \|v\|^2 + 2\|v\| \cdot \|w\| + \|w\|^2 = (\|v\| + \|w\|)^2,$$

with equality, $\|v + w\|^2 = (\|v\| + \|w\|)^2$, if, but only if, both inequalities in the proof become equalities, which occurs if, but only if, $\mathfrak{Re}(\langle v, w \rangle) = |\langle v, w \rangle|$, which means that $\langle v, w \rangle \geq 0$, and $|\langle v, w \rangle| = \|v\| \cdot \|w\|$, which means that $v = \frac{\langle v, w \rangle}{\langle w, w \rangle} \cdot w$. □

Table 4.4 Algebraic Properties of Norms

The following properties must hold for all $r \in \mathbb{F}$ and $v, w \in V$.

(1)	Nonnegativity of $\| \ \|$	$\|v\| \geq 0_{\mathbb{F}}$
(2)	Positivity of $\| \ \|$	If $\|v\| = 0_{\mathbb{F}}$ then $v = 0_V$
(3)	Homogeneity of $\| \ \|$	$\|rv\| = |r| \cdot \|v\|$
(4)	Triangle Inequality for $\| \ \|$	$\|v + w\| \leq \|v\| + \|w\|$

Definition 4.29 A **norm** defined on a linear space V over a field $\mathbb{F} \subseteq \mathbb{C}$ is a function

$$\| \ \| : V \to \mathbb{F}, \quad v \mapsto \|v\|,$$

which satisfies all the properties in Table 4.4. $\qquad\square$

Example 4.30 For each positive integer n and each number field $\mathbb{F} \subseteq \mathbb{C}$, the set $\mathbb{F}^n := \{(x_1, \ldots, x_n) : x_1, \ldots, x_n \in \mathbb{F}\}$ is a linear space over $\mathbb{F}$, with norm

$$\|(x_1, \ldots, x_n)\|_2 := \sqrt{\langle (x_1, \ldots, x_n), \ (x_1, \ldots, x_n) \rangle}$$

$$= \sqrt{x_1 \overline{x_1} + \cdots + x_n \overline{x_n}}. \qquad\square$$

Example 4.31 For each pair of real numbers $a < b$, the linear space

$$C^0 ([a, b], \mathbb{C})$$

of all continuous functions $f : [a, b] \to \mathbb{C}$, with $f(t) = u(t) + i v(t)$, has a norm defined by

$$\|f\|_2 := \sqrt{\langle f, f \rangle} = \sqrt{\int_a^b f(t) \overline{f(t)} \, dt}. \qquad\square$$

Not all norms arise from inner products, however.

Example 4.32 For each positive integer n and each number field $\mathbb{F} \subseteq \mathbb{C}$, the set $\mathbb{F}^n := \{(x_1, \ldots, x_n) : x_1, \ldots, x_n \in \mathbb{F}\}$ is a linear space over $\mathbb{F}$, with norm

$$\|(x_1, \ldots, x_n)\|_\infty := \max\{|x_1|, \ldots, |x_n|\}. \qquad\square$$

Example 4.33 For each positive integer n and each number field $\mathbb{F} \subseteq \mathbb{C}$, the set $\mathbb{F}^n := \{(x_1, \ldots, x_n) : x_1, \ldots, x_n \in \mathbb{F}\}$ is a linear space over $\mathbb{F}$, with norm

$$\|(x_1, \ldots, x_n)\|_1 := \sum_{k=1}^n |x_k| = |x_1| + \cdots + |x_n|. \qquad\square$$

Example 4.34 For each positive integer n and each number field $\mathbb{F} \subseteq \mathbb{C}$, and for each real $p \geq 1$, the set $\mathbb{F}^n$ is a linear space over $\mathbb{F}$, with norm

$$\|(x_1, \ldots, x_n)\|_p := \left(\sum_{k=1}^{n} |x_k|^p \right)^{\frac{1}{p}} = \left(|x_1|^p + \cdots + |x_n|^p \right)^{\frac{1}{p}}. \qquad \square$$

Theorem 4.35 *(Reverse Triangle Identity.) For all elements v and w in a linear space V with a norm $\| \ \|$,*

$$\|v - w\| \geq |\ \|v\| - \|w\|\ |.$$

PROOF: Apply the triangle inequality twice. First,

$$\|v\| = \|v + 0_V\| = \|v + (-w + w)\|$$
$$= \|(v - w) + w\| \leq \|v - w\| + \|w\|,$$
$$\|v\| - \|w\| \leq \|v - w\|.$$

Second,

$$\|w\| = \|w + 0_V\| = \|w + (-v + v)\|$$
$$= \|(w - v) + v\| \leq \|w - v\| + \|v\|,$$
$$\|w\| - \|v\| \leq \|w - v\|.$$

Yet $\|w - v\| = \|(-1)(v - w)\| = |-1| \cdot \|v - w\| = \|v - w\|$, whence changing signs gives

$$\|v\| - \|w\| \geq -\|v - w\|.$$

Thus,

$$-\|v - w\| \leq \|v\| - \|w\| \leq \|v - w\|,$$

which means that

$$\|v - w\| \geq |\ \|v\| - \|w\|\ |. \qquad \square$$

Theorem 4.36 *(Polar Identity.) For all vectors v and w in a linear space V over a field $\mathbb{F} \subseteq \mathbb{C}$ with an inner product $\langle \ , \ \rangle$ and with the corresponding norm defined by $\|u\| = \sqrt{\langle u, u \rangle}$, the following identities hold:*

$$\mathfrak{Re}(\langle v, w \rangle) = \tfrac{1}{4} \cdot \left(\|v + w\|^2 - \|v - w\|^2 \right),$$

$$\mathfrak{Im}(\langle v, w \rangle) = \tfrac{1}{4} \cdot \left(\|v + iw\|^2 - \|v - iw\|^2 \right).$$

In particular, if $\mathbb{F} \subseteq \mathbb{R}$, then $\mathfrak{Im}(\langle v, w \rangle) = 0$ and hence

$$\langle v, w \rangle = \tfrac{1}{4} \cdot \left(\|v + w\|^2 - \|v - w\|^2 \right).$$

PROOF: Apply the definition of the norm:

$$\|v + w\|^2 - \|v - w\|^2 = \langle v + w, v + w \rangle - \langle v - w, v - w \rangle$$
$$= \langle v, v \rangle + \langle v, w \rangle + \langle w, v \rangle + \langle w, w \rangle$$
$$- (\langle v, v \rangle - \langle v, w \rangle - \langle w, v \rangle + \langle w, w \rangle)$$
$$= 4\mathfrak{Re}(\langle v, w \rangle).$$

Similarly,

$$\|v + iw\|^2 - \|v - iw\|^2 = \langle v + iw, v + iw \rangle - \langle v - iw, v - iw \rangle$$
$$= \langle v, v \rangle + \langle v, iw \rangle + \langle iw, v \rangle + \langle iw, iw \rangle$$
$$- (\langle v, v \rangle - \langle v, iw \rangle - \langle iw, v \rangle + \langle iw, iw \rangle)$$
$$= -i\langle v, w \rangle + i\langle w, v \rangle - i\langle v, w \rangle + i\langle w, v \rangle$$
$$= 2i\left(\overline{\langle v, w \rangle} - \langle v, w \rangle\right)$$
$$= 2i(-2i)\mathfrak{Im}(\langle v, w \rangle). \qquad \square$$

Definition 4.37   In a linear space V with an inner product $\langle\ ,\ \rangle$, two vectors v and w are mutually **orthogonal,** or **perpendicular,** if, but only if, $\langle v, w \rangle = 0_{\mathbb{F}}$. The notation $v \perp w$ then means that $\langle v, w \rangle = 0_{\mathbb{F}}$. Also, v and w are mutually **orthonormal** if, but only if, $\langle v, w \rangle = 0_{\mathbb{F}}$ and $\|v\| = 1 = \|w\|$. $\qquad \square$

Theorem 4.38   *(Pythagorean Theorem.) In a linear space V over a field $\mathbb{F} \subseteq \mathbb{C}$ with an inner product $\langle\ ,\ \rangle$, for each pair of vectors v and w in V, the relation $\|v + w\|^2 = \|v\|^2 + \|w\|^2$ holds if, but only if, $\mathfrak{Re}(\langle v, w \rangle) = 0$.*

PROOF:

$$\|v + w\|^2 = \langle v + w, v + w \rangle$$
$$= \langle v, v \rangle + \langle v, w \rangle + \langle w, v \rangle + \langle w, w \rangle$$
$$= \|v\|^2 + \langle v, w \rangle + \overline{\langle v, w \rangle} + \|w\|^2$$
$$= \|v\|^2 + 2\mathfrak{Re}(\langle v, w \rangle) + \|w\|^2. \qquad \square$$

Theorem 4.39   *(Bessel's Inequality.) In a linear space V over a field $\mathbb{F} \subseteq \mathbb{C}$ with an inner product $\langle\ ,\ \rangle$, for each $v \in V$ and for each set of nonzero orthogonal vectors $\{w_1, w_2, \ldots, w_{m-1}, w_m\} \subseteq V$,*

$$\sum_{k=1}^{m} \frac{|\langle v, w_k \rangle|^2}{\|w_k\|^2} \leq \|v\|^2.$$

PROOF:

$$0 \leq \left\| v - \sum_{k=1}^{m} \frac{\langle v, w_k \rangle}{\|w_k\|^2} w_k \right\|^2$$

$$= \left\langle v - \sum_{k=1}^{m} \frac{\langle v, w_k \rangle}{\|w_k\|^2} w_k, \, v - \sum_{k=1}^{m} \frac{\langle v, w_k \rangle}{\|w_k\|^2} w_k \right\rangle$$

$$= \langle v, v \rangle - \sum_{k=1}^{m} \langle v, w_k \rangle \cdot \frac{\overline{\langle v, w_k \rangle}}{\|w_k\|^2} - \sum_{k=1}^{m} \langle w_k, v \rangle \cdot \frac{\langle v, w_k \rangle}{\|w_k\|^2}$$

$$+ \sum_{k=1}^{m} \sum_{\ell=1}^{m} \frac{\langle v, w_k \rangle}{\|w_k\|^2} \frac{\overline{\langle v, w_\ell \rangle}}{\|w_\ell\|^2} \cdot \langle w_k, w_\ell \rangle$$

$$= \|v\|^2 - \sum_{k=1}^{m} \frac{|\langle v, w_k \rangle|^2}{\|w_k\|^2} - \sum_{k=1}^{m} \frac{|\langle v, w_k \rangle|^2}{\|w_k\|^2} + \sum_{k=1}^{m} \frac{|\langle v, w_k \rangle|^2}{\|w_k\|^2}$$

$$= \|v\|^2 - \sum_{k=1}^{m} \frac{|\langle v, w_k \rangle|^2}{\|w_k\|^2}. \qquad \qquad \square$$

4.2.2 Gram–Schmidt Orthogonalization

Gram–Schmidt Orthogonalization provides a method to construct orthonormal bases, which will prove convenient in the calculation of orthogonal projections. In this context, a **basis** for a linear space V over a field $\mathbb{F}$ is a subset U of V that satisfies two conditions. Firstly, U **spans** V, which means that for each element $v \in V$ there exist a positive integer n, some elements $u_1, \ldots, u_n$ in U, and n coefficients $c_1, \ldots, c_n$ in $\mathbb{F}$ such that $v = c_1 u_1 + \cdots + c_n u_n$. Secondly, the set U is **linearly independent**, which means that if $v = 0_V$, so that $0_V = c_1 u_1 + \cdots + c_n u_n$, then all the coefficients equal zero: $c_1 = 0_{\mathbb{F}}, \ldots, c_n = 0_{\mathbb{F}}$. Also, for each subset W of V, the subspace Span W **spanned** by W is the set of all **linear combinations** $c_1 w_1 + \cdots + c_n w_n$ with any number n of elements $w_1, \ldots, w_n$ in W and coefficents $c_1, \ldots, c_n$ in $\mathbb{F}$. Moreover, if the linear space V has an inner product $\langle \, , \, \rangle$, then a basis or a subset Z of V is **orthogonal** if and only if all distinct elements x and z in Z are mutually orthogonal: if $x \neq z$ then $\langle x, z \rangle = 0_{\mathbb{F}}$. Finally, a basis or a subset Z of V is **orthonormal** if and only if it is orthogonal and every element z in Z has unit norm: $\langle z, z \rangle = 1_{\mathbb{F}}$.

Theorem 4.40 *(Gram–Schmidt Orthogonalization). For each linearly independent finite or infinite set $\{v_1, v_2, \ldots, v_k, \ldots\}$ in a linear space V with an inner product $\langle \, , \, \rangle$, there exists an orthonormal set $U = \{u_1, u_2, \ldots, u_k, \ldots\}$ such that for each index k,*

$$\langle u_k, v_k \rangle > 0,$$

$$\text{Span}\{u_1, \ldots, u_k\} = \text{Span}\{v_1, \ldots, v_k\}.$$

PROOF: The present proof first constructs an *orthogonal* set $W = \{w_1, w_2, \ldots, w_k, w_{k+1}, \ldots\}$ such that $\langle w_k, v_k \rangle > 0$ and Span$\{w_1, \ldots, w_k\}$ = Span$\{v_1, \ldots, v_k\}$. Divisions by the norms $\|w_k\| = \sqrt{\langle w_k, w_k \rangle}$ will then produce the orthonormal set U. To this end, let $w_1 := v_1$, so that $\langle w_1, v_1 \rangle = \langle v_1, v_1 \rangle > 0$

and Span $\{w_1\}$ = Span $\{v_1\}$. Proceeding by induction, assume that mutually orthogonal vectors $w_1, \ldots, w_m$ exist such that

$$\langle w_k, v_k \rangle > 0,$$

$$\text{Span}\{w_1, \ldots, w_k\} = \text{Span}\{v_1, \ldots, v_k\},$$

for each $k \in \{1, \ldots, m\}$. Then define w_{m+1} by

$$w_{m+1} := v_{m+1} - \sum_{k=1}^{m} \frac{\langle v_{m+1}, w_k \rangle}{\langle w_k, w_k \rangle} \cdot w_k.$$

Then the set $\{w_1, \ldots, w_m, w_{m+1}\}$ is again orthogonal:

$$\langle w_{m+1}, w_\ell \rangle = \left\langle v_{m+1} - \sum_{k=1}^{m} \frac{\langle v_{m+1}, w_k \rangle}{\langle w_k, w_k \rangle} \cdot w_k, w_\ell \right\rangle$$

$$= \langle v_{m+1}, w_\ell \rangle - \sum_{k=1}^{m} \frac{\langle v_{m+1}, w_k \rangle}{\langle w_k, w_k \rangle} \langle w_k, w_\ell \rangle$$

$$= \langle v_{m+1}, w_\ell \rangle - \frac{\langle v_{m+1}, w_\ell \rangle}{\langle w_\ell, w_\ell \rangle} \langle w_\ell, w_\ell \rangle$$

$$= 0,$$

because if $k \neq \ell$, then $\langle w_k, w_\ell \rangle = 0$. Moreover,

$$\text{Span}\{w_1, \ldots, w_m, w_{m+1}\}$$

$$= \text{Span}\left\{v_1, \ldots, v_m, v_{m+1} - \sum_{k=1}^{m} \frac{\langle v_{m+1}, w_k \rangle}{\langle w_k, w_k \rangle} \cdot w_k\right\}$$

$$= \text{Span}\{v_1, \ldots, v_m, v_{m+1}\}.$$

Furthermore, $\langle w_{m+1}, v_{m+1} \rangle > 0$ by Bessel's inequality:

$$\langle w_{m+1}, v_{m+1} \rangle = \left\langle v_{m+1} - \sum_{k=1}^{m} \frac{\langle v_{m+1}, w_k \rangle}{\langle w_k, w_k \rangle} \cdot w_k, v_{m+1} \right\rangle$$

$$= \langle v_{m+1}, v_{m+1} \rangle - \sum_{k=1}^{m} \frac{|\langle v_{m+1}, w_k \rangle|^2}{\langle w_k, w_k \rangle}$$

$$> 0.$$

Finally, define $u_k := \|w_k\|^{-1} \cdot w_k$. □

Remark 4.41 (Modified Gram–Schmidt Orthogonalization) With digital computations, a Modified Gram–Schmidt Orthogonalization yields greater accuracy than Gram–Schmidt Orthogonalization as just presented. For the Modified Gram–Schmidt Orthogonalization, subtract first the projection of w_1 not only from v_2, but from all the vectors $v_2, v_3, \ldots$, thus defining a sequence

$$w_1 = v_1,$$

$$w_2 = v_2 - \frac{\langle v_2, w_1 \rangle}{\langle w_1, w_1 \rangle} \cdot w_1,$$

$$w_3^{(2)} = v_3 - \frac{\langle v_3, w_1 \rangle}{\langle w_1, w_1 \rangle} \cdot w_1,$$

$$\vdots$$

$$w_k^{(2)} = v_k - \frac{\langle v_k, w_1 \rangle}{\langle w_1, w_1 \rangle} \cdot w_1,$$

$$\vdots$$

Then subtract the projection of w_2 not only from $w_3^{(1)}$, but from all the vectors $w_3^{(1)}, w_4^{(1)}, \ldots$:

$$w_1 = v_1,$$

$$w_2 = v_2 - \frac{\langle v_2, w_1 \rangle}{\langle w_1, w_1 \rangle} \cdot w_1,$$

$$w_3 = w_3^{(2)} - \frac{\langle w_3^{(2)}, w_2 \rangle}{\langle w_2, w_2 \rangle} \cdot w_2,$$

$$\vdots$$

$$w_k^{(3)} = w_k^{(2)} - \frac{\langle v_k, w_2 \rangle}{\langle w_2, w_2 \rangle} \cdot w_2,$$

$$\vdots$$

Then let $w_k := w_k^{(k)}$ and $u_k := \| w_k \|^{-1} \cdot w_k$. $\qquad \qquad \square$

4.2.3 Orthogonal Projections

The preceding results lead to methods and algorithms to calculate orthogonal projections. Consider a linear subspace $W \subseteq V$ of a linear space V over a field $\mathbb{F} \subseteq \mathbb{C}$ with inner product $\langle \, , \, \rangle$ and induced norm $\| \ \|$. Also, let $\vec{p} \perp W$ mean $\langle \vec{p}, \vec{q} \rangle = 0$ for every $\vec{q} \in W$.

Theorem 4.42 *If $\vec{w} \in W$, if $\vec{v} \in V$, and if $(\vec{v} - \vec{w}) \perp W$, then $\vec{w}$ is the member of W closest to $\vec{v}$.*

PROOF: For every other member $\vec{u} \in W$, $\vec{v} - \vec{w} \perp \vec{w} - \vec{u}$, and the Pythagorean theorem gives

$$\|\vec{v} - \vec{u}\|^2 = \|\vec{v} - \vec{w}\|^2 + \|\vec{w} - \vec{u}\|^2 \geq \|\vec{v} - \vec{w}\|^2.$$

Square roots then yield $\|\vec{v}-\vec{u}\| \geq \|\vec{v}-\vec{w}\|$, which means that $\vec{w}$ lies closer to $\vec{v}$ than $\vec{u}$ does. The features of the norm also guarantee that equality, $\|\vec{v} - \vec{u}\| = \|\vec{v} - \vec{w}\|$, occurs only at $\vec{u} := \vec{w}$, which means that in W there exists only one element $\vec{w}$ closest to $\vec{v}$.   □

Theorem 4.43 *If $(\vec{w}_1, \ldots, \vec{w}_n)$ is an orthogonal basis for W, then for each $\vec{v} \in V$ the element $\vec{w} \in W$ closest to $\vec{v}$ is*

$$\vec{w} = \sum_{i=1}^{n} \frac{\langle \vec{v}, \vec{w}_i \rangle}{\langle \vec{w}_i, \vec{w}_i \rangle} \vec{w}_i.$$

PROOF: Verify that $\langle \vec{v} - \vec{w}, \vec{u} \rangle = 0$ for each $\vec{u} \in W$, and invoke Theorem 4.42. Specifically, because $(\vec{w}_1, \ldots, \vec{w}_n)$ is a basis for W, for each $\vec{u} \in W$ there exist coefficients $u_1, \ldots, u_n \in \mathbb{F}$ such that

$$\vec{u} = \sum_{j=1}^{n} u_j \vec{w}_j.$$

Consequently, the linear (additive and multiplicative) properties of the inner product and the orthogonality of the basis yield

$$\langle \vec{v} - \vec{w}, \vec{u} \rangle = \left\langle \vec{v} - \sum_{i=1}^{n} \frac{\langle \vec{v}, \vec{w}_i \rangle}{\langle \vec{w}_i, \vec{w}_i \rangle} \vec{w}_i, \ \sum_{j=1}^{n} u_j \vec{w}_j \right\rangle$$

$$= \sum_{j=1}^{n} \overline{u_j} \left(\langle \vec{v}, \vec{w}_j \rangle - \sum_{i=1}^{n} \frac{\langle \vec{v}, \vec{w}_i \rangle}{\langle \vec{w}_i, \vec{w}_i \rangle} \langle \vec{w}_i, \vec{w}_j \rangle \right)$$

$$= \sum_{j=1}^{n} \overline{u_j} \left(\langle \vec{v}, \vec{w}_j \rangle - \frac{\langle \vec{v}, \vec{w}_j \rangle}{\langle \vec{w}_j, \vec{w}_j \rangle} \langle \vec{w}_j, \vec{w}_j \rangle \right)$$

$$= \sum_{j=1}^{n} \overline{u_j} \left(\langle \vec{v}, \vec{w}_j \rangle - \langle \vec{v}, \vec{w}_j \rangle \right)$$

$$= 0.   □$$

Theorem 4.44 *Let $V = \mathbb{R}^m$, $\mathbb{F} = \mathbb{R}$, and let $\langle \ , \ \rangle$ denote the usual dot product. If $Q = (\vec{q}_1, \ldots, \vec{q}_n) \in \mathbf{M}_{m \times n}(\mathbb{R})$ is a rectangular matrix with orthonormal columns, and if $\vec{v} \in \mathbb{R}^m$, then $Q^T \vec{v}$ is the vector of coordinates, with respect to the basis $(\vec{q}_1, \ldots, \vec{q}_n)$, of the orthogonal projection of $\vec{v}$ onto the subspace $W = \mathrm{Span}\{\vec{q}_1, \ldots, \vec{q}_n\}$.*

PROOF: Apply Theorem 4.43 to $(Q^T \vec{v})_i = \vec{q}_i^T \cdot \vec{v} = \langle \vec{q}_i, \vec{v} \rangle$. □

Theorem 4.45 *Let $V = \mathbb{R}^m$, $\mathbb{F} = \mathbb{R}$, and let $\langle \, , \, \rangle$ denote the usual dot product. A linear transformation $L : V \to V$, $L(\vec{x}) = Q \cdot \vec{x}$ preserves distances if, but only if, its matrix Q with respect to the canonical basis is orthogonal, that is, $Q^T Q = I$.*

PROOF: If $Q^T Q = I$, then

$$\|Q\vec{x}\|_2^2 = \langle Q\vec{x}, Q\vec{x} \rangle = (Q\vec{x})^T Q\vec{x} = \vec{x}^T Q^T Q\vec{x} = \vec{x}^T I \vec{x} = \|\vec{x}\|_2^2.$$

Conversely, if $\|Q\vec{x}\|_2 = \|\vec{x}\|_2$ for every $\vec{x} \in V$, then for all $\vec{u}, \vec{v} \in V$ the polar identity gives

$$\langle \vec{u}, \vec{v} \rangle = \tfrac{1}{4} \left(\|\vec{u} + \vec{v}\|_2^2 - \|\vec{u} - \vec{v}\|_2^2 \right)$$
$$= \tfrac{1}{4} \left(\|Q\vec{u} + Q\vec{v}\|_2^2 - \|Q\vec{u} - Q\vec{v}\|_2^2 \right)$$
$$= \langle Q\vec{u}, Q\vec{v} \rangle.$$

In particular, $\vec{u} = \vec{e}_i$ and $\vec{v} = \vec{e}_j$ yield $(Q^T Q)_{i,j} = \langle Q\vec{e}_i, Q\vec{e}_j \rangle = \langle \vec{e}_i, \vec{e}_j \rangle = \delta_{i,j} = I_{i,j}$. □

Theorem 4.46 *(Converse to Theorem 4.42.) If $\vec{w}$ is the member of W closest to $\vec{v}$, then $(\vec{v} - \vec{w}) \perp W$.*

PROOF: If W has a finite basis, then Theorem 4.43 gives a formula for a $\vec{w}$ with $(\vec{v} - \vec{w}) \perp W$, and then Theorem 4.42 guarantees that $\vec{w}$ is the only member of W closest to $\vec{v}$. If W does not have a finite basis, an alternative proof may proceed as follows. Since $\vec{w}$ is the element of W closest to $\vec{v} \in V$, it follows that $\|\vec{v} - \vec{u}\|^2 \geq \|\vec{v} - \vec{w}\|^2$ for every $\vec{u} \in W$. Hence

$$\|\vec{v} - \vec{u}\|^2 = \|(\vec{v} - \vec{w}) + (\vec{w} - \vec{u})\|^2$$
$$= \|\vec{v} - \vec{w}\|^2 + 2\mathfrak{Re}(\langle \vec{v} - \vec{w}, \vec{w} - \vec{u} \rangle) + \|\vec{w} - \vec{u}\|^2,$$

whence

$$\|\vec{w} - \vec{u}\|^2 + 2\mathfrak{Re}\langle \vec{v} - \vec{w}, \vec{w} - \vec{u} \rangle = \|\vec{v} - \vec{u}\|^2 - \|\vec{v} - \vec{w}\|^2 \geq 0,$$

and $\mathfrak{Re}(\langle \vec{v} - \vec{w}, \vec{w} - \vec{u} \rangle) \geq -\tfrac{1}{2}\|\vec{w} - \vec{u}\|^2$. Since this inequality holds for every $\vec{u} \in W$, it follows that it also holds for every $\vec{u}_\lambda := \lambda\vec{u} + (1 - \lambda)\vec{w}$, so that $\vec{w} - \vec{u}_\lambda = \lambda(\vec{w} - \vec{u})$. Thus, the inequality just established becomes $\lambda\mathfrak{Re}(\langle \vec{v} - \vec{w}, \vec{w} - \vec{u} \rangle) \geq -(\lambda^2/2)\|\vec{w} - \vec{u}\|^2$. Dividing by λ and letting λ tend to zero from either side yields

$$\lim_{\lambda \to 0-} \mathfrak{Re}(\langle \vec{v} - \vec{w}, \vec{w} - \vec{u} \rangle) \leq 0 \leq \lim_{\lambda \to 0+} \mathfrak{Re}(\langle \vec{v} - \vec{w}, \vec{w} - \vec{u} \rangle).$$

Consequently, $\mathfrak{Re}(\langle \vec{v} - \vec{w}, \vec{w} - \vec{u} \rangle) = 0$ for every $\vec{u} \in W$. Setting $\vec{u} := \vec{0} \in W$ yields $\mathfrak{Re}(\langle \vec{v} - \vec{w}, \vec{w} \rangle) = 0$, so that $\mathfrak{Re}(\langle \vec{v} - \vec{w}, \vec{u} \rangle) = 0$ for every $\vec{u} \in W$. For $\mathbb{F} = \mathbb{C}$, if $\vec{u} \in W$, then also $i\vec{u} \in W$, whence

$$0 = \mathfrak{Re}(\langle \vec{v} - \vec{w}, i\vec{u} \rangle) = \mathfrak{Re}(-i \langle \vec{v} - \vec{w}, \vec{u} \rangle) = \mathfrak{Im}(\langle \vec{v} - \vec{w}, \vec{u} \rangle).$$

Therefore, $\langle \vec{v} - \vec{w}, \vec{u} \rangle = 0$ for every $\vec{u} \in W$. $\square$

The matrices occurring in Theorems 4.44 and 4.45 bear a special name. Examples of such matrices will appear in the sequel.

Definition 4.47   A matrix $Q \in M_{n \times n}(\mathbb{F})$ is **orthogonal** (if $\mathbb{F} \subseteq \mathbb{R}$) or **unitary** (if $\mathbb{R} \subset \mathbb{F} \subseteq \mathbb{C}$) if, and only if, all its columns are mutually *orthonormal*. $\square$

EXERCISES

Exercise 4.3. Prove that if $P : V \rightarrow V$ is a linear transformation such that $P \circ P = P$, with $\circ$ denoting the composition of functions, then P restricts to the identity on its range: For every $z \in V$, if $z = P(x)$ for some $x \in V$, then $P(z) = z$.

Exercise 4.4. Prove that if $P : V \rightarrow V$ is an orthogonal projection, and if $I : V \rightarrow V$ represents the identity, then $Q := I - P$ is the orthogonal projection of V onto the kernel (null space) of P, and $P \circ Q = O = Q \circ P$.

4.3   APPLICATIONS OF ORTHOGONAL PROJECTIONS

4.3.1   Application to Three-Dimensional Computer Graphics

For three-dimensional graphics, the ambient three-dimensional space may correspond to the linear space $V := \mathbb{R}^3$, in which a two-dimensional subspace $W \subset V$ represents the plane of the screen. To produce a picture of a point $\vec{x} \in V$, a graphics procedure may then use an orthogonal projection $P : V \rightarrow W$ to map each point $\vec{x}$ to its image $P(\vec{x}) \in W$ on the screen. Computationally, such a graphics procedure may endow the screen W with an orthonormal basis $(\vec{u}, \vec{v})$, then calculate the coordinates on the screen of the image $P(\vec{x})$ by means of inner products,

$$P(\vec{x}) = \langle \vec{x}, \vec{u} \rangle \vec{u} + \langle \vec{x}, \vec{v} \rangle \vec{v},$$

and finally draw the image of $\vec{x}$ by plotting on the screen the point (p, q) with coordinates $p := \langle \vec{x}, \vec{u} \rangle$ and $q := \langle \vec{x}, \vec{v} \rangle$, as in Figure 4.1.

Example 4.48   The plane $W \subset \mathbb{R}^3$ that passes through the origin perpendicularly to the unit vector $\vec{w} := \left(\frac{6}{7}, \frac{2}{7}, -\frac{3}{7} \right)$ admits the orthonormal basis

$$\vec{u} := \left(\frac{2}{7}, \frac{3}{7}, \frac{6}{7} \right), \quad \vec{v} := \left(\frac{3}{7}, -\frac{6}{7}, \frac{2}{7} \right).$$

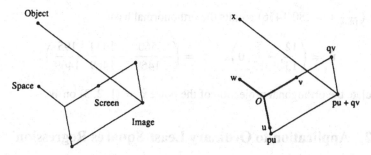

Figure 4.1 Three-dimensional graphics project objects in the three-dimensional ambient space onto a two-dimensional screen.

Plotting the image of the point $\vec{x} := (1, 2, 3)$ on the screen W then amounts to calculating the coordinates

$$p = \langle \vec{x}, \vec{u} \rangle = \left\langle (1, 2, 3), \left(\tfrac{2}{7}, \tfrac{3}{7}, \tfrac{6}{7} \right) \right\rangle = \tfrac{26}{7},$$

$$q = \langle \vec{x}, \vec{v} \rangle = \left\langle (1, 2, 3), \left(\tfrac{3}{7}, -\tfrac{6}{7}, \tfrac{2}{7} \right) \right\rangle = -\tfrac{3}{7},$$

and then marking the point $(\tfrac{26}{7}, -\tfrac{3}{7})$ on the screen. □

Remark 4.49 The foregoing outline in terms of inner products aims only at introducing the nature—as opposed to the details of computations— of three-dimensional computer graphics. Still, thanks to its simplicity, the approach just presented with inner products works well on very small machines, for instance, on pocket graphing calculators. In contrast, on larger machines, the use of projective geometry also allows for perspectives and for rotations and translations with matrices. □

EXERCISES

Exercise 4.5. In the three-dimensional space $V := \mathbb{R}^3$ with the usual inner product, the plane W passing through the origin perpendicularly to the unit vector $\vec{w} := (\tfrac{3}{13}, \tfrac{4}{13}, \tfrac{12}{13})$ admits the orthonormal basis

$$\vec{u} := \left(-\tfrac{4}{5}, \tfrac{3}{5}, 0 \right), \quad \vec{v} := \left(-\tfrac{36}{65}, -\tfrac{48}{65}, \tfrac{25}{65} \right).$$

Calculate the orthogonal projection of the point $\vec{x} := (1, 2, 3)$ on W.

Exercise 4.6. In the three-dimensional space $V := \mathbb{R}^3$ with the usual inner product, the plane W passing through the origin perpendicularly to the unit vector

$\vec{w} := (\frac{1}{1469})(75, 180, 1456)$ admits the orthonormal basis

$$\vec{u} := \left(\frac{12}{13}, -\frac{5}{13}, 0\right), \quad \vec{v} := \left(-\frac{560}{1469}, -\frac{1344}{1469}, \frac{195}{1469}\right).$$

Calculate the orthogonal projection of the point $\vec{x} := (1, 2, 3)$ on W.

4.3.2   Application to Ordinary Least-Squares Regression

The statistical method of Ordinary Least-Squares (OLS) Regression corresponds to an orthogonal projection of a vector of data on a linear subspace of specified vectors of coefficients.

Example 4.50   Consider the problem of fitting a straight line L with equation $c_1 x + c_0 = y$ to the data points

$$(2, 3), \quad (4, 7), \quad (5, 8), \quad (6, 9).$$

If all data points were on L, then they would satisfy the system

$$2c_1 + c_0 = 3,$$
$$4c_1 + c_0 = 7,$$
$$5c_1 + c_0 = 8,$$
$$6c_1 + c_0 = 9,$$

which, in terms of vectors, takes the equivalent form

$$c_1 \begin{pmatrix} 2 \\ 4 \\ 5 \\ 6 \end{pmatrix} + c_0 \begin{pmatrix} 1 \\ 1 \\ 1 \\ 1 \end{pmatrix} = \begin{pmatrix} 3 \\ 7 \\ 8 \\ 9 \end{pmatrix}.$$

Yet no solution exists, because the vector of data $\vec{y} := (3, 7, 8, 9) \in \mathbb{R}^4$ does not lie in the subspace $W := \mathrm{Span}\{\vec{x}, \vec{1}\}$ spanned by the vectors $\vec{x} := (2, 4, 5, 6)$ and $\vec{1} := (1, 1, 1, 1)$. Therefore, the method of Ordinary Least-Squares determines the linear combination $c_1 \vec{x} + c_0 \vec{1}$ closest to $\vec{y}$ by means of the orthogonal projection $P : \mathbb{R}^4 \rightarrow \mathrm{Span}\{\vec{x}, \vec{1}\}$; thus $c_1 \vec{x} + c_0 \vec{1} = P(\vec{y})$. For this example, the Gram–Schmidt algorithm followed by Theorem 4.43 gives the line displayed in Figure 4.2, with

$$c_1 = \tfrac{53}{35} \approx 1.514\,285\,714\,285\ldots,$$

$$c_0 = \tfrac{11}{35} \approx 0.314\,285\,714\,285\ldots,$$

so that the fitted line has equation $y = (\tfrac{53}{35})x + (\tfrac{11}{35})$. $\qquad\square$

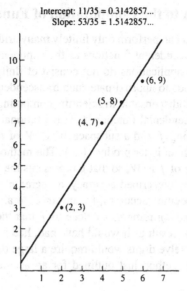

Figure 4.2 Ordinary Least-Squares Regression.

Remark 4.51 The foregoing example aims only at demonstrating that Ordinary Least-Squares Regression coincides with an orthogonal projection. In principle, the algorithms described in the preceding sections apply: Gram–Schmidt orthogonalization gives an orthogonal basis $(\vec{w}_1, \vec{w}_2)$ for Span $\{\vec{x}, \vec{1}\}$, whence inner products yield the projection $\langle \vec{y}, \vec{w}_1 \rangle \vec{w}_1 + \langle \vec{y}, \vec{w}_2 \rangle \vec{w}_2$, and then arithmetic converts the coefficients $\langle \vec{y}, \vec{w}_1 \rangle$ and $\langle \vec{y}, \vec{w}_1 \rangle$ back to c_1 and c_0. However, for Ordinary Least-Squares, specialized methods yield greater speed and accuracy with digital computers, for instance, by means of matrix factorizations [27, Ch. 5]. □

EXERCISES

Exercise 4.7. Calculate the OLS line for the data points

$$(1, 2), (2, 6), (6, 1).$$

Exercise 4.8. Calculate the OLS line for the data points

$$(-7, 1), (-1, 5), (1, 6), (7, 8).$$

Exercise 4.9. Calculate the OLS line for the data points

$$(30, 6), (1, 2), (10, 4), (13, 5).$$

Exercise 4.10. Calculate the OLS line for the data points

$$(1, 3), (3, 1), (4, 5), (5, 7), (7, 4).$$

4.3.3   Application to the Computation of Functions

While digital computers can perform only finitely many arithmetic operations and logical tests, such transcendental functions as the exponential and trigonometric functions that occur in applications do not consist of finitely many such operations. Therefore, methods to approximate such transcendental functions with rational functions prove indispensable in scientific computing. One of the methods to approximate a transcendental function f with a rational function g involves a linear space V containing f and a subspace $W \subset V$ of rational functions of a specified degree n, with an inner product $\langle\ ,\ \rangle$. The method then determines the orthogonal projection g of f in W, so that g represents the rational function from W closest to f. For any prescribed accuracy, a degree n exists for which g approximates f to the specified accuracy, for instance, to all displayed digits on a calculator. In other words, g remains so close to f that the calculator computes g and displays the same result as it would have had, had it computed f. Though a typical accuracy of twelve digits would require a large degree n, the following example illustrates the method just outlined for an accuracy of one significant digit.

Example 4.52   This example demonstrates how to determine a polynomial p of degree 1 in the design of a computer algorithm for the computation of the square-root function. Consider the field $\mathbb{R}$ and the linear space $V := C^0([\frac{1}{4}, 1], \mathbb{R})$ consisting of all real functions continuous on the interval $[\frac{1}{4}, 1]$, with the inner product

$$\langle f, g \rangle := \int_{\frac{1}{4}}^{1} f(x)g(x)\,dx.$$

The restriction of the square-root function to that interval, $f : [1/4, 1] \to \mathbb{R}$, $f(x) := \sqrt{x}$, lies in the linear space $V = C^0([1/4, 1], \mathbb{R})$.

Also, consider the linear subspace $W \subset V$ consisting of all polynomials of degree at most 1 on the same interval, so that every $p \in W$ has the form $p(x) = c_0 + c_1 x$ for some coefficients $c_0, c_1 \in \mathbb{R}$. The problem examined here then amounts to calculating the orthogonal projection p of the square root f in the space W of affine polynomials. To this end, the first task lies in finding an orthonormal or orthogonal basis for W, for example, by applying the Gram–Schmidt process to the basis $(v_1, v_2) := (1, x)$, which will utilize the following inner products:

$$\langle 1, 1 \rangle = \int_{\frac{1}{4}}^{1} 1 \cdot 1\,dx = \frac{3}{4},$$

$$\langle x, 1 \rangle = \int_{\frac{1}{4}}^{1} x \cdot 1\,dx = \left.\frac{x^2}{2}\right|_{\frac{1}{4}}^{1} = \frac{1^2 - (\frac{1}{4})^2}{2} = \frac{\frac{16}{16} - \frac{1}{16}}{2} = \frac{15}{32},$$

$$\langle x, x\rangle = \int_{\frac{1}{4}}^{1} x\cdot x\,dx = \frac{x^3}{3}\Bigg|_{\frac{1}{4}}^{1} = \frac{1^3 - (\frac{1}{4})^3}{3} = \frac{\frac{64}{64} - \frac{1}{64}}{3} = \frac{63}{252}.$$

Hence, Gram–Schmidt orthogonalization gives

$$w_1 = v_1 = 1,$$

$$\langle w_1, w_1\rangle = \langle 1, 1\rangle = \frac{3}{4};$$

$$w_2 = v_2 - \frac{\langle v_2, w_1\rangle}{\langle w_1, w_1\rangle}w_1 = x - \frac{\langle x, 1\rangle}{\langle 1, 1\rangle}1 = x - \frac{\frac{15}{32}}{\frac{3}{4}}1 = x - \frac{5}{8},$$

$$\langle w_2, w_2\rangle = \int_{\frac{1}{4}}^{1}\left(x - \frac{5}{8}\right)^2 dx = \frac{\left(x - \frac{5}{8}\right)^3}{3}\Bigg|_{\frac{1}{4}}^{1} = \frac{9}{256} = \frac{3^2}{2^8}.$$

The orthogonal projection of f on W then follows Theorem 4.43:

$$\langle f, w_1\rangle = \int_{\frac{1}{4}}^{1} \sqrt{x}\cdot 1\,dx = \frac{x^{\frac{3}{2}}}{\frac{3}{2}}\Bigg|_{\frac{1}{4}}^{1} = \frac{1 - \frac{1}{8}}{\frac{3}{2}} = \frac{7}{12} = \frac{7}{2^2\times 3},$$

$$\langle f, v_2\rangle = \int_{\frac{1}{4}}^{1} \sqrt{x}\cdot x\,dx = \frac{x^{\frac{5}{2}}}{\frac{5}{2}}\Bigg|_{\frac{1}{4}}^{1} = \frac{1 - \frac{1}{32}}{\frac{5}{2}} = \frac{31}{80} = \frac{31}{2^4\times 5},$$

$$\langle f, w_2\rangle = \langle f, v_2 - \frac{5}{8}w_1\rangle$$

$$= \langle f, v_2\rangle - \left(\frac{5}{8}\right)\langle f, w_1\rangle$$

$$= \frac{31}{80} - \left(\frac{5}{8}\right)\frac{7}{12} = \frac{11}{480} = \frac{11}{2^5\times 3\times 5}.$$

Hence, the affine polynomial p closest to the square root f on the interval $[\frac{1}{4}, 1]$ takes the form

$$p(x) = \frac{\langle f, w_1\rangle}{\langle w_1, w_1\rangle}w_1(x) + \frac{\langle f, w_2\rangle}{\langle w_2, w_2\rangle}w_2(x),$$

$$= \frac{\frac{7}{12}}{\frac{3}{4}}1 + \frac{\frac{11}{480}}{\frac{9}{256}}\left(x - \frac{5}{8}\right)$$

$$= \frac{88}{135}x + \frac{10}{27}.$$

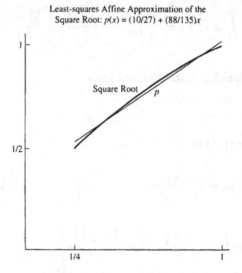

Figure 4.3   A least-squares affine approximation of the square root.

The result appears in Figure 4.3, with

$$c_0 = 10/27 \approx 0.370\,370\,370\,370\ldots,$$

$$c_1 = 88/135 \approx 0.651\,851\,851\,851\ldots.$$

An analysis of the discrepancy $D(x) := c_0 + c_1 x - \sqrt{x}$ may provide an estimate of the accuracy achieved by the approximation just obtained. On the open interval $]\frac{1}{4}, 1[$, where $\frac{1}{4} < x < 1$, calculus shows that the discrepancy D has a local minimum at $x_* := 1/(4 \cdot c_1^2) \approx 0.588\,358\,729\,339\ldots$, where $D(x_*) = c_0 - [1/(4c_1)] = -125/9504 \approx -0.013\,152\,357\ldots$, $D'(x_*) = 0$, and $D''(x_*) > 0$. However, at the endpoints, $D(1) = 1/45 = 0.022\,222\ldots$ and $D(\frac{1}{4}) = 1/30 = 0.033\,333\ldots$. Consequently, the absolute discrepancy reaches it maximum at $\frac{1}{4}$. Because the square root also has its minimum absolute value there, the relative discrepancy attains its maximum at the same endpoint. Thus,

$$\frac{|c_0 + c_1 x - \sqrt{x}|}{|\sqrt{x}|} \le \frac{|c_0 + c_1(\frac{1}{4}) - \sqrt{\frac{1}{4}}|}{\left|\sqrt{\frac{1}{4}}\right|} \le \frac{\frac{1}{30}}{\frac{1}{2}} = \frac{1}{15} = 0.0666\ldots. \qquad \square$$

The affine polynomial $p(x) = c_0 + c_1 x$ just obtained approximates the square root $\sqrt{}$ on the interval $[\frac{1}{4}, 1]$.

Example 4.53   If $x := \frac{9}{25}$, then $\sqrt{\frac{9}{25}} = \frac{3}{5} = 0.6$ and $p(\frac{9}{25}) = c_0 + c_1 \frac{9}{25} \approx 0.605\,037\,037\,037\ldots \approx 0.6$. $\qquad \square$

The affine polynomial $p(x) = c_0 + c_1 x$ can also be used to approximate the square root $\sqrt{}$ everywhere on the positive real line $\mathbb{R}_+^* =]0, \infty[$.

Example 4.54 If $z := 9$, then $\sqrt{9} = 3$. Moreover, $z = 9 = 25 \cdot \frac{9}{25}$ with $\frac{9}{25} \in [\frac{1}{4}, 1]$. Hence $p(\frac{9}{25}) = c_0 + c_1 \frac{9}{25} \approx 0.605\,037\,037\,037\ldots$ and

$$\sqrt{9} = \sqrt{25 \cdot \frac{9}{25}} = \sqrt{25} \cdot \sqrt{\frac{9}{25}} \approx \sqrt{25} \cdot p\left(\frac{9}{25}\right)$$

$$\approx 5 * 0.605\,037\,037\,037\ldots$$

$$\approx 3.025\,185\,185\ldots. \qquad\qquad \square$$

Remark 4.55 Many other methods exist to approximate functions to any degree of accuracy, for example, splines, and Chebyshev's least absolute value approximation. Algorithms for computers may follow such approximation by Newton's method to produce greater accuracy [27], [34], [39]. $\qquad \square$

EXERCISES

Exercise 4.11. Consider the field $\mathbb{R}$ and the linear space $V := C^0([\frac{1}{8}, 1], \mathbb{R})$ consisting of all real functions continuous on the interval $[\frac{1}{8}, 1]$, with the inner product

$$\langle f, g \rangle := \int_{\frac{1}{8}}^{1} f(x)g(x)\,dx.$$

(a) Apply the Gram–Schmidt process to the subset $\{1, x\}$.

(b) In the subspace $W := \mathrm{Span}\,\{1, x\}$, determine the affine function q of the form $q(x) = c_0 + c_1 x$ closest to the cube-root function $\sqrt[3]{} \in V$.

(c) Compare $\sqrt[3]{\frac{64}{125}}$ with its approximation $q(\frac{64}{125})$.

(d) Explain how to use q to approximate the cube root of each real number, and verify your procedure with $\sqrt[3]{27}$.

(e) Estimate the maximum relative discrepancy between the cube root and its approximation q on $[\frac{1}{8}, 1]$.

Exercise 4.12. This exercise demonstrates a method to design an algorithm to compute the exponential function. Consider the linear space $V := C^0([-1, 0], \mathbb{R})$ of all continuous real-valued functions defined on the closed interval $[-1, 0]$, with the inner product

$$\langle f, g \rangle := \int_{-1}^{0} f(x) \cdot g(x)\,dx,$$

and with the function $\exp_2 : [-1, 0] \to \mathbb{R}$ defined by $\exp_2(x) := 2^x$.

(a) Apply the Gram–Schmidt process to the subset $\{1, x\}$.

(b) Calculate the orthogonal projection, denoted by g, of the function $\exp_2$ just defined on the linear subspace $W := \mathrm{Span}\,\{1, x\}$.

(c) Compare $\exp_2(-\frac{1}{2})$ with its approximation $g(-\frac{1}{2})$.

(d) Explain how to utilize g to approximate e^x for each $x \in \mathbb{R}$, and test your procedure with e^1.

(e) Estimate the maximum relative discrepancy between $\exp_2$ and its approximation g.

4.3.4   Applications to Wavelets

In a manner similar to the approximation of functions by mutually orthogonal polynomials explained in the preceding subsection, orthogonal projections produce approximations of signals by combinations of mutually orthogonal wavelets. The general context of wavelets includes a linear space V of functions with an inner product $\langle\ ,\ \rangle$ and a linear subspace $W \subseteq V$ consisting of combinations of mutually orthogonal wavelets

$$w_0, w_1, w_2, \ldots, w_k, w_{k+1}, \ldots.$$

For each function (signal) f in V, the representation of f by wavelets consists of the orthogonal projection $\tilde{f}$ of f on the linear subspace W spanned by the wavelets:

$$\tilde{f} = \sum_k \frac{\langle f, w_k \rangle}{\langle w_k, w_k \rangle}\, w_k,$$

with wavelet coefficients

$$c_k = \frac{\langle f, w_k \rangle}{\langle w_k, w_k \rangle}.$$

If $W = V$, then $\tilde{f} = f$. If $W \not\subseteq V$, however, then the wavelet representation $\tilde{f}$ does not reconstruct exactly, but only approximates, the signal f. For instance, with the Haar wavelets, consider the linear space

$$V := C_I^0\,([0, 1], \mathbb{R})$$

of all piecewise continuous functions defined on the closed unit interval $[0, 1]$, with inner product

$$\langle f, g \rangle = \int_0^1 f(t)\, g(t)\, dt.$$

The following proposition verifies that step functions over disjoint intervals are mutually orthogonal.

Proposition 4.56   *For each positive integer n, and for all integers k, the step functions*

$$\varphi_{[0,2^{-n}[}, \quad \varphi_{[2^{-n},2*2^{-n}[}, \quad \varphi_{[2*2^{-n},3*2^{-n}[}, \cdots, \varphi_{[(2^n-1)*2^{-n},(2^n)*2^{-n}[},$$

defined by

$$\varphi_{[k2^{-n},(k+1)2^{-n}[}(r) := \begin{cases} 1 & \text{if } k2^{-n} \le r < (k+1)2^{-n}, \\ 0 & \text{if } r < k2^{-n} \text{ or } (k+1)2^{-n} \le r, \end{cases}$$

are mutually orthogonal functions in the linear space $V := C_I^0\,([0,1],\mathbb{R})$ *with the inner product* $\langle f, g \rangle = \int_0^1 f(t)\,g(t)\,dt.$

PROOF: For all integers k and ℓ,

$$\langle \varphi_{[k2^{-n},(k+1)2^{-n}[}, \quad \varphi_{[\ell2^{-n},(\ell+1)2^{-n}[} \rangle$$

$$= \int_0^1 \varphi_{[k2^{-n},(k+1)2^{-n}[}(r) \cdot \varphi_{[\ell2^{-n},(\ell+1)2^{-n}[}(r)\,dr.$$

If $k \ne \ell$, for instance if $k < \ell$, then for each real number r, either $r < \ell$ and then $\varphi_{[\ell2^{-n},(\ell+1)2^{-n}[}(r) = 0$, or $\ell \le r$ and then $\varphi_{[k2^{-n},(k+1)2^{-n}[}(r) = 0$. Consequently, if $k < \ell$, then

$$\varphi_{[k2^{-n},(k+1)2^{-n}[}(r) \cdot \varphi_{[\ell2^{-n},(\ell+1)2^{-n}[}(r) = 0$$

for every r, whence

$$\langle \varphi_{[k2^{-n},(k+1)2^{-n}[}, \quad \varphi_{[\ell2^{-n},(\ell+1)2^{-n}[} \rangle$$

$$= \int_0^1 \varphi_{[k2^{-n},(k+1)2^{-n}[}(r) \cdot \varphi_{[\ell2^{-n},(\ell+1)2^{-n}[}(r)\,dr = \int_0^1 0\,dr = 0.$$

Switching the roles of k and ℓ confirms the same result if $\ell < k$. Therefore, the functions $\varphi_{[k2^{-n},(k+1)2^{-n}[}$ are mutually orthogonal.
   In contrast, if $k = \ell$, then

$$\varphi_{[k2^{-n},(k+1)2^{-n}[}(r) \cdot \varphi_{[k2^{-n},(k+1)2^{-n}[}(r) = \begin{cases} 1 & \text{if } k2^{-n} \le r < (k+1)2^{-n}, \\ 0 & \text{otherwise,} \end{cases}$$

whence

$$\langle \varphi_{[k2^{-n},(k+1)2^{-n}[}, \quad \varphi_{[k2^{-n},(k+1)2^{-n}[} \rangle$$

$$= \int_0^1 \varphi_{[k2^{-n},(k+1)2^{-n}[}(r) \cdot \varphi_{[k2^{-n},(k+1)2^{-n}[}(r)\,dr$$

$$= \int_{k2^{-n}}^{(k+1)2^{-n}} 1\,dr = 2^{-n}. \qquad\qquad \square$$

Thus the step functions considered here span the linear subspace

$$W := \mathrm{Span}\{\varphi_{[k2^{-n},(k+1)2^{-n}[} : k \in \{0, 1, \ldots, 2^n - 2, 2^n - 1\}\}.$$

Such functions approximate a continuous signal $f \in V = C_I^0([0, 1], \mathbb{R})$ by the combination of step functions

$$\tilde{f} := \sum_{k=0}^{2^n-1} \frac{\langle f, \varphi_{[k2^{-n},(k+1)2^{-n}[} \rangle}{\langle \varphi_{[k2^{-n},(k+1)2^{-n}[}, \varphi_{[k2^{-n},(k+1)2^{-n}[} \rangle} \varphi_{[k2^{-n},(k+1)2^{-n}[}.$$

Approximating the integral inner products by a Riemann sum through the left-hand endpoint gives

$$\langle f, \varphi_{[k2^{-n},(k+1)2^{-n}[} \rangle$$

$$= \int_{k2^{-n}}^{(k+1)2^{-n}} f(r)\, \varphi_{[k2^{-n},(k+1)2^{-n}[}(r)\, dr \approx f(k2^{-n})2^{-n},$$

$$\langle \varphi_{[k2^{-n},(k+1)2^{-n}[}, \varphi_{[k2^{-n},(k+1)2^{-n}[} \rangle$$

$$= \int_{k2^{-n}}^{(k+1)2^{-n}} \varphi_{[k2^{-n},(k+1)2^{-n}[}(r)\, \varphi_{[k2^{-n},(k+1)2^{-n}[}(r)\, dr \approx 2^{-n},$$

$$\frac{\langle f, \varphi_{[k2^{-n},(k+1)2^{-n}[} \rangle}{\langle \varphi_{[k2^{-n},(k+1)2^{-n}[}, \varphi_{[k2^{-n},(k+1)2^{-n}[} \rangle}$$

$$= \frac{\int_{k2^{-n}}^{(k+1)2^{-n}} f(r)\, \varphi_{[k2^{-n},(k+1)2^{-n}[}(r)\, dr}{\int_{k2^{-n}}^{(k+1)2^{-n}} \varphi_{[k2^{-n},(k+1)2^{-n}[}(r)\, \varphi_{[k2^{-n},(k+1)2^{-n}[}(r)\, dr}$$

$$\approx \frac{f(k2^{-n})2^{-n}}{2^{-n}} = f(k2^{-n}).$$

The approximation just obtained explains why the values of the sample $s_k = f(k2^{-n})$ can serve as the initial coefficients for the approximation of the signal f by the combination of wavelets $\tilde{f}$.

The orthogonality of the Haar wavelets also corresponds to the orthogonality of the following matrix, as verified in the exercises.

Definition 4.57   For each nonnegative integer n, let $N := 2^n$, and set

$$_H^N\Omega := \begin{pmatrix} 1 & 1 & & & & & \\ 1 & -1 & & & & & \\ & & 1 & 1 & & & \\ & & 1 & -1 & & & \\ & & & & \ddots & & \\ & & & & & 1 & 1 \\ & & & & & 1 & -1 \end{pmatrix}$$

with N rows and N columns, where all blank entries equal zero. $\square$

Daubechies wavelets lend themselves to a similar explanation. However, the derivation of the existence, continuity, and orthogonality of Daubechies wavelets requires further mathematics, to be presented in the following chapters.

EXERCISES

This exercise demonstrates that the Haar Wavelet Transform corresponds to a rotation of coordinates in a linear space of functions.

Exercise 4.13. Define $_H^N\Omega$ as in Definition 4.57.

(a) Verify that $(_H^N\Omega)(_H^N\Omega)^T = 2I$.

(b) Deduce that $(_H^N\Omega)^T$ is the matrix inverse of $(\frac{1}{2})(_H^N\Omega)$.

(c) Conclude that $(1/\sqrt{2})(_H^N\Omega)$ is an orthogonal matrix:

$$\left[(1/\sqrt{2})(_H^N\Omega)\right]\left[(1/\sqrt{2})(_H^N\Omega)\right]^T = I.$$

Exercise 4.14. Define $_H^N\Omega$ as in Definition 4.57.

(a) Verify that for each array of $N = 2^n$ numbers

$$\vec{s}^{(n)} = \left(a_0^{(n)}, a_1^{(n)}, \ldots, a_{2^n-2}^{(n)}, a_{2^n-1}^{(n)}\right),$$

a multiplication by the matrix $(\frac{1}{2})_H^N\Omega$ produces the first sweep of the In-Place Haar Wavelet Transform:

$$\begin{pmatrix} a_0^{(n-1)} \\ c_0^{(n-1)} \\ a_1^{(n-1)} \\ c_1^{(n-1)} \\ \vdots \\ a_{2^{(n-1)}-1}^{(n-1)} \\ c_{2^{(n-1)}-1}^{(n-1)} \end{pmatrix} = \frac{1}{2} \begin{pmatrix} 1 & 1 & & & & \\ 1 & -1 & & & & \\ & & 1 & 1 & & \\ & & 1 & -1 & & \\ & & & & \ddots & \\ & & & & 1 & 1 \\ & & & & 1 & -1 \end{pmatrix} \begin{pmatrix} a_0^{(n)} \\ a_1^{(n)} \\ a_2^{(n)} \\ a_3^{(n)} \\ \vdots \\ a_{2^n-2}^{(n)} \\ a_{2^n-1}^{(n)} \end{pmatrix}.$$

(b) Determine the matrix that yields one sweep of the *inverse* transform.

Exercise 4.15. Verify that for each positive integer n, and for all integers k, the 2^{n-1} step functions

$$\varphi_{[0,2^{-(n-1)}[}, \cdots, \varphi_{[(2^{n-1}-1)*2^{-(n-1)},(2^{n-1})*2^{-(n-1)}[},$$

defined for each $k \in \{0, \ldots, 2^{n-1} - 1\}$ by

$$\varphi_{[k2^{-(n-1)},(k+1)2^{-(n-1)}[}(r) := \begin{cases} 1 & \text{if } k2^{-(n-1)} \le r < (k+1)2^{-(n-1)}, \\ 0 & \text{if } r < k2^{-(n-1)} \text{ or } (k+1)2^{-(n-1)} \le r, \end{cases}$$

and the 2^{n-1} wavelets

$$\psi_{[0,2^{-(n-1)}[}, \ldots, \psi_{[(2^{n-1}-1)*2^{-(n-1)},(2^{n-1})*2^{-(n-1)}[},$$

defined for each $k \in \{0, \ldots, 2^{n-1} - 1\}$ by

$$\psi_{[k2^{-(n-1)},(k+1)2^{-(n-1)}[}(r)$$

$$:= \begin{cases} 1 & \text{if } k2^{-(n-1)} \leq r < (k + \frac{1}{2})2^{-(n-1)}, \\ -1 & \text{if } (k + \frac{1}{2})2^{-(n-1)} \leq r < (k + 1)2^{-(n-1)}, \\ 0 & \text{if } r < k2^{-(n-1)} \text{ or } (k + 1)2^{-(n-1)} \leq r, \end{cases}$$

are all mutually orthogonal functions in the linear space $C_I^0([0, 1], \mathbb{R})$ with the inner product $\langle f, g \rangle = \int_0^1 f(t) g(t) dt$.

Exercise 4.16. Verify that the linear subspace W spanned by the 2^n scaling functions

$$\varphi_{[0,2^{-n}[}, \quad \varphi_{[2^{-n},2*2^{-n}[}, \ldots, \varphi_{[(2^n-1)*2^{-n},(2^n)*2^{-n}[},$$

coincides with the linear subspace spanned by the 2^{n-1} scaling functions and 2^{n-1} wavelets

$$\varphi_{[0,2^{-(n-1)}[}, \ldots, \varphi_{[(2^{n-1}-1)*2^{-(n-1)},(2^{n-1})*2^{-(n-1)}[},$$
$$\psi_{[0,2^{-(n-1)}[}, \ldots, \psi_{[(2^{n-1}-1)*2^{-(n-1)},(2^{n-1})*2^{-(n-1)}[}.$$

CHAPTER 5

Discrete and Fast Fourier Transforms

5.0 INTRODUCTION

This chapter introduces a few features of the Discrete Fourier Transform (DFT) and its computation through the Fast Fourier Transform (FFT), which interpolates finite sequences of numbers by Fourier Series. The Fast Fourier Transform provides a comparison and contrast with the Fast Haar and Daubechies Wavelet Transforms in a context that involves complex exponentials and some linear algebra but that does not yet require calculus. Fourier Series and wavelets share the following common features:

(W1) Orthonormal bases allow for simple formulae for coefficients.

(W2) Algebraic relations allow for fast transforms.

(W3) Complete bases allow for approximations to any accuracy.

Yet Fourier series and wavelets differ by one essential feature.

(W4) Wavelets have compact support, but Fourier series do not.

In other words, every wavelet equals zero outside of an interval with finite length, whereas the complex exponential functions in Fourier series have nonzero values over the entire extent of the sample.

5.1 THE DISCRETE FOURIER TRANSFORM (DFT)

The Discrete Fourier Transform amounts to a change of coordinates in space, which yields information by consideration of the same data from a different point of view.

5.1.1   Definition and Inversion

The Discrete Fourier Transform will express arrays of real or complex data as linear combinations of other arrays ordered by frequencies instead of the initial values of the data. Every data set consisting of a finite sequence of complex numbers $f_0, \ldots, f_{N-1}$ corresponds to the coordinates of a vector $\vec{f} = (f_0, \ldots, f_{N-1}) \in \mathbb{C}^N$ with respect to the canonical basis

$$\vec{e}_0 = (1, 0, \ldots, 0, 0),$$

$$\vdots$$

$$\vec{e}_{N-1} = (0, 0, \ldots, 0, 1),$$

so that

$$\begin{pmatrix} f_0 \\ f_1 \\ \vdots \\ f_{N-2} \\ f_{N-1} \end{pmatrix} = f_0 \begin{pmatrix} 1 \\ 0 \\ \vdots \\ 0 \\ 0 \end{pmatrix} + f_1 \begin{pmatrix} 0 \\ 1 \\ \vdots \\ 0 \\ 0 \end{pmatrix} + \cdots + f_{N-2} \begin{pmatrix} 0 \\ 0 \\ \vdots \\ 1 \\ 0 \end{pmatrix} + f_{N-1} \begin{pmatrix} 0 \\ 0 \\ \vdots \\ 0 \\ 1 \end{pmatrix}.$$

$$= f_0 \vec{e}_0 + f_1 \vec{e}_1 + \cdots + f_{N-2} \vec{e}_{N-2} + f_{N-1} \vec{e}_{N-1}.$$

Every such array $\vec{f} = (f_0, \ldots, f_{N-1})$ can, but need not, be a sample from a function $f : \mathbb{R} \to \mathbb{C}$ sampled at N points

$$x_0 := 0, \ldots, \quad x_\ell := \ell \cdot (2\pi/N), \ldots, \quad x_{N-1} := (N-1) \cdot (2\pi/N)$$

in $[0, 2\pi[$, with values

$$\vec{f} = (f(x_0), f(x_1), \ldots, f(x_\ell), \ldots, f(x_{N-1})).$$

The Discrete Fourier Transform then expresses such an array $\vec{f}$ not with its values f_ℓ but with linear combinations of arrays of the type

$$\vec{w}_k := \left(e^{ikx_\ell} \right)_{\ell=0}^{N-1} = \left(1, e^{ik2\pi/N}, \ldots, e^{ik\ell \cdot 2\pi/N}, \ldots, e^{ik(N-1) \cdot 2\pi/N} \right),$$

where the coordinate with index ℓ in the array $\vec{w}_k$ equals the Nth root of unity, $\omega_N := e^{i \cdot 2\pi/N}$, raised to the power $k\ell$:

$$(\vec{w}_k)_\ell = \left(e^{i \cdot 2\pi/N} \right)^{k\ell} = \omega_N^{k\ell}.$$

Example 5.1   If $N = 1$, then $\omega_N = \omega_1 = e^{i2\pi/1} = 1$ and $\vec{w}_0 = (\omega_1^0) = (1)$. If $N = 2$, then $\omega_N = \omega_2 = e^{i2\pi/2} = -1$ and

$$\vec{w}_0 = \left([\omega_2^0]^0, [\omega_2^0]^1 \right) = (1, 1),$$

$$\vec{w}_1 = \left([\omega_2^1]^0, [\omega_2^1]^1 \right) = (1, -1).$$

If $N = 4$, then $\omega_N = \omega_4 = e^{i2\pi/4} = i$ and

$$\vec{\mathbf{w}}_0 = \left([\omega_4^0]^0, [\omega_4^0]^1, [\omega_4^0]^2, [\omega_4^0]^3\right) = (1, 1, 1, 1),$$

$$\vec{\mathbf{w}}_1 = \left([\omega_4^1]^0, [\omega_4^1]^1, [\omega_4^1]^2, [\omega_4^1]^3\right) = (1, i, -1, -i),$$

$$\vec{\mathbf{w}}_2 = \left([\omega_4^2]^0, [\omega_4^2]^1, [\omega_4^2]^2, [\omega_4^2]^3\right) = (1, -1, 1, -1),$$

$$\vec{\mathbf{w}}_3 = \left([\omega_4^3]^0, [\omega_4^3]^1, [\omega_4^3]^2, [\omega_4^3]^3\right) = (1, -i, -1, i). \qquad \square$$

Several features of the basis $(\vec{\mathbf{w}}_0, \ldots, \vec{\mathbf{w}}_{N-1})$ will allow for fast computations. For instance, the arrays $\vec{\mathbf{w}}_k$ are mutually orthonormal with respect to a variant of the usual inner product.

Definition 5.2 For each positive integer N, index the coordinates of vectors in $\mathbb{C}^N$ with indices from 0 through $N - 1$, and define an inner product $\langle\,,\,\rangle_N$ on $\mathbb{C}^N$ by

$$\langle\vec{\mathbf{z}}, \vec{\mathbf{w}}\rangle_N := \frac{1}{N}\sum_{m=0}^{N-1} z_m \cdot \overline{w_m}. \qquad \square$$

Remark 5.3 Multiplicative constants other than $1/N$ also occur in this context. For instance, some texts and computer programs use the constant $1/\sqrt{N}$, which will then yield the same constant for the inverse transform, but with the additional computation of $\sqrt{N}$ if N is not an even power of two. Alternatively, other texts and computer programs omit such a constant and use $\langle\vec{\mathbf{z}}, \vec{\mathbf{w}}\rangle_N := \sum_{m=0}^{N-1} z_m \overline{w_m}$, which then requires $1/N$ for the inverse transform but also allows users to choose their own constant. Still other texts and computer programs call the transform just introduced the "Inverse Discrete Fourier Transform," thus swapping the roles of the forward and inverse transforms. The constant selected here, $1/N$, will later prove consistent with complex Fourier series. $\qquad \square$

The following formula for "geometric series" allows for the simplification of sums of consecutive integral powers.

Lemma 5.4 (Geometric Series.) *For each $w \in \mathbb{C} \setminus \{1\}$ and for each $N \in \mathbb{N}$,*

$$\sum_{k=0}^{N} w^k = (1 - w)^{-1}(1 - w^{N+1}).$$

PROOF: Let $S_N := \sum_{k=0}^{N} w^k$ and calculate $S_N - w \cdot S_N$:

$$S_N = 1 + w + \cdots + w^{N-1} + w^N,$$

$$w \cdot S_N = w + w^2 + \cdots + w^N + w^{N+1},$$

$$S_N - w \cdot S_N = 1 \qquad\qquad - w^{N+1};$$

then $(1 - w)S_N = 1 - w^{N+1}$ and $S_N = (1 - w)^{-1}(1 - w^{N+1})$. $\qquad \square$

Remark 5.5 As just stated, $\sum_{k=0}^{N} w^k = (1 - w)^{-1}(1 - w^{N+1})$, the formula for the geometric series, holds not only for complex numbers $w \neq 1$, but also for matrices w such that the matrix $I - w$ is invertible, with I denoting the identity matrix [39, p. 256, # 9]. The same formula also holds for certain linear operators w between linear spaces [45, p. 260]. $\square$

Lemma 5.6 *For each positive integer N, the set*

$$\left\{ \vec{w}_k : k \in \{0, \ldots, N - 1\} \right\}$$

is orthonormal with respect to the inner product $\langle \, , \, \rangle_N$.

PROOF: Apply the definitions of $\vec{w}_k$ and $\langle \vec{z}, \vec{w} \rangle_N$. For all integers $k, \ell \in \mathbb{Z}$ such that N divides $k - \ell$, so that $\ell = k + JN$ for some integer J,

$$
\begin{aligned}
\langle \vec{w}_k, \vec{w}_\ell \rangle_N &= \frac{1}{N} \sum_{m=0}^{N-1} (\vec{w}_k)_m \overline{(\vec{w}_\ell)_m} = \frac{1}{N} \sum_{m=0}^{N-1} e^{ikm \cdot 2\pi/N} \overline{e^{i\ell m \cdot 2\pi/N}} \\
&= \frac{1}{N} \sum_{m=0}^{N-1} e^{ikm \cdot 2\pi/N} e^{-i\ell m \cdot 2\pi/N} \\
&= \frac{1}{N} \sum_{m=0}^{N-1} e^{ikm \cdot 2\pi/N} e^{-i(k+JN)m \cdot 2\pi/N} \\
&= \frac{1}{N} \sum_{m=0}^{N-1} e^{ikm \cdot 2\pi/N} e^{-ikm \cdot 2\pi/N} e^{-iJm \cdot 2\pi} = \frac{1}{N} \sum_{m=0}^{N-1} 1 = 1.
\end{aligned}
$$

In contrast, for all integers $k, \ell \in \mathbb{Z}$ such that N does *not* divide $k - \ell$, the formula for the sum of the geometric series,

$$\sum_{m=0}^{N-1} w^m = \frac{1 - w^N}{1 - w},$$

with $w := e^{i \cdot [k-\ell] \cdot 2\pi/N} \neq 1$ gives

$$
\begin{aligned}
\langle \vec{w}_k, \vec{w}_\ell \rangle_N &= \frac{1}{N} \sum_{m=0}^{N-1} (\vec{w}_k)_m \overline{(\vec{w}_\ell)_m} = \frac{1}{N} \sum_{m=0}^{N-1} e^{ikm \cdot 2\pi/N} \overline{e^{i\ell m \cdot 2\pi/N}} \\
&= \frac{1}{N} \sum_{m=0}^{N-1} e^{ikm \cdot 2\pi/N} e^{-i\ell m \cdot 2\pi/N} = \frac{1}{N} \sum_{m=0}^{N-1} e^{i[k-\ell]m \cdot 2\pi/N} \\
&= \frac{1}{N} \sum_{m=0}^{N-1} \left(e^{i[k-\ell] \cdot 2\pi/N} \right)^m = \frac{1}{N} \frac{1 - \left(e^{i[k-\ell] \cdot 2\pi/N} \right)^N}{1 - e^{i[k-\ell] \cdot 2\pi/N}} \\
&= \frac{1}{N} \frac{1 - \left(e^{2\pi i} \right)^{[k-\ell]}}{1 - e^{i[k-\ell] \cdot 2\pi/N}} = \frac{1}{N} \frac{1 - 1}{1 - e^{i[k-\ell] \cdot 2\pi/N}} = 0. \quad \square
\end{aligned}
$$

Definition 5.7 For each positive integer N, define the Fourier matrix $\underset{F}{N}\Omega$ so that the ℓth column of $\underset{F}{N}\Omega$ contains $\vec{w}_\ell$:

$$\underset{F}{N}\Omega_{k,\ell} := (\vec{w}_\ell)_k = e^{ik\ell 2\pi/N} = \omega_N^{k\ell}. \qquad \square$$

Example 5.8 If $N = 1$, then $\omega_N = \omega_1 = e^{i2\pi/1} = 1$, and $\underset{F}{1}\Omega = ((\omega_1^0)^0) = 1$.
If $N = 2$, then $\omega_N = \omega_2 = e^{i2\pi/2} = -1$, and

$$\underset{F}{2}\Omega = (\vec{w}_0, \quad \vec{w}_1) = \begin{pmatrix} (\omega_2^0)^0 & (\omega_2^1)^0 \\ (\omega_2^0)^1 & (\omega_2^1)^1 \end{pmatrix} = \begin{pmatrix} 1 & 1 \\ 1 & -1 \end{pmatrix}.$$

Thus, for $N = 2$, the Fourier matrix coincides with the Haar matrix.

If $N = 4$, then $\omega_N = \omega_4 = e^{i2\pi/4} = i$, and

$$\underset{F}{4}\Omega = (\vec{w}_0, \quad \vec{w}_1 \quad \vec{w}_2, \quad \vec{w}_3)$$

$$= \begin{pmatrix} 1 & (i^1)^0 & (i^2)^0 & (i^3)^0 \\ 1 & (i^1)^1 & (i^2)^1 & (i^3)^1 \\ 1 & (i^1)^2 & (i^2)^2 & (i^3)^2 \\ 1 & (i^1)^3 & (i^2)^3 & (i^3)^3 \end{pmatrix} = \begin{pmatrix} 1 & 1 & 1 & 1 \\ 1 & i & -1 & -i \\ 1 & -1 & 1 & -1 \\ 1 & -i & -1 & i \end{pmatrix}. \qquad \square$$

Remark 5.9 Because the linear space $\mathbb{C}^N$ has dimension N over the field $\mathbb{C}$, it follows from linear algebra that the N mutually orthonormal vectors $\vec{w}_0, \ldots, \vec{w}_{N-1}$ form a basis for $\mathbb{C}^N$, so that each vector $\vec{f} \in \mathbb{C}^N$ admits a representation as a linear combination of $\vec{w}_0, \ldots, \vec{w}_{N-1}$. The following definition gives a name to the coefficients of such combinations. $\qquad \square$

Definition 5.10 For each positive integer N, and for each array $\vec{f} \in \mathbb{C}^N$, the **Discrete Fourier Transform** of $\vec{f}$ is the array $\hat{f}$ defined by

$$\hat{f}_k = \langle \vec{f}, \vec{w}_k \rangle_N = \frac{1}{N} \sum_{m=0}^{N-1} f_m \cdot e^{-ikm \cdot 2\pi/N}. \qquad \square$$

Remark 5.11 Because the Discrete Fourier Transform (abbreviated DFT) consists of the coefficients of the same vector $\vec{f}$ with respect to the new orthonormal basis $(\vec{w}_k)_{k=0}^{N-1}$, the Discrete Fourier Transform is a rotation of coordinates in the space $\mathbb{C}^N$. $\qquad \square$

Proposition 5.12 *The Discrete Fourier Transform corresponds to a multiplication by the transposed conjugate matrix* $(1/N)\overline{\underset{F}{N}\Omega^T}$.

PROOF:

$$\hat{f}_{N,k} = \langle \vec{f}, \vec{w}_k \rangle_N = \frac{1}{N} \cdot \sum_{\ell=0}^{N-1} f_\ell e^{-i \cdot k \cdot \ell \cdot 2\pi/N}$$

$$= \frac{1}{N} \cdot \sum_{\ell=0}^{N-1} \overline{\underset{F}{N}\Omega^T}_{k,\ell} \cdot f_\ell = \frac{1}{N} \cdot (\overline{\underset{F}{N}\Omega^T} \cdot \vec{f})_k. \qquad \square$$

As with every orthonormal basis, an elementary sum provides the **Inverse Discrete Fourier Transform,** which expresses each vector $\vec{\mathbf{f}} \in \mathbb{C}^N$ in terms of its Discrete Fourier Transform.

Proposition 5.13 *For each positive integer N, and for each array $\vec{\mathbf{f}} \in \mathbb{C}^N$, the following inversion formula holds:*

$$\vec{\mathbf{f}} = \sum_{k=0}^{N-1} \hat{f}_k \vec{\mathbf{w}}_k; \qquad \hat{f}_k = \langle \vec{\mathbf{f}}, \vec{\mathbf{w}}_k \rangle_N = \frac{1}{N} \sum_{m=0}^{N-1} f_m \cdot e^{-ikm \cdot 2\pi/N}.$$

PROOF: Several proofs exist, among them the following three proofs: a short but abstract one, one with matrix algebra, and a longer but merely computational one.

First proof. Thanks to the orthonormality of $(\vec{\mathbf{w}}_k)_{k=0}^{N-1}$ and to the fact that $(\vec{\mathbf{w}}_k)_{k=0}^{N-1}$ forms a basis for $\mathbb{C}^N$, the formula

$$\vec{\mathbf{f}} = \sum_{k=0}^{N-1} \hat{f}_k \vec{\mathbf{w}}_k = \sum_{k=0}^{N-1} \langle \vec{\mathbf{f}}, \vec{\mathbf{w}}_k \rangle_N \vec{\mathbf{w}}_k$$

represents the orthogonal projection of $\vec{\mathbf{f}}$ on $\mathbb{C}^N$, which projection is $\vec{\mathbf{f}}$ because $\vec{\mathbf{f}}$ already lies in the range $\mathbb{C}^N$ of the projection.

Second proof. In terms of matrix algebra, with I denoting the identity matrix, the orthogonality of $(\vec{\mathbf{w}}_k)_{k=0}^{N-1}$ means that

$$\overset{N}{\underset{F}{}}\Omega \cdot \frac{1}{N} \overline{\overset{N}{\underset{F}{}}\Omega^T} = I.$$

Therefore,

$$\vec{\mathbf{f}} = I\vec{\mathbf{f}} = \left[\overset{N}{\underset{F}{}}\Omega \cdot \frac{1}{N} \overline{\overset{N}{\underset{F}{}}\Omega^T} \right] \vec{\mathbf{f}} = \overset{N}{\underset{F}{}}\Omega \cdot \left[\frac{1}{N} \overline{\overset{N}{\underset{F}{}}\Omega^T} \cdot \vec{\mathbf{f}} \right] = \left(\overset{N}{\underset{F}{}}\Omega \right) \hat{\mathbf{f}}.$$

Third proof. Direct calculations give

$$\left[\sum_{k=0}^{N-1} \hat{f}_k \vec{\mathbf{w}}_k \right]_m = \sum_{k=0}^{N-1} \left(\frac{1}{N} \sum_{\ell=0}^{N-1} f_\ell e^{-ik\ell 2\pi/N} \right) e^{ikm2\pi/N}$$

$$= \sum_{\ell=0}^{N-1} f_\ell \frac{1}{N} \sum_{k=0}^{N-1} e^{-ik\ell 2\pi/N} e^{ikm2\pi/N} = \sum_{\ell=0}^{N-1} f_\ell \langle \vec{\mathbf{w}}_m, \vec{\mathbf{w}}_\ell \rangle_N = f_m,$$

because $\langle \vec{\mathbf{w}}_m, \vec{\mathbf{w}}_m \rangle_N = 1$ but $\langle \vec{\mathbf{w}}_m, \vec{\mathbf{w}}_\ell \rangle_N = 0$ for every $m \neq \ell$.   □

The next examples illustrate the concepts just defined. However, the Discrete Fourier Transform becomes more effective in the form of the Fast Fourier Transform, as explained in the next section.

Example 5.14   If $N := 1$, then $\omega_1 = 1$, and arrays have only $N = 1$ coordinate: $\vec{w}_0 = \omega_1^0 = 1$, whence $\vec{f} = (f_0)$ and $\hat{f} = (\hat{f}_0)$ with $\hat{f}_0 = f_0\overline{\omega_1} = f_0$. Thus, $\hat{f} = \vec{f}$ for $N = 1$.

With $N := 2$, the square root of unity becomes

$$\omega_N = \omega_2 = e^{i \cdot 2\pi/N} = e^{i \cdot 2\pi/2} = e^{i\pi} = -1,$$

and the N arrays $\vec{w}_k$ take the form

$$\vec{w}_0 = \left([e^{i \cdot 2\pi/2}]^{0 \cdot 0}, [e^{i \cdot 2\pi/2}]^{0 \cdot 1}\right) = \left([-1]^{0 \cdot 0}, [-1]^{0 \cdot 1}\right) = (1, 1),$$

$$\vec{w}_1 = \left([e^{i \cdot 2\pi/2}]^{1 \cdot 0}, [e^{i \cdot 2\pi/2}]^{1 \cdot 1}\right) = \left([-1]^{1 \cdot 0}, [-1]^{1 \cdot 1}\right) = (1, -1).$$

Hence, for each array $\vec{f} = (f_0, f_1)$, inner products give the Discrete Fourier Transform $\hat{f} = (\hat{f}_0, \hat{f}_1)$ given by

$$\hat{f}_0 = \langle \vec{f}, \vec{w}_0 \rangle_2 = \langle (f_0, f_1), (1, 1) \rangle_2 = \frac{1}{2} \cdot (f_0 \cdot \overline{1} + f_1 \cdot \overline{1})$$

$$= \frac{f_0 + f_1}{2},$$

$$\hat{f}_1 = \langle \vec{f}, \vec{w}_1 \rangle_2 = \langle (f_0, f_1), (1, -1) \rangle_2 = \frac{1}{2} \cdot (f_0 \cdot \overline{1} + f_1 \cdot \overline{-1})$$

$$= \frac{f_0 - f_1}{2}.$$

Thus, for $N = 2$ the Discrete Fourier Transform coincides with the Haar Wavelet Transform and yields multiples of the sum and difference of the values of the data. The inversion formula then recovers the initial data:

$$\vec{f} = \hat{f}_0 \vec{w}_0 + \hat{f}_1 \vec{w}_1,$$

$$(f_0, f_1) = \frac{f_0 + f_1}{2} \cdot (1, 1) + \frac{f_0 - f_1}{2} \cdot (1, -1),$$

which expresses the initial array $\vec{f} = (f_0, f_1)$ as a linear combination of a constant array $(1, 1)$ and an array $(1, -1)$ with one period.   □

Example 5.15   With $N := 2$, consider the array of data

$$\vec{f} = (f_0, f_1) := (7, 5).$$

Inner products give the Discrete Fourier Transform

$$\hat{f}_0 = \langle \vec{f}, \vec{w}_0 \rangle_2 = \langle (7, 5), (1, 1) \rangle_2 = \frac{1}{2} \cdot (7 \cdot \overline{1} + 5 \cdot \overline{1}) = 6,$$

$$\hat{f}_1 = \langle \vec{f}, \vec{w}_1 \rangle_2 = \langle (7, 5), (1, -1) \rangle_2 = \frac{1}{2} \cdot (7 \cdot \overline{1} + 5 \cdot \overline{-1}) = 1.$$

As a verification,

$$\vec{f} = \hat{f}_0 \vec{w}_0 + \hat{f}_1 \vec{w}_1,$$

$$(7, 5) = 6 \cdot (1, 1) + 1 \cdot (1, -1),$$

which expresses the initial array $\vec{f} = (7, 5)$ as the sum of a constant array $6 \cdot (1, 1)$ and an array $1 \cdot (1, -1)$ with one period.   □

Example 5.16   With $N := 4$, the fourth root of unity becomes

$$\omega_N = \omega_4 = e^{i \cdot 2\pi/N} = e^{i \cdot 2\pi/4} = e^{i\pi/2} = i.$$

The N arrays $\vec{w}_k$ become

$$\vec{w}_0 = \left([i]^{0 \cdot 0}, [i]^{0 \cdot 1}, [i]^{0 \cdot 2}, [i]^{0 \cdot 3} \right) = (1, 1, 1, 1),$$

$$\vec{w}_1 = \left([i]^{1 \cdot 0}, [i]^{1 \cdot 1}, [i]^{1 \cdot 2}, [i]^{1 \cdot 3} \right) = (1, i, -1, -i),$$

$$\vec{w}_2 = \left([i]^{2 \cdot 0}, [i]^{2 \cdot 1}, [i]^{2 \cdot 2}, [v]^{2 \cdot 3} \right) = (1, -1, 1, -1),$$

$$\vec{w}_3 = \left([i]^{3 \cdot 0}, [i]^{3 \cdot 1}, [i]^{3 \cdot 2}, [i]^{3 \cdot 3} \right) = (1, -i, -1, i).$$

As a numerical example, consider the array of data

$$\vec{f} = (f_0, f_1, f_2, f_3) := (9, 7, 5, 7).$$

Hence, inner products give the Discrete Fourier Transform

$$\hat{f}_0 = \langle \vec{f}, \vec{w}_0 \rangle_4 = \langle (9, 7, 5, 7), (1, 1, 1, 1) \rangle_4$$

$$= \frac{1}{4} \cdot (9 \cdot \overline{1} + 7 \cdot \overline{1} + 5 \cdot \overline{1} + 7 \cdot \overline{1}) = 7,$$

$$\hat{f}_1 = \langle \vec{f}, \vec{w}_1 \rangle_4 = \langle (9, 7, 5, 7), (1, i, -1, -i) \rangle_4$$

$$= \frac{1}{4} \cdot (9 \cdot \overline{1} + 7 \cdot \overline{i} + 5 \cdot \overline{-1} + 7 \cdot \overline{-i}) = 1,$$

$$\hat{f}_2 = \langle \vec{f}, \vec{w}_2 \rangle_4 = \langle (9, 7, 5, 7), (1, -1, 1, -1) \rangle_4$$

$$= \frac{1}{4} \cdot (9 \cdot \overline{1} + 7 \cdot \overline{-1} + 5 \cdot \overline{1} + 7 \cdot \overline{-1}) = 0,$$

$$\hat{f}_3 = \langle \vec{f}, \vec{w}_1 \rangle_4 = \langle (9, 7, 5, 7), (1, -i, -1, i) \rangle_4$$

$$= \frac{1}{4} \cdot (9 \cdot \overline{1} + 7 \cdot \overline{-i} + 5 \cdot \overline{-1} + 7 \cdot \overline{i}) = 1.$$

Thus,

$$\vec{f} = \hat{f}_0 \vec{w}_0 + \hat{f}_1 \vec{w}_1 + \hat{f}_2 \vec{w}_2 + \hat{f}_3 \vec{w}_3,$$

$$(9,7,5,7) = 7 \cdot (1,1,1,1) + 1 \cdot (1,i,-1,-i)$$

$$+ 0 \cdot (1,-1,1,-1) + 1 \cdot (1,-i,-1,i). \qquad \square$$

EXERCISES

The following four exercises establish algebraic relations to convert the Discrete Fourier Transform as just defined with the constant $1/N$ to and from similar transforms with other constants.

Exercise 5.1. Determine the complex constants r_k for which the vectors $r_k \cdot \vec{w}_k$ become orthonormal with the inner product

$$\langle \vec{z}, \vec{w} \rangle_{\sqrt{N}} := \frac{1}{\sqrt{N}} \sum_{m=0}^{N-1} z_m \cdot \overline{w_m}.$$

Exercise 5.2. Determine the complex constants s_k for which the vectors $s_k \cdot \vec{w}_k$ become orthonormal with the inner product

$$\langle \vec{z}, \vec{w} \rangle := \sum_{m=0}^{N-1} z_m \cdot \overline{w_m}.$$

Exercise 5.3. With the notation of Exercise 1, determine a formula for the coefficients $\sqrt{N} \hat{f}_k$ such that

$$\vec{f} = \sum_{k=0}^{N-1} (\sqrt{N} \hat{f}_k) \cdot (r_k \cdot \vec{w}_k).$$

Exercise 5.4. With the notation of Exercise 2, determine a formula for the coefficients $_1 \hat{f}_k$ such that

$$\vec{f} = \sum_{k=0}^{N-1} (_1 \hat{f}_k) \cdot (s_k \cdot \vec{w}_k).$$

5.1.2   Unitary Operators

The foregoing results show that the Discrete Fourier Transform is a linear operator

$$\mathrm{DFT} : \mathbb{C}^N \to \mathbb{C}^N, \qquad \vec{f} \mapsto \hat{f} = (1/N)_F^{\overline{N}} \Omega^T \vec{f},$$

because it corresponds to a multiplication by a matrix, $(1/N)\overline{{}_F^N\Omega^T}$. Moreover, it has an inverse linear operator,

$$\text{DFT}^{-1} : \mathbb{C}^N \to \mathbb{C}^N, \qquad \hat{\vec{f}} \mapsto \vec{f} = {}_F^N\Omega\hat{\vec{f}},$$

which also corresponds to a multiplication by a matrix, ${}_F^N\Omega$. Furthermore, with the multiplicative constant $1/\sqrt{N}$ instead of N, the Discrete Fourier Transform corresponds to a multiplication by the matrix ${}_F^N U := (1/\sqrt{N})\overline{{}_F^N\Omega^T}$, so that $({}_F^N U)^{-1} = (\overline{{}_F^N U})^T$, as verified in the exercises. Such operators have a special name.

Definition 5.17   For each linear space V with the number field $\mathbb{F} \subseteq \mathbb{C}$ and with an inner product $\langle \, , \, \rangle$, a linear operator $L : V \to V$ is **unitary** if L **preserves the inner product,** which means that for all elements v and w in V,

$$\langle Lv, Lw \rangle = \langle v, w \rangle.$$

Similarly, if V has a norm $\| \ \|$, then L **preserves the norm** if for every element v in V,

$$\|Lv\| = \|v\|. \qquad \qquad \square$$

With the multiplicative constant $1/\sqrt{N}$, the Discrete Fourier Transform is a unitary operator. Another but related unitary operator, the Fourier Transform on the real line, will prove crucial in designing wavelets, as explained in subsequent chapters.

EXERCISES

Exercise 5.5. Verify that the following matrix ${}_H^N Q$ (Haar's matrix multiplied by $1/\sqrt{2}$) is an orthogonal matrix ($[{}_H^N Q]^T [{}_H^N Q] = I$):

$$
{}_H^N Q := \frac{1}{\sqrt{2}} {}_H^N\Omega := \frac{1}{\sqrt{2}}
\begin{pmatrix}
1 & 1 & & & & & \\
1 & -1 & & & & & \\
& & 1 & 1 & & & \\
& & 1 & -1 & & & \\
& & & & \ddots & & \\
& & & & & 1 & 1 \\
& & & & & 1 & -1
\end{pmatrix}.
$$

Exercise 5.6. Verify that ${}_F^N U := (1/\sqrt{N})\overline{{}_F^N\Omega^T}$ is unitary.

Exercise 5.7. Verify that for each orthogonal $Q \in \mathbb{M}_{n \times n}(\mathbb{R})$, so that $Q^T Q = I$, the linear operator $L : \mathbb{R}^n \to \mathbb{R}^n$ defined by a multiplication by Q, so that $L(v) := Qv$, is a unitary operator.

Exercise 5.8. Verify that for each unitary matrix $U \in M_{n \times n}(\mathbb{C})$, so that $\overline{U}^T U = I$, the linear operator $L : \mathbb{C}^n \to \mathbb{C}^n$ defined by a multiplication by U, so that $L(v) := Uv$, is a unitary operator.

Exercise 5.9. Prove that if the norm arises from the inner product, which means that $\|v\| = \sqrt{\langle v, v \rangle}$, and if L preserves the inner product, then L also preserves the norm.

Exercise 5.10. Use the polar identity to prove that if the norm arises from the inner product, which means that $\|v\| = \sqrt{\langle v, v \rangle}$, and if L preserves the norm, then L also preserves the inner product.

5.2   THE FAST FOURIER TRANSFORM (FFT)

The exponential property that $e^{w+z} = e^w \cdot e^z$ for all real or complex numbers w and z leads to a reduction in the number of arithmetic operations necessary to calculate the Discrete Fourier Transform (DFT), through an algorithm published in 1965 by James W. Cooley and John W. Tukey [5] and known as the Fast Fourier Transform (FFT). *The Fast Fourier Transform produces the same coefficients as the Discrete Fourier Transform but with fewer arithmetic operations.*

5.2.1   The Forward Fast Fourier Transform

The following result derives a recursive transform from the exponential rule $e^{w+z} = e^w \cdot e^z$.

Definition 5.18   For each even positive integer N, for each array $\vec{f} \in \mathbb{C}^N$, define corresponding arrays $_{\text{even}}\vec{f}$ and $_{\text{odd}}\vec{f}$ of one-half the length: for each index $m \in \{0, \dots, (N/2) - 1\}$, let

$$_{\text{even}}x_m := x_{2m} = 2m \cdot 2\pi/N = m \cdot 2\pi/(N/2),$$

$$_{\text{odd}}x_m := x_{2m+1} = [2m + 1] \cdot 2\pi/N = m \cdot 2\pi/(N/2) + 2\pi/N,$$

$$(_{\text{even}}\vec{f})_m := \vec{f}_{2m} = f(2m \cdot 2\pi/N),$$

$$(_{\text{odd}}\vec{f})_m := \vec{f}_{2m+1} = f([2m + 1] \cdot 2\pi/N).$$

Moreover, let $\hat{\vec{f}}$ denote the Discrete Fourier Transform of $\vec{f}$, let $_{\text{even}}\hat{\vec{f}}$ denote the Discrete Fourier Transform of $_{\text{even}}\vec{f}$, and let $_{\text{odd}}\hat{\vec{f}}$ denote the Discrete Fourier Transform of $_{\text{odd}}\vec{f}$.   □

Lemma 5.19   *For each $k \in \{0, \dots, (N/2) - 1\}$,*

$$\hat{f}_k = \frac{1}{2} \cdot \left(_{\text{even}}\hat{f}_k + [e^{-i \cdot 2\pi/N}]^k [_{\text{odd}}\hat{f}_k] \right),$$

$$\hat{f}_{k+(N/2)} = \frac{1}{2} \cdot \left(_{\text{even}}\hat{f}_k - [e^{-i \cdot 2\pi/N}]^k [_{\text{odd}}\hat{f}_k] \right).$$

PROOF: Consider two cases. In each case, split the sum defining the Discrete

Fourier Transform into a sum with even indices and a sum with odd indices. In the first case, for each index $k \in \{0, \ldots, \frac{N}{2} - 1\}$,

$$\hat{f}_k = \langle \vec{f}, \vec{w}_k \rangle_N = \frac{1}{N} \sum_{m=0}^{N-1} f_m e^{-ikm2\pi/N}$$

$$= \frac{1}{N} \sum_{m=0}^{(N/2)-1} f_{2m} e^{-ik[2m]2\pi/N} + \frac{1}{N} \sum_{m=0}^{(N/2)-1} f_{2m+1} e^{-ik[2m+1]2\pi/N}$$

$$= \frac{1}{2} \cdot \frac{1}{N/2} \sum_{m=0}^{(N/2)-1} f_{2m} e^{-ikm2\pi/(N/2)}$$

$$+ \frac{1}{2} \cdot \frac{1}{N/2} \sum_{m=0}^{(N/2)-1} f_{2m+1} e^{-ikm2\pi/(N/2)} \cdot e^{-ik\pi/(N/2)}$$

$$= \frac{1}{2} \cdot \left({}_{\text{even}}\hat{f}_k + [e^{-i \cdot 2\pi/N}]^k [{}_{\text{odd}}\hat{f}_k] \right).$$

Similarly, for each index $k \in \{0, \ldots, \frac{N}{2} - 1\}$,

$$\hat{f}_{k+(N/2)} = \langle \vec{f}, \vec{w}_{k+(N/2)} \rangle_N$$

$$= \frac{1}{N} \sum_{m=0}^{N-1} f_m e^{-i[k+(N/2)]m2\pi/N}$$

$$= \frac{1}{N} \sum_{m=0}^{(N/2)-1} f_{2m} e^{-i[k+(N/2)][2m]2\pi/N}$$

$$+ \frac{1}{N} \sum_{m=0}^{(N/2)-1} f_{2m+1} e^{-i[k+(N/2)][2m+1]2\pi/N}$$

$$= \frac{1}{2} \cdot \frac{1}{N/2} \sum_{m=0}^{(N/2)-1} f_{2m} e^{-i[k+(N/2)]m2\pi/(N/2)}$$

$$+ \frac{1}{2} \cdot \frac{1}{N/2} \sum_{m=0}^{(N/2)-1} f_{2m+1} e^{-i[k+(N/2)]m2\pi/(N/2)} \cdot e^{-i[k+(N/2)]\pi/(N/2)}$$

$$= \frac{1}{2} \cdot \frac{1}{N/2} \sum_{m=0}^{(N/2)-1} f_{2m} e^{-ikm2\pi/(N/2)} \cdot e^{-im2\pi}$$

$$+ \frac{1}{2} \cdot \frac{1}{N/2} \sum_{m=0}^{(N/2)-1} f_{2m+1} e^{-ikm2\pi/(N/2)} \cdot e^{-im2\pi} \cdot e^{-ik\pi/(N/2)} \cdot e^{-i\pi}$$

$$= \frac{1}{2} \cdot \left({}_{\text{even}}\hat{f}_k - [e^{-i \cdot 2\pi/N}]^k [{}_{\text{odd}}\hat{f}_k] \right). \qquad \square$$

Example 5.20   For $N := 1$, each array $\vec{\mathbf{f}} = (f_0)$ consists of only one number $f_0 \in \mathbb{C}$, and there exists only one array $\vec{\mathbf{w}}_0 = (1)$. Thus, for $N = 1$, the Fast Fourier Transform merely copies $\vec{\mathbf{f}}$ into $\hat{\mathbf{f}}$ without change:

$$\hat{f}_0 = \langle \vec{\mathbf{f}}, \vec{\mathbf{w}}_0 \rangle_1 = \frac{1}{1} \cdot f_0 \cdot \bar{1} = f_0. \qquad \square$$

Example 5.21   With $N := 2$, each array $\vec{\mathbf{f}} = (f_0, f_1)$ splits into two arrays $_{\text{even}}\vec{\mathbf{f}} = (f_0)$ and $_{\text{odd}}\vec{\mathbf{f}} = (f_1)$. Example 5.20 then gives

$$_{\text{even}}\hat{f}_0 = {}_{\text{even}} f_0 = f_{2 \cdot 0} = f_0,$$

$$_{\text{odd}}\hat{f}_0 = {}_{\text{odd}} f_0 = f_{2 \cdot 0 + 1} = f_1.$$

Hence, with $e^{-i \cdot 0 \cdot \pi / (N/2)} = 1$, the recursion yields

$$\hat{f}_0 = \frac{1}{2} \cdot \left({}_{\text{even}}\hat{f}_0 + 1 \cdot {}_{\text{odd}} \hat{f}_0 \right) = \frac{f_0 + f_1}{2},$$

$$\hat{f}_{1+0} = \frac{1}{2} \cdot \left({}_{\text{even}}\hat{f}_0 - 1 \cdot {}_{\text{odd}} \hat{f}_0 \right) = \frac{f_0 - f_1}{2},$$

confirming the same result from the Discrete Fourier Transform. $\qquad \square$

Example 5.22   With $N := 4$, consider the array

$$\vec{\mathbf{f}} = (f_0, f_1, f_2, f_3) := (5, 1, 2, 8).$$

A first recursion splits the initial array $\vec{\mathbf{f}}$ into two subarrays, here

$$_{\text{even}}\vec{\mathbf{f}} = (f_0, f_2) = (5, 2),$$

$$_{\text{odd}}\vec{\mathbf{f}} = (f_1, f_3) = (1, 8).$$

For $_{\text{even}}\vec{\mathbf{f}}$, Example 5.21 gives

$$_{\text{even}}\hat{f}_0 = \frac{{}_{\text{even}} f_0 + {}_{\text{even}} f_1}{2} = \frac{f_0 + f_2}{2} = \frac{5 + 2}{2} = \frac{7}{2},$$

$$_{\text{even}}\hat{f}_1 = \frac{{}_{\text{even}} f_0 - {}_{\text{even}} f_1}{2} = \frac{f_0 - f_2}{2} = \frac{5 - 2}{2} = \frac{3}{2}.$$

Thus $_{\text{even}}\vec{\mathbf{f}} = (5, 2) = (\frac{7}{2}) \cdot (1, 1) + (\frac{3}{2}) \cdot (1, -1)$. For $_{\text{odd}}\vec{\mathbf{f}}$,

$$_{\text{odd}}\hat{f}_0 = \frac{{}_{\text{odd}} f_0 + {}_{\text{odd}} f_1}{2} = \frac{f_1 + f_3}{2} = \frac{1 + 8}{2} = \frac{9}{2},$$

$$_{\text{odd}}\hat{f}_1 = \frac{{}_{\text{odd}} f_0 - {}_{\text{odd}} f_1}{2} = \frac{f_1 - f_3}{2} = \frac{1 - 8}{2} = -\frac{7}{2}.$$

This means that $_{\text{odd}}\vec{\mathbf{f}} = (1, 8) = (\frac{9}{2}) \cdot (1, 1) + (-\frac{7}{2}) \cdot (1, -1)$. Hence, with $N = 4$

and $N/2 = 2$, the recursion applies to the two arrays

$$\text{even}\hat{f} = \left(\frac{7}{2}, \frac{3}{2}\right), \qquad \text{odd}\hat{f} = \left(\frac{9}{2}, -\frac{7}{2}\right),$$

which gives, with $e^{-ik\pi/(N/2)} = \left(e^{-i\pi/(4/2)}\right)^k = (-i)^k$,

$$\hat{f}_0 = \frac{1}{2} \cdot \left(\text{even}\hat{f}_0 + (-i)^0 (\text{odd}\hat{f})_0\right)$$

$$= \frac{1}{2} \cdot \left(\frac{7}{2} + \frac{9}{2}\right) = 4,$$

$$\hat{f}_{2+0} = \frac{1}{2} \cdot \left(\text{even}\hat{f}_0 - (-i)^0 (\text{odd}\hat{f})_0\right)$$

$$= \frac{1}{2} \cdot \left(\frac{7}{2} - \frac{9}{2}\right) = -\frac{1}{2},$$

$$\hat{f}_1 = \frac{1}{2} \cdot \left(\text{even}\hat{f}_1 + (-i)^1 (\text{odd}\hat{f})_1\right)$$

$$= \frac{1}{2} \cdot \left(\frac{3}{2} + (-i)\frac{-7}{2}\right) = \frac{3 + 7i}{4},$$

$$\hat{f}_{2+1} = \frac{1}{2} \cdot \left(\text{even}\hat{f}_0 - (-i)^1 (\text{odd}\hat{f})_0\right)$$

$$= \frac{1}{2} \cdot \left(\frac{3}{2} - (-i)\frac{-7}{2}\right) = \frac{3 - 7i}{4}.$$

Thus, the initial array $\vec{f} = (5, 1, 2, 8)$ has the Discrete Fourier Transform

$$\hat{\mathbf{f}} = (\hat{f}_0, \hat{f}_1, \hat{f}_2, \hat{f}_3) = \left(4, \frac{3 + 7i}{4}, -\frac{1}{2}, \frac{3 - 7i}{4}\right),$$

which means that

$$\vec{f} = (5, 1, 2, 8) = 4 \cdot (1, 1, 1, 1) + \frac{3 + 7i}{4} \cdot (1, i, -1, -i)$$

$$+ \frac{-1}{2} \cdot (1, -1, 1, -1) + \frac{3 - 7i}{4} \cdot (1, -i, -1, i).$$

Alternatively, if the array $\vec{f}$ represents values of a function f sampled at $x_0 = 0$, $x_1 = 2\pi/4$, $x_2 = 2 \cdot 2\pi/4$, $x_3 = 3 \cdot 2\pi/4$, then the results just obtained interpolate the function f by the "exponential polynomial" $\hat{f}$ defined by

$$\hat{f}(x) = \hat{f}_0 \cdot 1 + \hat{f}_1 \cdot e^{ix} + \hat{f}_2 \cdot e^{2ix} + \hat{f}_3 \cdot e^{3ix}$$

$$= 4 \cdot 1 + \frac{3 + 7i}{4} \cdot e^{ix} + \frac{-1}{2} \cdot e^{2ix} + \frac{3 - 7i}{4} \cdot e^{3ix}.$$

Indeed, $\hat{f}(x_\ell) = f(x_\ell)$ for every $\ell \in \{0, \dots, N-1\}$, by the Inverse Discrete Fourier Transform. □

5.2.2 The Inverse Fast Fourier Transform

For each positive integer N, the Fast Fourier Transform admits an inverse mapping, called the Inverse Fast Fourier Transform (IFFT), which starts from an array $\hat{\vec{f}} \in \mathbb{C}^N$ and constructs the array of data $\vec{f} \in \mathbb{C}^N$ such that $\hat{\vec{f}}$ is the Discrete Fourier Transform of $\vec{f}$. By the formula for the Inverse Discrete Fourier Transform, established in Proposition 5.13,

$$\vec{f} = \hat{f}_0 \vec{w}_0 + \cdots + \hat{f}_k \vec{w}_k + \cdots + \hat{f}_{N-1} \vec{w}_{N-1}.$$

For the coordinates with index ℓ this means that

$$f_\ell = \sum_{k=0}^{N-1} \hat{f}_k (\vec{w}_k)_\ell = \sum_{k=0}^{N-1} \hat{f}_k \cdot e^{i \cdot k \cdot \ell \cdot 2\pi / N}.$$

However, the commutativity $k \cdot \ell = \ell \cdot k$ and two consecutive complex conjugations $\overline{\overline{w}} = w$ represent the result just obtained—the Inverse Discrete Fourier Transform—as yet another Fast Fourier Transform:

$$f_\ell = \sum_{k=0}^{N-1} \hat{f}_k \cdot e^{i \cdot k \cdot \ell \cdot 2\pi / N} = \overline{\sum_{k=0}^{N-1} \overline{\hat{f}_k \cdot e^{i \cdot k \cdot \ell \cdot 2\pi / N}}}$$

$$= N \cdot \overline{\frac{1}{N} \sum_{k=0}^{N-1} \overline{\hat{f}_k} \cdot e^{-i \cdot \ell \cdot k \cdot 2\pi / N}} = N \cdot \overline{\hat{\overline{\hat{f}}}_\ell}.$$

Thus, to calculate the data $\vec{f}$ from $\hat{\vec{f}}$, it suffices to form the complex conjugate $\overline{\hat{\vec{f}}}$, then to calculate its Fast Fourier Transform $\hat{\overline{\hat{\vec{f}}}}$ multiplied by N, and to take the complex conjugate again:

$$\vec{f} = N \cdot \overline{\hat{\overline{\hat{\vec{f}}}}} = N \cdot \left(\widehat{\overline{\hat{\vec{f}}}} \right)^{\overline{}}.$$

5.2.3 Interpolation by the Inverse Fast Fourier Transform

The Inverse Fast Fourier Transform also provides an algorithm to interpolate a function f at points x other than the points $x_\ell = \ell \cdot 2\pi / N$ in the sample. At each point x_ℓ in the sample, by definition of the Discrete Fourier Transform,

$$f(x_\ell) = \sum_{k=0}^{N-1} \hat{f}_k \cdot e^{i \cdot k \cdot x_\ell}.$$

Consequently, one method to interpolate f at x consists in approximating $f(x)$ by

$$f(x) \approx \sum_{k=0}^{N-1} \hat{f}_k \cdot e^{i \cdot k \cdot x}.$$

In situations that involve several values of x spaced by multiples of $2\pi/N$, the Fast Fourier Transform provides a more economical algorithm. To this end, for any ℓ, let $u := x - x_\ell$, so that $x = u + x_\ell$. Thus,

$$f(x) \approx \sum_{k=0}^{N-1} \hat{f}_k \cdot e^{i \cdot k \cdot x} = \sum_{k=0}^{N-1} \hat{f}_k \cdot e^{i \cdot k \cdot [u + x_\ell]}$$

$$= \sum_{k=0}^{N-1} \left(\hat{f}_k \cdot e^{i \cdot k \cdot u} \right) \cdot e^{i \cdot k \cdot x_\ell} = \sum_{k=0}^{N-1} \left(\hat{f}_k \cdot e^{i \cdot k \cdot u} \right) \cdot e^{i \cdot k \cdot \ell \cdot 2\pi/N},$$

but the sums just obtained amount to the Inverse Fast Fourier Transform applied to the coefficients $(\hat{f}_k \cdot e^{i \cdot k \cdot u})_{k=0}^{N-1}$. Thus, to interpolate f at $x = u + \ell \cdot 2\pi/N$ for each $\ell \in \{0, \dots, N-1\}$, it suffices to multiply each Fourier coefficient $\hat{f}_k$ by $e^{i \cdot k \cdot u} = (e^{i \cdot u})^k$, and then to calculate the Inverse Fast Fourier Transform starting from the array $(\hat{f}_k \cdot e^{i \cdot k \cdot u})_{k=0}^{N-1}$.

EXERCISES

Exercise 5.11. Interpolate $v \mapsto f(v - x_\ell)$ with the FFT.

The following two exercises demonstrate a method to interpolate functions at points that are not necessarily equally spaced.

Exercise 5.12. For each sequence $(z_0, \dots, z_{2m})$ of $2m + 1$ *distinct and nonzero* complex numbers, and for each sequence $(w_0, \dots, w_{2m})$ of *any* $2m + 1$ complex numbers, prove that $2m + 1$ complex numbers $c_{-m}, \dots, c_{-1}, c_0, c_1, \dots, c_m$ exist such that the function $g : \mathbb{C} \setminus \{0\} \to \mathbb{C}$ defined by

$$g(z) := \sum_{k=-m}^{m} c_k \cdot z^k$$

takes the value $g(z_\ell) = w_\ell$ for each $\ell \in \{0, \dots, 2m\}$.

Exercise 5.13. For each sequence $0 \le x_0 < x_1 < \cdots < x_{2m-1} < x_{2m} < 2\pi$ of $2m + 1$ *distinct* real numbers, define a sequence $(z_0, \dots, z_{2m})$ of $2m + 1$ complex numbers by $z_\ell := e^{ix_\ell}$. For each sequence $(w_0, \dots, w_{2m})$ of any *real* numbers, verify that the function g defined in the preceding exercise has coefficients such that $c_{-k} = \overline{c_k}$ for each $k \in \{0, \dots, m\}$.

Exercise 5.14. With the notation and hypotheses of the preceding exercise, prove that *real* numbers $a_0, a_1, \ldots, a_m$ and $b_1, \ldots, b_m$ exist such that

$$g(e^{ix}) = \frac{a_0}{2} + \sum_{k=1}^{m} [a_k \cdot \cos(k \cdot x) + b_k \cdot \sin(k \cdot x)].$$

5.2.4 Bit Reversal

The foregoing subsections have demonstrated the Fast Fourier Transform recursively, but in this instance recursion does not yet yield a convenient algorithm, because it requires some record-keeping with indices. Several methods exist to simplify such recursions into an induction with less record-keeping, for instance, the method of "bit reversal" presented here [25], [39, pp. 81–88]. Essentially, bit reversal rearranges the initial sequence so that the Fast Fourier Transform operates on adjacent pairs of sequences at each step.

Definition 5.23 Bit reversal transforms each finite sequence $(p_{k-1}, p_{k-2}, \ldots, p_1, p_0)$ of k binary integers $p_{k-1}, \ldots, p_0 \in \{0, 1\}$ into the **bit-reversed** sequence $(p_0, p_1, \ldots, p_{k-2}, p_{k-1})$. Equivalently, bit reversal transforms each binary integer

$$p = \sum_{\ell=0}^{k-1} p_\ell \cdot 2^\ell = p_{k-1} 2^{k-1} + p_{k-2} 2^{k-2} + \cdots + p_1 2 + p_0$$

into the binary integer

$$q = B(p) := \sum_{\ell=0}^{k-1} p_\ell \cdot 2^{(k-1)-\ell} = p_{k-1} + p_{k-2} 2 + \cdots + p_1 2^{k-2} + p_0 2^{k-1}. \quad \square$$

Bit reversal allows for an arrangement of the data so that the Fast Fourier Transform transforms *adjacent* pairs of sequences.

Proposition 5.24 *For each integer power of two* $N = 2^n$, *the Fast Fourier Transform amounts to the following operations:*

(0) *For each binary index* $p \in \{0, \ldots, N-1\}$, *calculate the bit-reversed index* $q = B(p)$, *and arrange the data* $\vec{z} = (z_0, \ldots, z_{N-1})$ *in the order* $\vec{z}_B := (z_{B(0)}, \ldots, z_{B(N-1)})$;

(1) *For each* $k \in \{0, \ldots, n-1\}$, *perform one step of the Fast Fourier Transform on each of the* $2^{(n-k)-1}$ *pairs of adjacent sequences of* 2^k *elements.*

PROOF: Proceed by induction. For each index $p \in \{0, \ldots, N-1\}$,

• either p is even, hence ends with $p_0 = 0$, and $q = B(p)$ begins with $q_{N-1} = p_0 = 0$,

- or p is odd, hence ends with $p_0 = 1$, and $q = B(p)$ begins with $q_{N-1} = p_0 = 1$.

Consequently, in the bit-reversed sequence of indices q, the first $N/2$ bit-reversed indices come from all the even indices p, while the last $N/2$ bit-reversed indices come from all the odd indices p. This proves the proposition for the last step of the Fast Fourier Transform.

Yet after the initial $q_0 = 0$ in the first $N/2$ bit-reversed indices come the bit-reversed indices of the order of the even indices, and similarly after the initial $q_0 = 1$ for the odd indices.   □

Thus, the Fast Fourier Transform may proceed as follows.

(0) For each index $p \in \{0, \dots, N-1\}$, calculate the bit-reversed index $q = B(p)$, and arrange the data $\vec{z} = (z_0, \dots, z_{N-1})$ in the order

$$\vec{z}_B := (z_{B(0)}, \dots, z_{B(N-1)}).$$

(1) Compute the $N/2$ Fast Fourier Transform steps from one to two points of each of the $N/2$ pairs of two numbers:

$$([z_{B(0)}, z_{B(1)}], [z_{B(1)}, z_{B(2)}], \dots,$$
$$[z_{B(N-4)}, z_{B(N-3)}], [z_{B(N-2)}, z_{B(N-1)}]),$$

which gives

$$([\hat{z}_{B(0)}, \hat{z}_{B(1)}], [\hat{z}_{B(1)}, \hat{z}_{B(2)}], \dots,$$
$$[\hat{z}_{B(N-4)}, \hat{z}_{B(N-3)}], [\hat{z}_{B(N-2)}, \hat{z}_{B(N-1)}]).$$

(2) Compute the $N/4$ Fast Fourier Transform steps from two to four points of each of the $N/4$ sequences of 4 numbers:

$$([\hat{z}_{B(0)}, \hat{z}_{B(1)}, \hat{z}_{B(1)}, \hat{z}_{B(2)}], \dots,$$
$$[\hat{z}_{B(N-4)}, \hat{z}_{B(N-3)}, \hat{z}_{B(N-2)}, \hat{z}_{B(N-1)}]).$$

$(n-1)$ Compute the 1 Fast Fourier Transform step from $N/2$ to N points of each of the 1 sequence of N numbers:

$$([\hat{z}_{B(0)}, \hat{z}_{B(1)}, \hat{z}_{B(1)}, \hat{z}_{B(2)} \dots,$$
$$\hat{z}_{B(N/2-1)}], [\hat{z}_{B(N/2)}, \hat{z}_{B(N/2+1)}, \hat{z}_{B(N/2+2)},$$
$$\dots \hat{z}_{B(N/2+N/2-1)}]).$$

5.3   APPLICATIONS OF THE FAST FOURIER TRANSFORM

5.3.1   Noise Reduction Through the Fast Fourier Transform

Given a periodic signal with a larger amplitude added to noise from various sources with smaller amplitudes, the Fast Fourier Transform can decompose the sum of the signal and noise into a linear combination of terms with different frequencies, identify and keep the signal's contribution from the larger coefficients, and reject the smaller coefficients from the noise, thus restoring the initial signal.

Example 5.25   This example demonstrates the use of the Discrete Fourier Transform to remove "noise" from a discrete set of data by retaining only a linear combination with the largest coefficients and discarding the other ones.

The top panel of Figure 5.1 displays a sequence $\vec{z} = (z_0, \ldots, z_{255})$ of $N = 256$ numbers, plotted in the form (k, z_k), and joined by segments for visual convenience.

The middle panel of Figure 5.1 (next page) shows the size of the Fourier coefficients $|\hat{z}| = (|\hat{z}_0|, \ldots, |\hat{z}_{255}|)$, plotted in the same fashion. Besides many coefficients of size less than 0.1, the plot reveals a few coefficients of size about 0.25 near $k = 100$ and near $k = 155$. An automated search identifies these largest coefficients as the following:

$$\hat{z}_{99} = -0.232517 - 0.163612i,$$

$$\hat{z}_{101} = +0.183198 + 0.136200i,$$

$$\hat{z}_{155} = +0.183198 - 0.136200i,$$

$$\hat{z}_{157} = -0.232517 + 0.163612i.$$

The bottom panel of Figure 5.1 shows the inverse transform obtained from only the four largest coefficients just listed, with all the other coefficients annihilated:

$$\tilde{z}_k := \hat{z}_{99}e^{i\cdot 99\cdot k\cdot 2\pi/256} + \hat{z}_{101}e^{i\cdot 101\cdot k\cdot 2\pi/256}$$
$$+ \hat{z}_{155}e^{i\cdot 155\cdot k\cdot 2\pi/256} + \hat{z}_{157}e^{i\cdot 157\cdot k\cdot 2\pi/256}.$$

The identity $e^{ik\cdot\ell\cdot 2\pi/256} = e^{i[k-256]\cdot\ell\cdot 2\pi/256}$ leads to alternative trigonometric expressions of the result, with $\Re$ and $\Im$ denoting the real and imaginary parts of complex numbers:

$$\hat{f}_{99}e^{i99\ell\cdot 2\pi/256} + \hat{f}_{101}e^{i101\ell\cdot 2\pi/256} + \hat{f}_{155}e^{i155\ell\cdot 2\pi/256} + \hat{f}_{157}e^{i157\ell\cdot 2\pi/256}$$

$$= \hat{f}_{99}e^{i99\ell\cdot 2\pi/256} + \overline{\hat{f}_{99}}e^{-i99\ell\cdot 2\pi/256}$$

$$+ \hat{f}_{101}e^{i101\ell\cdot 2\pi/256} + \overline{\hat{f}_{101}}e^{-i101\ell\cdot 2\pi/256}$$

$$= 2\Re\left(\hat{f}_{99}e^{i99\ell\cdot 2\pi/256} + \hat{f}_{101}e^{i101\ell\cdot 2\pi/256}\right)$$

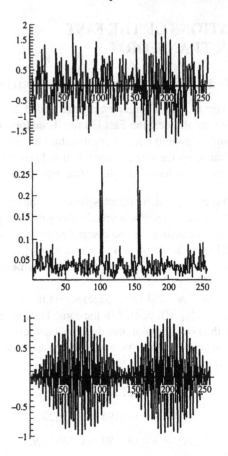

Figure 5.1 *Top:* 256 noisy amplitude-modulated data points. *Middle:* the DFT shows four coefficients with larger magnitudes. *Bottom:* reconstruction from only those four frequencies.

$$= 2\Re(\hat{f}_{99})\cos{(99\ell \cdot 2\pi/256)} - 2\Im(\hat{f}_{99})\sin{(99\ell \cdot 2\pi/256)}$$

$$+ 2\Re(\hat{f}_{101})\cos{(101\ell \cdot 2\pi/256)} - 2\Im(\hat{f}_{101})\sin{(101\ell \cdot 2\pi/256)}$$

$$\vdots$$

$$= 2\left[\Re(\hat{f}_{99}) - \Re(\hat{f}_{101})\right][\sin{(100\ell \cdot 2\pi/256)}\sin{(1\ell \cdot 2\pi/256)}] + \cdots$$

$$\approx 0.970 \cdot [\sin{(100\ell \cdot 2\pi/256)}\sin{(1\ell \cdot 2\pi/256)}] + \cdots,$$

where the ellipses [···] stand for terms with smaller amplitudes. Thus, the signal under consideration appears to be an amplitude-modulated signal with noise. Indeed, the simulated data considered here consisted of a superposition of a sample from an amplitude-modulated signal $\sin(100t) \cdot \sin(t)$, similar to a tone transmitted by amplitude-modulated (AM) radio, and a random noise of unit amplitude.

Thus, even though the noise had an amplitude equal to that of the signal, the Forward and Inverse Discrete Fourier Transform successfully extracted about 97% of the intensity of the signal and attenuated most of the random noise, because the large amplitude of such random noise consisted of a sum of many components with high frequencies, which the Discrete Fourier Transform detected.   □

5.3.2   Convolution and Fast Multiplication

The Forward and Inverse Fast Fourier Transforms provide an algorithm to multiply two polynomials, and hence two sums of any type of powers, for instance, two integers, with fewer multiplications than involved in the definition of such multiplications, but with the same result, according to a general pattern called "convolution."

Definition 5.26   For each pair of vectors $\vec{w}, \vec{z} \in \mathbb{C}^N$ of length N, their **convolution** is the vector of length $2N$ denoted by $\vec{w} * \vec{z}$ and defined for each $k \in \{0, \ldots, 2N - 1\}$ by

$$(\vec{w} * \vec{z})_k := \frac{1}{N} \sum_{\ell = \max\{0, k-(N-1)\}}^{\min\{k, N-1\}} w_{k-\ell} \cdot z_\ell .$$

Alternatively, but equivalently, extend each vector $\vec{w}$ and $\vec{z}$ by infinitely many zeros on each side,

$$\vec{w} = (\ldots, 0, w_0, w_1, \ldots, w_{N-2}, w_{N-1}, 0, \ldots),$$

$$\vec{z} = (\ldots, 0, z_0, z_1, \ldots, z_{N-2}, z_{N-1}, 0, \ldots),$$

and set

$$(\vec{w} * \vec{z})_k := \frac{1}{N} \sum_{\ell=0}^{2N-1} w_{k-\ell} \cdot z_\ell .$$ □

The following lemmas establish intermediate results leading to the fact that the Fourier Transform of a convolution coincides with the "coordinatewise" product of Fourier Transforms, which forms the basis for the algorithm for faster multiplication.

Lemma 5.27   *The product of convolution, $*$, commutes.*

PROOF: Set $h := k - \ell$, $p := \max\{0, k - (N - 1)\}$, and $q := \min\{k, N - 1\}$:

$$(\vec{w} * \vec{z})_k = \frac{1}{N} \sum_{\ell=p}^{q} w_{k-\ell} \cdot z_\ell = \frac{1}{N} \sum_{h=q}^{p} w_h \cdot z_{k-h}$$

$$= \frac{1}{N} \sum_{h=p}^{q} z_{k-h} \cdot w_h = (\vec{z} * \vec{w})_k .$$ □

Lemma 5.28 *The product of convolution $*$ is linear with respect to each variable: for all $\vec{w}, \vec{u}, \vec{v} \in \mathbb{C}^N$ and all $r, s, \in \mathbb{C}$,*

$$\vec{w} * (r \cdot \vec{u} + s \cdot \vec{v}) = r \cdot (\vec{w} * \vec{u}) + s \cdot (\vec{w} * \vec{v}),$$

$$(r \cdot \vec{u} + s \cdot \vec{v}) * \vec{w} = r \cdot (\vec{u} * \vec{w}) + s \cdot (\vec{v} * \vec{w}).$$

PROOF: For each index $k \in \{0, \ldots, 2N - 1\}$,

$$[\vec{w} * (r \cdot \vec{u} + s \cdot \vec{v})]_k = \frac{1}{N} \sum_{\ell=0}^{2N-1} w_{k-\ell} \cdot (r \cdot \vec{u} + s \cdot \vec{v})_\ell$$

$$= \frac{1}{N} \sum_{\ell=0}^{2N-1} w_{k-\ell}(ru_\ell + sv_\ell)$$

$$= \frac{1}{N} \sum_{\ell=0}^{2N-1} [(w_{k-\ell}ru_\ell) + (w_{k-\ell}sv_\ell)]$$

$$= r \cdot \frac{1}{N} \sum_{\ell=0}^{2N-1} w_{k-\ell} \cdot u_\ell + s \cdot \frac{1}{N} \sum_{\ell=0}^{2N-1} w_{k-\ell} \cdot v_\ell$$

$$= r \cdot (\vec{w} * \vec{u})_k + s \cdot (\vec{w} * \vec{v})_k$$

$$= [r \cdot (\vec{w} * \vec{u}) + s \cdot (\vec{w} * \vec{v})]_k.$$

Consequently, $\vec{w} * (r \cdot \vec{u} + s \cdot \vec{v}) = r \cdot (\vec{w} * \vec{u}) + s \cdot (\vec{w} * \vec{v})$. The equality $(r \cdot \vec{u} + s \cdot \vec{v}) * \vec{w} = r \cdot (\vec{u} * \vec{w}) + s \cdot (\vec{v} * \vec{w})$ follows similarly, or from the equality just obtained and commutativity.   □

Lemma 5.29 *For each $\vec{w}$ and each basis vector $\vec{e}_h$ in $\mathbb{C}^N$,*

$$(\widehat{\vec{w} * \vec{e}_h})_k = \hat{w}_k \cdot (\hat{e}_h)_k.$$

PROOF: For the Discrete Fourier Transform of the basis vector,

$$(\hat{e}_h)_k = \frac{1}{N} \sum_{\ell=0}^{2N-1} (e_h)_\ell \cdot e^{-ik\ell 2\pi/(2N)} = \frac{1}{N} \cdot 1 \cdot e^{-ikh\pi/N}.$$

Consequently,

$$\hat{w}_k \cdot (\hat{e}_h)_k = \frac{1}{N} \cdot \hat{w}_k \cdot e^{-ikh\pi/N},$$

whereas for the convolution,

$$(\vec{w} * \vec{e}_h)_\ell = \frac{1}{N} \cdot \sum_{k=0}^{2N-1} w_{\ell-k} \cdot (e_h)_k = \frac{1}{N} \cdot w_{\ell-h} \cdot 1.$$

Hence, the Discrete Fourier Transform of the convolution becomes

$$
\begin{aligned}
\widehat{(\vec{w} * \vec{e}_h)}_k &= \frac{1}{N} \sum_{\ell=0}^{2N-1} (\vec{w} * \vec{e}_h)_\ell \cdot e^{-ik\ell 2\pi/(2N)} \\
&= \frac{1}{N} \sum_{\ell=0}^{2N-1} \frac{1}{N} \cdot w_{\ell-h} \cdot e^{-ik\ell 2\pi/(2N)} \\
&= \frac{1}{N} \sum_{\ell=0}^{2N-1} w_{\ell-h} \cdot e^{-ik[\ell-h]2\pi/(2N)} \cdot \frac{1}{N} \cdot e^{-ikh2\pi/(2N)} \\
&= \frac{1}{N} \sum_{m=0}^{2N-1} w_m \cdot e^{-ikm2\pi/(2N)} \cdot \frac{1}{N} \cdot e^{-ikh2\pi/(2N)} \\
&= \hat{w}_k \cdot (\hat{e}_h)_k. \qquad\qquad\qquad\qquad\qquad \square
\end{aligned}
$$

Theorem 5.30   *For each pair of arrays $\vec{w}, \vec{z} \in \mathbb{C}^N$ of length N, the Discrete Fourier Transform of the convolution $\vec{w} * \vec{z}$ of length $2N$ equals the product of the Discrete Fourier Transforms of length $2N$ of the extensions by zeros $(w_0, \ldots, w_{N-1}, 0, \ldots, 0)$ and $(z_0, \ldots, z_{N-1}, 0, \ldots, 0)$:*

$$
\widehat{(\vec{w} * \vec{z})}_k = \hat{w}_k \cdot \hat{z}_k.
$$

PROOF: Apply the linearity of the convolution and of the Discrete Fourier Transform, and the foregoing result for the particular case of bases vectors: Let

$$
\vec{z} = \sum_{h=0}^{N-1} z_h \cdot \vec{e}_h,
$$

with $z_h = 0$ for every $h \in \{N, \ldots, 2N - 1\}$. Hence,

$$
\widehat{(\vec{w} * \vec{z})}_k = \left(\vec{w} * \overbrace{\sum_{h=0}^{N-1} z_h \vec{e}_h} \right)_k = \sum_{h=0}^{N-1} z_h \widehat{\left(\vec{w} * \vec{e}_h\right)}_k
$$

$$
= \sum_{h=0}^{N-1} z_h \left(\hat{w}_k (\hat{e}_h)_k \right) = \hat{w}_k \cdot \left(\overbrace{\sum_{h=0}^{N-1} z_h \vec{e}_h} \right)_k = \hat{w}_k \cdot \hat{z}_k \qquad \square
$$

Corollary 5.31   *For each pair of arrays $\vec{p}, \vec{q} \in \mathbb{C}^N$, and for each array $\vec{r} \in \mathbb{C}^{2N}$, consider the polynomials P, Q, and R defined by*

$$
P(X) := \sum_{k=0}^{N-1} p_k X^k,
$$

$$Q(X) := \sum_{k=0}^{N-1} q_k X^k,$$

$$R(X) := \sum_{k=0}^{2N-1} r_k X^k.$$

If $R := P \cdot Q$, then $\vec{r}$ is the Inverse Discrete Fourier Transform of the "coordinatewise" product $(\hat{p}_0 \cdot \hat{q}_0, \ldots, \hat{p}_{2N-1} \cdot \hat{q}_{2N-1})$.

PROOF: The entries $(\vec{p} * \vec{q})_k$ of the convolution $\vec{p} * \vec{q}$ are the coefficients of R, for

$$R(X) = P(X) \cdot Q(X) = \left(\sum_{k=0}^{N-1} p_k X^k \right) \cdot \left(\sum_{\ell=0}^{N-1} q_\ell X^\ell \right)$$

$$= \sum_{h=0}^{2N-1} \sum_{k+\ell=h} p_k q_\ell X^h = \sum_{h=0}^{2N-1} (\vec{p} * \vec{q})_h X^h,$$

where $\widehat{(\vec{p} * \vec{q})}_k = \hat{p}_k \cdot \hat{q}_k$. $\qquad\qquad\qquad\qquad\qquad\qquad\qquad\qquad$ □

The foregoing result means that to compute all the coefficients r_h of the product of two polynomials, it suffices to compute the Fast Fourier Transforms of the extensions of $\vec{p}$ and $\vec{q}$ of lengths $2N$, then to multiply the $2N$ pairs $\hat{p}_k \hat{q}_k$, and finally to compute the Inverse Fast Fourier Transform of the sequence $(\hat{p}_0 \hat{q}_0, \ldots, \hat{p}_{2N-1} \hat{q}_{2N-1})$.

Applying the result just obtained to $X := 2$ or $X := 10$, for instance, yields an algorithm to multiply two integers in binary or decimal expansions.

Example 5.32   Consider the two polynomials

$$P(X) := 4 - 4X, \qquad Q(X) := 6 + 2X,$$

with extended vectors of coefficients

$$\vec{p} = (4, -4, 0, 0), \qquad \vec{q} = (6, 2, 0, 0).$$

The Fast Fourier Transform with four points gives

$$\hat{p} = (0, 1 + i, 2, 1 - i), \qquad \hat{q} = (2, (3 - i)/2, 1, (3 + i)/2).$$

Hence, multiplication of corresponding pairs of coordinates produces

$$\hat{p}\hat{q} = (0 \cdot 2, (1 + i) \cdot (3 - i)/2, 2 \cdot 1, (1 - i) \cdot (3 + i)/2)$$

$$= (0, 2 + i, 2, 2 - i),$$

and the Inverse Fourier Transform of $(0, 2 + i, 2, 2 - i)$ yields

$$\vec{p} * \vec{q} = (24, -16, -8, 0),$$

which means that $P(X) \cdot Q(X) = 24 - 16X - 8X^2$. □

Remark 5.33 In practice, for fast multiplication of integers, the computation and multiplication of powers of the complex root ω_N already consumes a substantial amount of time. Therefore, for practical fast multiplication, a yet more abstract Fourier Transform proves useful. See, for instance, [28, p. 502].

5.4 MULTIDIMENSIONAL DISCRETE AND FAST FOURIER TRANSFORMS

The multidimensional Discrete Fourier Transform consists of consecutive one-dimensional Discrete Fourier Transforms in each dimension, each of which can proceed with the one-dimensional Fast Fourier Transform for the same result. As a result, the multidimensional Discrete Fourier Transform yields the coefficients of tensor products of one-dimensional complex exponentials.

Example 5.34 Consider the following rectangular array of $2 \times 4 = 8$ values on a rectangular grid,

$$\vec{f} := \begin{pmatrix} 9 & 7 & 6 & 2 \\ 8 & 2 & 4 & 0 \end{pmatrix}.$$

The first step of the two-dimensional Discrete or Fast Fourier Transform consists of a one-dimensional Discrete or Fast Fourier Transform for each row. An optional preparation can perform a bit reversal in each row:

$$\begin{matrix} 00 & 01 & 10 & 11 \\ \begin{pmatrix} 9 & 7 & 6 & 2 \\ 8 & 2 & 4 & 0 \end{pmatrix} \end{matrix} \begin{matrix} \to \\ \to \end{matrix} \begin{matrix} 00 & 10 & 01 & 11 \\ \begin{pmatrix} 9 & 6 & 7 & 2 \\ 8 & 4 & 2 & 0 \end{pmatrix} \end{matrix}.$$

In each row, the Fast Fourier Transform first operates on consecutive pairs,

$$\begin{pmatrix} 9 & 6 & 7 & 2 \\ 8 & 4 & 2 & 0 \end{pmatrix} \begin{matrix} \to \\ \to \end{matrix} \begin{pmatrix} \frac{9+6}{2} & \frac{9-6}{2} & \frac{7+2}{2} & \frac{7-2}{2} \\ \frac{8+4}{2} & \frac{8-4}{2} & \frac{2+0}{2} & \frac{2-0}{2} \end{pmatrix} = \begin{pmatrix} \frac{15}{2} & \frac{3}{2} & \frac{9}{2} & \frac{5}{2} \\ \frac{12}{2} & \frac{4}{2} & \frac{2}{2} & \frac{2}{2} \end{pmatrix},$$

and then on consecutive blocks of four entries,

$$\begin{pmatrix} \frac{15}{2} & \frac{3}{2} & \frac{9}{2} & \frac{5}{2} \\ \frac{12}{2} & \frac{4}{2} & \frac{2}{2} & \frac{2}{2} \end{pmatrix} \begin{matrix} \to \\ \to \end{matrix} \begin{pmatrix} \frac{\frac{15}{2}+\frac{9}{2}}{2} & \frac{\frac{3}{2}-i\frac{5}{2}}{2} & \frac{\frac{15}{2}-\frac{9}{2}}{2} & \frac{\frac{3}{2}+i\frac{5}{2}}{2} \\ \frac{\frac{12}{2}+\frac{2}{2}}{2} & \frac{\frac{4}{2}-i\frac{2}{2}}{2} & \frac{\frac{12}{2}-\frac{2}{2}}{2} & \frac{\frac{4}{2}+i\frac{2}{2}}{2} \end{pmatrix}$$

$$= \frac{1}{4}\begin{pmatrix} 24 & 3-5i & 6 & 3+5i \\ 14 & 4-2i & 10 & 4+2i \end{pmatrix}.$$

The result obtained so far shows the one-dimensional Discrete Fourier Transform of each row. The two-dimensional Discrete Fourier Transform then continues with one-dimensional Discrete (or Fast) Fourier Transforms in each column of the result just obtained (in situations with more than two rows, an optional preparation could perform a bit reversal in each column):

$$\frac{1}{4} \begin{pmatrix} 24 & 3-5i & 6 & 3+5i \\ 14 & 4-2i & 10 & 4+2i \end{pmatrix}$$

$$\downarrow \qquad \downarrow \qquad \downarrow \qquad \downarrow$$

$$\frac{1}{8} \begin{pmatrix} 38 & 7-7i & 16 & 7+7i \\ 10 & -1-3i & -4 & -1+3i \end{pmatrix}$$

$$= \frac{1}{8} \begin{pmatrix} 24+14 & (3-5i)+(4-2i) & 6+10 & (3+5i)+(4+2i) \\ 24-14 & (3-5i)-(4-2i) & 6-10 & (3+5i)-(4+2i) \end{pmatrix}.$$

The Fourier coefficient $\hat{f}_{k,\ell}$ in row k and column ℓ corresponds to the tensor product $\vec{w}_k \otimes \vec{w}_\ell$, where

$$(\vec{w}_k \otimes \vec{w}_\ell)_{p,q} = (\vec{w}_k)_p (\vec{w}_\ell)_q = e^{ikp2\pi/M} e^{i\ell q 2\pi/N}.$$

With $M = 2$ rows and $N = 4$ columns, the particular example just computed means that

$$\vec{f}_{p,q} = \frac{1}{8} \Big(38 \cdot e^{i0p2\pi/2} e^{i0q2\pi/4} + (7-7i) \cdot e^{i0p2\pi/2} e^{i1q2\pi/4}$$

$$+ 16 \cdot e^{i0p2\pi/2} e^{i2q2\pi/N} + (7+7i) \cdot e^{i0p2\pi/2} e^{i3q2\pi/4}$$

$$+ 10 \cdot e^{i1p2\pi/2} e^{i0q2\pi/4} + (-1-3i) \cdot e^{i1p2\pi/2} e^{i1q2\pi/4}$$

$$+ (-4) \cdot e^{i1p2\pi/2} e^{i2q2\pi/4} + (-1+3i) \cdot e^{i1p2\pi/2} e^{i3q2\pi/4} \Big). \qquad \square$$

The verification of the assertions just made proceeds along the following outline.

Definition 5.35 On the set $\mathbb{M}_{M \times N}(\mathbb{C}) = \mathbb{C}^{MN}$ of all rectangular arrays with M rows and N columns of complex entries, define the inner product

$$\langle \vec{f}, \vec{g} \rangle_{M,N} := \frac{1}{MN} \sum_{k=0}^{M-1} \sum_{\ell=0}^{M-1} f_{k,\ell} \overline{g_{k,\ell}}.$$

Similarly, for arrays with three (or more) dimensions, define the inner product

$$\langle \vec{f}, \vec{g} \rangle_{L,M,N,\dots} := \frac{1}{LMN \cdots} \sum_{h=0}^{L-1} \sum_{k=0}^{M-1} \sum_{\ell=0}^{N-1} \cdots f_{h,k,\ell,\dots} \overline{g_{h,k,\ell,\dots}}. \qquad \square$$

The innermost sum, here with respect to the index ℓ, represents the one-dimensional inner product along each row, and then the next sum, here with respect to the index k, represents the one-dimensional inner product along each column. Consequently, results for the one-dimensional Discrete Fourier Transform also hold for the multidimensional Discrete Fourier Transform, with verification proceeding along each consecutive dimension.

Lemma 5.36 *For all positive integers M and N, and for all indices $m \in \{0, \ldots, M-1\}$ and $n \in \{0, \ldots, N-1\}$, the arrays*

$$\vec{w}_m \otimes \vec{w}_n = \left(e^{imk2\pi/M} e^{in\ell2\pi/N} \right)_{k=0,\ell=0}^{k=M-1,\ell=N-1}$$

are mutually orthonormal.

PROOF: The following proof holds for all tensor products of orthonormal arrays:

$$\langle \vec{w}_m \otimes \vec{w}_n, \vec{w}_p \otimes \vec{w}_q \rangle_{M,N} \tag{5.1}$$

$$= \frac{1}{MN} \sum_{k=0}^{M-1} \sum_{\ell=0}^{N-1} (\vec{w}_m \otimes \vec{w}_n)_{k,\ell} \overline{(\vec{w}_p \otimes \vec{w}_q)_{k,\ell}}$$

$$= \frac{1}{MN} \sum_{k=0}^{M-1} \sum_{\ell=0}^{N-1} e^{imk2\pi/M} e^{in\ell2\pi/N} e^{-ipk2\pi/M} e^{-iq\ell2\pi/N}$$

$$= \frac{1}{M} \sum_{k=0}^{M-1} e^{imk2\pi/M} e^{-ipk2\pi/M} \frac{1}{N} \sum_{\ell=0}^{N-1} e^{in\ell2\pi/N} e^{-iq\ell2\pi/N}$$

$$= \langle \vec{w}_m, \vec{w}_p \rangle_M \cdot \langle \vec{w}_n, \vec{w}_q \rangle_N$$

$$= \begin{cases} 1 & \text{if } m = p \text{ and } n = q, \\ 0 & \text{if } m \neq p \text{ or } n \neq q. \end{cases} \qquad \square$$

Definition 5.37 For each array $\vec{f} \in \mathbb{M}_{M \times N}(\mathbb{C}) = \mathbb{C}^{MN}$, the two-dimensional Discrete Fourier Transform of $\vec{f}$ consists of the array $\hat{\vec{f}}$ of the coordinates of $\vec{f}$ with respect to the basis

$$\left(\vec{w}_m \otimes \vec{w}_n \right)_{m=0,n=0}^{m=M-1,n=N-1}. \qquad \square$$

Proposition 5.38 *For each array $\vec{f} \in \mathbb{M}_{M \times N}(\mathbb{C}) = \mathbb{C}^{MN}$, the two-dimensional Discrete Fourier Transform of $\vec{f}$ consists of the one-dimensional Discrete Fourier Transform of each row of $\vec{f}$ followed by the one-dimensional Discrete Fourier Transform of each column.*

PROOF: The following proof holds for all tensor products of orthonormal arrays:

$$\hat{\mathbf{f}}_{m,n} = \langle \vec{\mathbf{f}}, \vec{\mathbf{w}}_m \otimes \vec{\mathbf{w}}_n, \rangle_{M,N} = \frac{1}{MN} \sum_{k=0}^{M-1} \sum_{\ell=0}^{N-1} f_{k,\ell} \overline{(\vec{\mathbf{w}}_m \otimes \vec{\mathbf{w}}_n)_{k,\ell}}$$

$$= \frac{1}{MN} \sum_{k=0}^{M-1} \sum_{\ell=0}^{N-1} f_{k,\ell} e^{-imk2\pi/M} e^{-in\ell2\pi/N}$$

$$= \frac{1}{M} \sum_{k=0}^{M-1} e^{-imk2\pi/M} \left(\frac{1}{N} \sum_{\ell=0}^{N-1} f_{k,\ell} e^{-in\ell2\pi/N} \right). \qquad \square$$

CHAPTER 6

Fourier Series for Periodic Functions

6.0 INTRODUCTION

This chapter introduces a few features of Fourier series. In contrast to the Discrete Fourier Transform, which pertains to finite-dimensional spaces, Fourier series provide an approach to the concept of approximation of functions in linear spaces that do not admit finite bases, in a context simpler than that of Daubechies wavelets.

Moreover, Fourier series allow for detailed analyses of various aspects of approximations—such as accuracy, rate of convergence, and edge effects—by means that require only calculus and linear algebra, whereas for the same topics Daubechies wavelets would demand further mathematics.

Finally, for comparison and contrast, the Fourier series of periodic functions outline the state of signal analysis before wavelets, and thus demonstrate some common features and some differences unique to wavelets. Consistent with the literature [31, p. 94], the following notation will prove convenient.

Definition 6.1 For all numbers u and w, the notation $[u, w]$ represents the interval of all numbers from u (included) to w (included):

$$[u, w] = \{r : u \leq r \leq w\}.$$

Also, the notation $[u, w[$ represents the interval of all numbers from u (included) to w (*excluded*):

$$[u, w[= \{r : u \leq r < w\},$$

and the notation $]u, w]$ represents the interval of all numbers from u (*excluded*) to w (included):

$$]u, w] = \{r : u < r \leq w\}.$$

Finally, the notation $]u, w[$ represents the interval of all numbers from u (*excluded*) to w (*excluded*):

$$]u, w[= \{r : u < r < w\}. \qquad \square$$

Such a notation as $]u, w[$ for an interval also avoids confusions with the notation (u, w) for the coordinates u and w of a point.

Definition 6.2   An interval $I \subset \mathbb{R}$ is **compact** if I contains both of its endpoints: $I = [a, b]$ for some $a, b \in \mathbb{R}$. Also, a function $f : \mathbb{R} \to \mathbb{C}$ has **compact support** if there is a compact interval $I = [a, b]$ such that $f(x) = 0$ for every $x \notin I$. $\square$

6.1   FOURIER SERIES

Fourier series approximate any integrable periodic function f globally over entire periods, relating f to series of the form

$$f(x) \approx \sum_{k=-n}^{n} c_{f,k} \cdot e^{ikx},$$

with

$$c_{f,k} := \frac{1}{2T} \int_{-T}^{T} f(t) \cdot e^{-ikt}\, dt.$$

This section specifies the significance of the approximation $\approx$ for functions that are continuous everywhere except possibly at finitely many points, as defined in a subsequent section.

6.1.1   Orthonormal Complex Trigonometric Functions

Orthonormal complex trigonometric functions provide computational tools to extract information from periodic functions.

Definition 6.3   A function f has a **period** P, or is **periodic** with period P, if $f(t + P) = f(t)$ for every t. $\square$

Example 6.4   The trigonometric functions cos and sin have period 2π. $\square$

Example 6.5   The complex exponential function has period $2\pi i$. $\square$

Definition 6.6   On the space $C^0_{I,2T}(\mathbb{R}, \mathbb{C})$ of all complex-valued piecewise continuous functions with period $2T$, define the inner product

$$\langle f, g \rangle := \frac{1}{2T} \cdot \int_{-T}^{T} f(s) \cdot \overline{g(s)}\, ds. \qquad \square$$

The following lemma demonstrates the calculation of the inner product just defined and verifies the orthonormality of a set of "complex trigonometric" exponential functions.

Lemma 6.7   *The functions $w_k : \mathbb{R} \to \mathbb{C}$ defined by $w_k(t) := e^{ikt\pi/T}$ for each $k \in \mathbb{Z}$ are orthonormal in $C^0_{1,2T}(\mathbb{R}, \mathbb{C})$ with respect to the inner product $\langle \, , \, \rangle$.*

PROOF: Direct calculation of the integrals for the inner product confirms the result:

$$\langle w_k, w_\ell \rangle = \frac{1}{2T} \int_{-T}^{T} w_k(s) \cdot \overline{w_\ell(s)} \, ds = \frac{1}{2T} \int_{-T}^{T} e^{iks\pi/T} \cdot e^{-i\ell s\pi/T} \, ds$$

$$= \frac{1}{2T} \int_{-T}^{T} e^{i(k-\ell)s\pi/T} \, ds.$$

If $k = \ell$, then

$$\frac{1}{2T} \int_{-T}^{T} e^{i(k-\ell)s\pi/T} \, ds = \frac{1}{2T} \int_{-T}^{T} 1 \, ds = \frac{1}{2T} \cdot [T - (-T)] = 1.$$

If $k \neq \ell$, then

$$\frac{1}{2T} \int_{-T}^{T} e^{i(k-\ell)s\pi/T} \, ds = \frac{1}{2} \cdot \frac{1}{i(k-\ell)\pi/T} \cdot e^{i(k-\ell)s\pi/T} \Big|_{-T}^{T}$$

$$= \frac{1}{2} \cdot \frac{1}{i(k-\ell)\pi} \cdot \left(e^{i(k-\ell)\pi} - e^{-i(k-\ell)\pi} \right)$$

$$= \frac{\sin((k-\ell)\pi)}{(k-\ell)\pi} = 0. \qquad \square$$

The theory of orthogonal projections immediately identifies which linear combination of the functions w_k best approximates a signal, as explained in the following subsections.

6.1.2   Definition and Examples of Fourier Series

The following results will show that the subspace spanned by the orthonormal complex trigonometric functions just introduced approximates every continuous periodic function to any accuracy.

Definition 6.8   For every function $f \in C^0_{1,2T}(\mathbb{R}, \mathbb{C})$, define the **complex Fourier coefficients** $c_{f,k}$ of f by

$$c_{f,k} := \langle f, w_k \rangle = \frac{1}{2T} \int_{-T}^{T} f(s) \cdot \overline{w_k(s)} \, ds = \frac{1}{2T} \int_{-T}^{T} f(s) \cdot e^{-iks\pi/T} \, ds. \quad \square$$

Remark 6.9   The Fourier coefficients just defined represent the limit of the Discrete Fourier Coefficients as the number of sample points increases. Indeed, through Riemann sums the integral inner product also represents the limit of the inner products $\langle\,,\,\rangle_N$:

$$\langle f, w_k \rangle = \frac{1}{2T} \int_{-T}^{T} f(t) \cdot \overline{w_k(t)} \, dt$$

$$= \frac{1}{2T} \lim_{N \to \infty} \sum_{m=0}^{N-1} f(t_m) \overline{w_k(t_m)} (2T/N) = \lim_{N \to \infty} \langle \vec{f}, \vec{w}_k \rangle_N. \qquad \Box$$

Definition 6.10   For every function $f \in C^0_{I,2T}(\mathbb{R}, \mathbb{C})$, and for each $N \in \mathbb{N}$, define the **partial sum** $S_N(f)$ by

$$S_N(f) = \langle f, w_k \rangle \cdot w_k = \sum_{k=-N}^{N} c_{f,k} w_k,$$

$$S_N(f)(t) = \sum_{k=-N}^{N} \left[\left(\frac{1}{2T} \int_{-T}^{T} f(s) \cdot e^{-iks\pi/T} \, ds \right) \cdot e^{ikt\pi/T} \right]. \qquad \Box$$

Thus, S_N represents the orthogonal projection of $f \in C^0_{I,2T}(\mathbb{R}, \mathbb{C})$ on the subspace W_N defined by

$$W_N := \text{Span}\,\{w_{-N}, \ldots, w_0, \ldots, w_N\}.$$

Example 6.11   For any pair of distinct real numbers $a < b$ in $[-\pi, \pi]$, consider the function $g := \chi_{[a,b]}$ defined by

$$g(t) := \begin{cases} 1 & \text{if } t \in [a, b], \\ 0 & \text{if } t \notin [a, b]. \end{cases}$$

Then the complex Fourier coefficients of g become

$$c_{g,k} = \frac{1}{2\pi} \int_{-\pi}^{\pi} g(t) e^{-ikt} \, dt = \frac{1}{2\pi} \int_{a}^{b} 1 \cdot e^{-ikt} \, dt = \frac{1}{2\pi} \cdot \frac{1}{-ik} \cdot e^{-ikt} \Big|_{a}^{b}$$

$$= \frac{e^{-ikb} - e^{-ika}}{-2\pi ik} = \frac{e^{-ik(a+b)/2}}{2\pi k} \cdot 2 \frac{e^{-ik(a-b)/2} - e^{-ik(b-a)/2}}{2i}$$

$$= \frac{e^{-ik(a+b)/2}}{\pi k} \cdot \sin(k[b - a]/2).$$

$$= \frac{\cos(k[a + b]/2) - i \sin(k[a + b]/2)}{\pi k} \cdot \sin(k[b - a]/2),$$

except for $k = 0$, for which

$$c_{g,0} = \frac{1}{2\pi} \int_{-\pi}^{\pi} g(t) \, dt = \frac{1}{2\pi} \int_{a}^{b} g(t) \, dt = \frac{b - a}{2\pi}. \qquad \Box$$

Example 6.12 In Example 6.11, if $a := -\pi/2$ and $b := \pi/2$, then

$$g(t) := \begin{cases} 1 & \text{if } t \in [-\pi/2, \pi/2], \\ 0 & \text{if } t \notin [-\pi/2, \pi/2], \end{cases}$$

and

$$c_{g,k} = \frac{\cos(k[a+b]/2) - i\sin(k[a+b]/2)}{\pi k} \cdot \sin(k[b-a]/2).$$

$$= \frac{\cos(k[0]/2) - i\sin(k[0]/2)}{\pi k} \cdot \sin(k[\pi]/2). = \frac{\sin(k\pi/2)}{\pi k}$$

$$= \begin{cases} 0 & \text{if } k \equiv 0 \text{ modulo } 4, \\ \frac{1}{\pi k} & \text{if } k \equiv 1 \text{ modulo } 4, \\ 0 & \text{if } k \equiv 2 \text{ modulo } 4, \\ \frac{-1}{\pi k} & \text{if } k \equiv 3 \text{ modulo } 4, \end{cases}$$

except for $k = 0$, for which

$$c_{g,0} = \frac{1}{2\pi} \int_{-\pi}^{\pi} g(t)\, dt = \frac{1}{2\pi} \int_{-\pi/2}^{\pi/2} 1\, dt = \frac{\pi}{2\pi} = \frac{1}{2}.$$

Thus, the complex Fourier series of g takes the form

$$\sum_{k=-\infty}^{\infty} c_{g,k} \cdot e^{ikt} = \frac{1}{\pi} \cdot \left[\cdots + \frac{1}{-7} \cdot e^{i\cdot(-7)\cdot t} + \frac{-1}{-5} \cdot e^{i\cdot(-5)\cdot t} \right.$$

$$+ \frac{1}{-3} \cdot e^{i\cdot(-3)\cdot t} + \frac{-1}{-1} \cdot e^{i\cdot(-1)\cdot t} + \frac{\pi}{2}$$

$$\left. + \frac{1}{1} \cdot e^{i\cdot 1\cdot t} + \frac{-1}{3} \cdot e^{i\cdot 3\cdot t} + \frac{1}{5} \cdot e^{i\cdot 5\cdot t} + \frac{-1}{7} \cdot e^{i\cdot 7\cdot t} + \cdots \right]$$

$$= \frac{1}{2} + \frac{2}{\pi} \cdot \left[\frac{1}{1} \cdot \cos(1 \cdot t) + \frac{-1}{3} \cdot \cos(3 \cdot t) \right.$$

$$\left. + \frac{1}{5} \cdot \cos(5 \cdot t) + \cdots \right].$$

Partial sums of the series appear in Figure 6.1.   □

The following lemmas lead to an alternative expression for the orthogonal projections $S_N(f)$, in terms of only one integral, of the product of f and another function called an integrating "kernel." Though not always computationally efficient, such integrating kernels can prove theoretically powerful because they can yield information more easily than other methods.

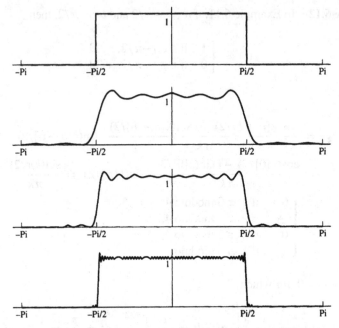

Figure 6.1   From top to bottom: characteristic function of $[-\pi/2, \pi/2]$ (the vertical edges serve only as visual guidelines); first 10 terms of the series; first 20 terms of the series; first 100 terms of the series.

Lemma 6.13   (Geometric series with negative and positive powers.)   *For every $N \in \mathbb{N}$, and for every $w \in \mathbb{C} \setminus \{1\}$,*

$$\sum_{k=-N}^{N} w^k = \frac{w^{-(N+\frac{1}{2})} - w^{(N+\frac{1}{2})}}{w^{-\frac{1}{2}} - w^{\frac{1}{2}}}.$$

In particular, for every $N \in \mathbb{N}$, and for every $z \in \mathbb{C}$ not equal to an integral multiple of 2π,

$$\sum_{k=-N}^{N} e^{ikz} = \frac{\sin([N + \frac{1}{2}]z)}{\sin(z/2)}. \qquad \qquad \Box$$

PROOF: Apply the formula for the geometric series, or mimic the proof for the geometric series: Let

$$S := \sum_{k=-N}^{N} w^k,$$

whence

$$S := \sum_{k=-N}^{N} w^k,$$

$$wS := \sum_{k=-N}^{N} w^{k+1} = \sum_{k=-N+1}^{N} w^k + w^{N+1},$$

$$(1-w)S = w^{-N} - w^{N+1},$$

$$S = (1-w)^{-1}(w^{-N} - w^{N+1}).$$

For either square root $w^{\frac{1}{2}}$, but with the same branch of the complex square root throughout,

$$S = \frac{w^{-N} - w^{N+1}}{1-w} = \frac{w^{-\frac{1}{2}}}{w^{-\frac{1}{2}}} \cdot \frac{w^{-N} - w^{N+1}}{1-w} = \frac{w^{(-N-\frac{1}{2})} - w^{(N+\frac{1}{2})}}{w^{(-\frac{1}{2})} - w^{(\frac{1}{2})}}.$$

In particular, if $z \neq 2m\pi$, then $w = e^{iz} \neq 1$, and

$$\sum_{k=-N}^{N} e^{ikz} = \frac{[e^{iz}]^{-N-1/2} - [e^{iz}]^{N+1/2}}{[e^{iz}]^{-1/2} - [e^{iz}]^{1/2}} = \frac{\sin([N+\frac{1}{2}]z)}{\sin(z/2)}. \qquad \Box$$

Lemma 6.14 (Integral expression for partial sums.) *For every $f \in C^0_{1,2T}$ $(\mathbb{R}, \mathbb{C})$, and for every $N \in \mathbb{N}$,*

$$S_N(f)(t) = \frac{1}{2T} \int_{-T}^{T} f(s) \cdot \frac{\sin([N+\frac{1}{2}](t-s)\pi/T)}{\sin((t-s)\pi/(2T))} \, ds.$$

PROOF: The function $t \mapsto \sin(([N+\frac{1}{2}](t-s))\pi/T)/\sin((t-s)\pi/(2T))$ extends to a continuous function at $t = s$, because $\lim_{x \to 0} \sin(x)/x = 1$; hence,

$$S_N(f)(t) = \sum_{k=-N}^{N} \left\{ \left[\frac{1}{2T} \int_{-T}^{T} f(s) \cdot e^{-iks\pi/T} \, ds \right] \cdot e^{ikt\pi/T} \right\}$$

$$= \frac{1}{2T} \int_{-T}^{T} f(s) \cdot \left\{ \sum_{k=-N}^{N} \left[e^{ik(t-s)\pi/T} \right] \right\} ds$$

$$= \frac{1}{2T} \int_{-T}^{T} f(s) \cdot \frac{\sin([N+\frac{1}{2}](t-s)\pi/T)}{\sin((t-s)\pi/(2T))} \, ds.$$

The last equality results from the choice of either branch of the complex square root for $(t\pi/T) - \pi < (t-s)\pi/T < (t\pi/T) + \pi$. $\qquad \Box$

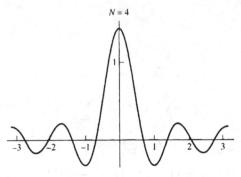

Figure 6.2 Dirichlet's kernel with $N = 4$: $s \mapsto D_N(s) := \{\sin([N + \frac{1}{2}]s\pi/T)\}/ \{\sin(s\pi/[2T])\}$.

Definition 6.15   The function $D_N : \mathbb{R} \to \mathbb{R}$ defined by

$$D_N(s) := \frac{\sin([N + \frac{1}{2}]s\pi/T)}{\sin(s\pi/[2T])}$$

is called the **Dirichlet kernel.** (See Figure 6.2.)   □

The expression defining the partial sums $S_N(f)$ by the Fourier coefficients $c_{f,k}$ will later allow for comparisons between discrete and continuous Fourier coefficients (Shannon's sampling theorem), while the expression of the same partial sums $S_N(f)$ with the Dirichlet kernel will provide a means to verify the convergence of the partial sums to the function f, as demonstrated in the following sections.

6.1.3   Relation Between Series and Discrete Transforms

The following result reveals a relation between the Discrete Fourier Transform $\hat{z}$ of a sample $\vec{z}$ from a function,

$$\vec{z} := (f(0), f(2\pi/N), \dots, f(k \cdot 2\pi/N), \dots, f([N-1] \cdot 2\pi/N)),$$

and the Fourier series of the same function:

$$\hat{z}_k = \langle \vec{z}, \vec{w}_k \rangle_N = \frac{1}{N} \cdot \sum_{m=0}^{N-1} z_m e^{-imk2\pi/N}$$

$$= \frac{1}{N} \cdot \sum_{m=0}^{N-1} f(m \cdot 2\pi/N) e^{-imk2\pi/N}$$

$$= \frac{1}{N} \cdot \sum_{m=0}^{N-1} \sum_{\ell \in \mathbb{Z}} c_{f,\ell} e^{i\ell(m \cdot 2\pi/N)} e^{-imk2\pi/N}$$

$$= \sum_{\ell \in \mathbb{Z}} c_{f,\ell} \frac{1}{N} \cdot \sum_{m=0}^{N-1} e^{i(\ell-k)(m \cdot 2\pi/N)},$$

where

$$\frac{1}{N} \cdot \sum_{m=0}^{N-1} e^{i(\ell-k)(m \cdot 2\pi/N)} = \begin{cases} 1 & \text{if } k \equiv \ell \text{ (modulo } N\text{)}, \\ 0 & \text{if } k \not\equiv \ell \text{ (modulo } N\text{)}. \end{cases}$$

Consequently,

$$\hat{z}_k = \sum_{n \in \mathbb{Z}} c_{f,k+nN}.$$

This means that each Discrete Fourier Coefficient $\hat{z}_k$ of a sample $\vec{z}$ of f consists not only of the corresponding "continuous" Fourier coefficient c_{fk}, but of the sum of all continuous Fourier coefficients $c_{f,k+nN}$ for frequencies $k+nN$ at an integer multiple of N apart. Such a phenomenon—called **aliasing**—becomes less severe if f has more derivatives, as proved in a subsequent section.

6.1.4 Multidimensional Fourier Series

Multidimensional Fourier series express periodic functions of several variables with tensor products of complex exponentials, which amounts to one-dimensional Fourier series in each dimension.

Example 6.16 Consider the characteristic function χ of the square $S = [-\pi/2, \pi/2] \times [-\pi/2, \pi/2]$ extended by a period 2π in each dimension. Thus,

$$\chi = \chi_{[-\pi/2, \pi/2]} \otimes \chi_{[-\pi/2, \pi/2]},$$

whence, with m and n representing any integers,

$$\chi(x, y) := \begin{cases} 1 & \text{if } -\pi/2 \leq x + 2m\pi \leq \pi/2 \text{ and } -\pi/2 \leq y + 2m\pi \leq \pi/2, \\ 0 & \text{otherwise.} \end{cases}$$

Consequently, the two-dimensional Fourier coefficients of χ become

$$c_{\chi;k,\ell} = \langle \chi, w_k \otimes w_\ell \rangle$$

$$= \frac{1}{2\pi} \int_{-\pi}^{\pi} \frac{1}{2\pi} \int_{-\pi}^{\pi} \chi(x, y) (w_k \otimes w_\ell)(x, y) \, dx \, dy$$

$$= \frac{1}{2\pi} \int_{-\pi}^{\pi} \frac{1}{2\pi} \int_{-\pi}^{\pi} \chi_{[-\pi/2, \pi/2]}(x) \chi_{[-\pi/2, \pi/2]}(y) \, w_k(x) w_\ell(y) \, dx \, dy$$

$$= \frac{1}{2\pi} \int_{-\pi}^{\pi} \chi_{[-\pi/2,\pi/2]}(x) w_k(x)\, dx \frac{1}{2\pi} \int_{-\pi}^{\pi} \chi_{[-\pi/2,\pi/2]}(y) w_\ell(y)\, dy$$

$$= c_{\chi;k} c_{\chi;\ell} = \frac{\sin(k\pi/2)}{k\pi} \frac{\sin(\ell\pi/2)}{\ell\pi},$$

except for $k = 0$ or $\ell = 0$, for which $c_{\chi;k,0} = [\sin(k\pi/2)]/(2k\pi)$ and $c_{\chi;0,\ell} = [\sin(\ell\pi/2)]/(2\ell\pi)$. Therefore, the two-dimensional Fourier series of χ takes the following form (see also Figure 6.3):

$$\sum_{k=-\infty}^{\infty} \sum_{\ell=-\infty}^{\infty} \frac{\sin(k\pi/2)}{k\pi} \frac{\sin(\ell\pi/2)}{\ell\pi} e^{iku} e^{i\ell v}. \qquad \qquad \Box$$

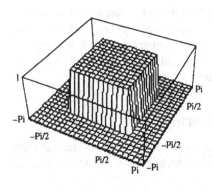

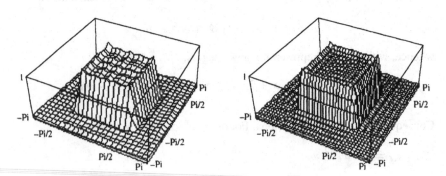

Figure 6.3   *Top.* Characteristic function of a square. *Bottom left.* 19 × 19 terms of the Fourier series. *Bottom right.* 39 × 39 terms of the Fourier series. Unlike the flat top border of the characteristic function, the ridge and gutter along the top border of the partial sums illustrate the Gibbs–Wilbraham phenomenon described in the next section.

EXERCISES

Comparison of Figure 6.1 with plots obtained in the following exercises demonstrates by examples how the accuracy of the partial sums of the Fourier series increases as the degree of differentiability of the function increases: the smoother the function, the more accurate its approximation by partial sums of its Fourier series.

Exercise 6.1. Calculate the Fourier coefficients of the function $g : \mathbb{R} \to \mathbb{R}$ defined by

$$g(x) := \begin{cases} x + \pi & \text{if } -\pi \le x < -\pi/2, \\ -x & \text{if } -\pi/2 \le x < \pi/2, \\ x - \pi & \text{if } \pi/2 \le x < \pi \end{cases}$$

and extended to have period 2π. Then plot a few partial sums.

Exercise 6.2. Calculate the Fourier coefficients of the function $h : \mathbb{R} \to \mathbb{R}$ defined by

$$h(x) := \begin{cases} (\frac{1}{2})(x + \pi)^2 & \text{if } -\pi \le x < -\pi/2, \\ (\pi^2/4) - (x^2/2) & \text{if } -\pi/2 \le x < \pi/2, \\ (\frac{1}{2})(x - \pi)^2 & \text{if } \pi/2 \le x < \pi, \end{cases}$$

and extended to have period 2π. Then plot a few partial sums.

6.2 CONVERGENCE AND INVERSION OF FOURIER SERIES

The foregoing examples suggest that as the number of terms increases, the partial sums converge at a rate depending upon the differentiability of the function under consideration, as investigated further by the following considerations.

6.2.1 The Gibbs–Wilbraham Phenomenon

This subsection illustrates with an example some peculiar features of the "pointwise" convergence of Fourier series at discontinuities. Such an illustration will justify the need for the theory explained in the following subsections.

Example 6.17 The characteristic function g of the interval $[-\pi/2, \pi/2]$, defined by

$$g(t) := \begin{cases} 1 & \text{if } t \in [-\pi/2, \pi/2], \\ 0 & \text{if } t \notin [-\pi/2, \pi/2], \end{cases}$$

has the Fourier series calculated in Example 6.12:

$$\sum_{k=-\infty}^{\infty} c_{g,k} \cdot e^{ikt} = \frac{1}{2} + \frac{2}{\pi} \cdot \left[\sum_{m=0}^{\infty} \frac{(-1)^m}{m} \cos([2m+1]t) \right]$$

$$= \frac{1}{2} + \frac{2}{\pi} \cdot \left[\frac{1}{1} \cdot \cos(t) + \frac{-1}{3} \cdot \cos(3t) + \frac{1}{5} \cdot \cos(5t) + \cdots \right].$$

Figure 6.4 displays two partial sums,

$$(S_{100}g)(t) = \sum_{k=-100}^{100} c_{g,k} e^{ikt} = \frac{1}{2} + \frac{2}{\pi} \left[\sum_{m=0}^{49} \frac{(-1)^m}{m} \cos([2m+1]t) \right],$$

$$(S_{250}g)(t) = \sum_{k=-250}^{250} c_{g,k} e^{ikt} = \frac{1}{2} + \frac{2}{\pi} \left[\sum_{m=0}^{124} \frac{(-1)^m}{m} \cos([2m+1]t) \right].$$

Figure 6.4 reveals that as the number of terms increases, the accuracy of the convergence increases, and ripples move closer and closer toward the discontinuity.
However, the *size* of the ripples does not decrease. Indeed, regardless of the number of terms in the partial sums of the Fourier series, the highest ripple "over-
shoots" the function by about 9% *and* "undershoots" the function by about 5%
on either side of the discontinuity. Figure 6.5 illustrates the constant magnitude of
"overshoot" and "undershoot" through a magnification of the plots of the partial
sums

$$(S_{250}g)(t) = \sum_{k=-250}^{250} c_{g,k} e^{ikt} = \frac{1}{2} + \frac{2}{\pi} \left[\sum_{m=0}^{124} \frac{(-1)^m}{m} \cos([2m+1]t) \right],$$

$$(S_{1000}g)(t) = \sum_{k=-1000}^{1000} c_{g,k} e^{ikt} = \frac{1}{2} + \frac{2}{\pi} \left[\sum_{m=0}^{499} \frac{(-1)^m}{m} \cos([2m+1]t) \right]. \quad \square$$

The simultaneous "overshoot" and "undershoot" of partial sums of Fourier
series near discontinuities is called the **Gibbs–Wilbraham phenomenon.** A the-

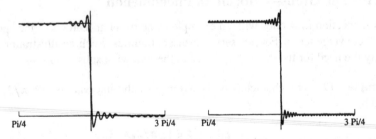

Pi/4   3 Pi/4   Pi/4   3 Pi/4

Figure 6.4   Illustration of the Gibbs–Wilbraham phenomenon: simultaneous "overshoot"
and "undershoot" at discontinuities. *Left.* Sum of the first 100 terms. *Right.* Sum of the first
250 terms.

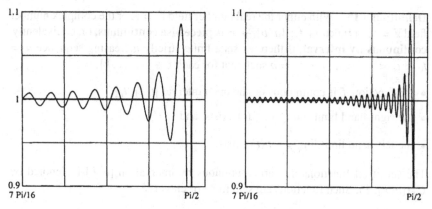

Figure 6.5   The Gibbs–Wilbraham phenomenon. Simultaneous "overshoot" by about 9% *and* "undershoot" by about 5% at a discontinuity, regardless of the number of terms in the partial sum. *Left.* Sum of the first 250 terms. *Right.* Sum of the first 1000 terms.

ory and history of the Gibbs–Wilbraham phenomenon was published in 1979 by Edwin Hewitt and Robert E. Hewitt [22], and in 1997 by Gottlieb and Shu [18]. Before them, Maxime Bôcher had already proved an observation made independently by by Josiah Willard Gibbs and Henry Wilbraham in the nineteenth century: at a jump discontinuity, the partial sums of the Fourier series "overshoot" a function by

$$\frac{2}{\pi} \int_0^\pi \frac{\sin(t)}{t}\, dt \approx \frac{2}{\pi} 1.851\,937\,0 \approx 1.178\,779\,7$$

multiplied by the size of the jump. This means that the partial sums of the Fourier series "overshoot" the function by one-half of this amount on each side of the jump: $(1.178\,779\,7 - 1)/2 \approx 0.09 = 9\%$, as corroborated on the left-hand side of the jump by Figure 6.5. Still, Hewitt and Hewitt also proved that for every function with a jump discontinuity, the partial sums of the Fourier series not only "overshoot" but also "undershoot" the function by

$$\frac{2}{\pi} \int_0^{2\pi} \frac{\sin(t)}{t}\, dt \approx \frac{2}{\pi} 1.418\,151\,6 \approx 0.902\,823\,3$$

multiplied by the size of the jump. This means that the partial sums of the Fourier series "undershoot" the function by one half of this amount on each side of the jump: $(1 - 0.902\,823\,3)/2 \approx 0.05 = 5\%$, as corroborated on the left-hand side of the jump by Figure 6.5.

6.2.2   Piecewise Continuous Functions

This section introduces several classes of functions suited for continuous—as opposed to discrete—periodic phenomena.

Definition 6.18   With either the real number field $\mathbb{F} = \mathbb{R}$ or the complex number field $\mathbb{F} = \mathbb{C}$, a function $f : [a, b] \to \mathbb{F}$ is **piecewise continuous**, or, equivalently, **continuous by interval,** if there exists a finite strictly increasing sequence $a = t_0 < t_1 < \cdots < t_{k-1} < t_k = b$ such that for each $\ell \in \{1, \ldots, k\}$,

- the function f is continuous on the open interval $]t_{\ell-1}, t_\ell[$,

- the right-hand limit $\lim_{t \searrow t_{\ell-1}^+} f(t)$ exists, and

- the left-hand limit $\lim_{t \nearrow t_\ell^-} f(t)$ exists.

The set of all functions that are continuous by interval on $[a, b]$ is denoted by $C_I^0([a, b], \mathbb{F})$. Such functions are said to be of class C_I^0. $\square$

Example 6.19   For each pair of real numbers $u < w$ consider the **characteristic function** of the interval $[u, w[$, denoted by $\chi_{[u,w[}$ and defined by

$$\chi_{[u,w[}(x) := \begin{cases} 1 & \text{if } u \leq x < w, \\ 0 & \text{if } x < u \text{ or if } w \leq x. \end{cases}$$

The characteristic function $\chi_{[u,w[} : \mathbb{R} \to \mathbb{R}$ is then continuous by interval, indeed constant (1) on the open interval $]u, w[$ and constant (0) outside the closed interval $[u, w]$, with the one-sided limits

$$\lim_{x \nearrow u^-} \chi_{[u,w[}(x) = \lim_{x \nearrow u^-} 0 = 0,$$

$$\lim_{x \searrow u^+} \chi_{[u,w[}(x) = \lim_{x \searrow u^+} 1 = 1,$$

$$\lim_{x \nearrow w^-} \chi_{[u,w[}(x) = \lim_{x \nearrow w^-} 1 = 1,$$

$$\lim_{x \searrow w^+} \chi_{[u,w[}(x) = \lim_{x \searrow w^+} 0 = 0. \qquad \square$$

Counterexample 6.20   The reciprocal function $x \mapsto 1/x$ is *not* continuous by interval, because one-sided limits do not exist at 0. $\square$

The following technical results will allow for logical verification of the convergence of Fourier series.

Lemma 6.21   *A function $f : [a, b] \to \mathbb{C}$ is piecewise continuous if there exists a finite strictly increasing sequence $a = t_0 < t_1 < \cdots < t_{k-1} < t_k = b$ such that for each $j \in \{1, \ldots, k\}$ on the open interval $I_j :=]t_{j-1}, t_j[$ the function f coincides with the restriction of a function $\tilde{f}_j : [t_{j-1}, t_j] \to \mathbb{C}$ that is continuous on the closed interval $J_j := [t_{j-1}, t_j]$, so that $f|_{I_j} = \tilde{f}|_{I_j}$.*

PROOF: If such continuous functions $\tilde{f}_j$ exist, then f is continuous by interval, because

$$\lim_{t \searrow t_{j-1}^+} f(t) = \lim_{t \searrow t_{j-1}^+} \tilde{f}_j(t) = \tilde{f}_j(t_{j-1}),$$

$$\lim_{t \nearrow t_j^-} f(t) = \lim_{t \nearrow t_j^-} \tilde{f}_j(t) = \tilde{f}_j(t_j).$$

Conversely, if f is continuous by interval, define each function $\tilde{f}_j$ by

$$\tilde{f}_j(x) := \begin{cases} \lim_{t \searrow t_{j-1}^+} f(t) & \text{if } x \leq t_{j-1}, \\ f(x) & \text{if } t_{j-1} < x < t_j, \\ \lim_{t \nearrow t_j^-} f(t) & \text{if } t_j < x. \end{cases}$$

Then each such function $\tilde{f}_j$ is continuous on $[t_{j-1}, t_j]$ by definition, and $\tilde{f}_j$ coincides with f on $]t_{j-1}, t_j[$.   □

Definition 6.22   A function $f: I \to \mathbb{C}$ is **uniformly continuous** on an interval $I \subseteq \mathbb{R}$ if

for each real $\varepsilon > 0$,
there exists a real $\delta > 0$,
such that $|f(u) - f(v)| < \varepsilon$ for all $u, v \in I$ with $|u - v| < \delta$.   □

Remark 6.23   With a uniformly continuous function, the same tolerances ε and δ hold everywhere in the interval I. In contrast, with a continuous but not uniformly continuous function, the tolerances will need adjustments depending upon the location in I: the definition of the continuity of a function f at a point u states that

for each point $u \in I$,
for each real $\varepsilon > 0$,
there exists a real $\delta > 0$,
such that $|f(u) - f(v)| < \varepsilon$ for *each* $v \in I$ with $|u - v| < \delta$,

so that the order of the foregoing specifications allows ε and δ to depend upon u. However, results from advanced calculus or from real analysis guarantee that if a function is continuous on an interval $I = [a, b]$ that contains both of its endpoints, then such a function is also uniformly continuous [21, p. 88 (7.18)]   □

Corollary 6.24   *If a function f is continuous by interval, then it is uniformly continuous on each bounded interval.*

PROOF: Each bounded interval I intersects only finitely many (m) subintervals $I_j = [t_{j-1}, t_j]$, where the corresponding function $\tilde{f}_j$ is uniformly continuous. This means that for each index $j \in \{1, \ldots, m\}$ and each real $\varepsilon > 0$, there exists a real $\delta_{j,\varepsilon} > 0$ such that $|\tilde{f}_j(r) - \tilde{f}_j(s)| < \varepsilon$ for all reals r and s such that $|r - s| < \delta_{j,\varepsilon}$. Consequently, $|f(r) - f(s)| < \varepsilon$ for all reals r and s in I such that $|r - s| < \delta_\varepsilon := \min_{1 \leq j \leq m} \delta_{j,\varepsilon}$.   □

Proposition 6.25 *For each function f continuous by interval, for each bounded interval $I \subsetneq \mathbb{R}$, and for each real $\varepsilon > 0$, there exists a simple function $\tilde{f}$ (sum of finitely many multiples of characteristic functions) such that $|f(t) - \tilde{f}(t)| < \varepsilon$ for all real $r, s \in I$.*

PROOF: If f is continuous by interval, then it is uniformly continuous on each bounded interval I. Consequently, for each real $\varepsilon > 0$, there exists a real $\delta_\varepsilon > 0$ such that $|f(r) - f(s)| < \varepsilon$ for all reals r and s in I such that $|r - s| < \delta_\varepsilon$. Hence, subdivide I into finitely many subintervals $[x_{k-1}, x_k[$ such that $|x_k - x_{k-1}| < 2\delta_\varepsilon$ for every $k \in \{1, \dots, N\}$, and approximate f on $[x_{k-1}, x_k[$ by its value at the midpoint z_k: Define

$$z_k := \frac{x_{k-1} + x_k}{2}, \quad \tilde{f} := \sum_{k=1}^{N} f(z_k) \cdot \chi_{[x_{k-1}, x_k[}.$$

Then each $x \in I$ lies in exactly one subinterval $[x_{k-1}, x_k[$, where $|x - z_k| < \delta_\varepsilon$, whence

$$|f(x) - \tilde{f}(x)| = |f(x) - f(z_k)| < \varepsilon. \qquad \square$$

Definition 6.26   A function $f : I \to \mathbb{C}$ defined on an *open* interval $I \subseteq \mathbb{R}$ is of class C^m if all the derivatives $f', f'', \dots, f^{(m-1)}, f^{(m)}$ exist and are continuous on I. $\qquad\square$

Definition 6.27   A function $f : [a, b] \to \mathbb{C}$ is of class C^m if there exists an extension $F : I \to \mathbb{R}$ of f to an open interval $I \supset [a, b]$ with F of class C^m on I. $\qquad\square$

Definition 6.28   A function $f : [a, b] \to \mathbb{C}$ is of class C_I^m if f is of class $C^{(m-1)}$ on $[a, b]$ and $f^{(m)}$ is of class C_I^0 on $[a, b]$. $\qquad\square$

Definition 6.29   A function $f : \mathbb{R} \to \mathbb{C}$ is of class $C_{I,T}^m$ if f has period T and is of class C_I^m on $[0, T]$. $\qquad\square$

EXERCISES

Exercise 6.3. Identify a class of functions that contains all of Haar's wavelets.

Exercise 6.4. Prove that every sum f of finitely many multiples of characteristic functions is continuous by interval:

$$f = \sum_{k=1}^{N} c_k \chi_{[u_k, w_k[}.$$

6.2.3 Convergence and Inversion of Fourier Series

This subsection demonstrates that the orthogonal projections $S_N(f)$ converge to the function f at all points of differentiability.

Lemma 6.30 (Riemann–Lebesgue.) *For each piecewise continuous $2T$-periodic function g,*

$$\lim_{w \to \infty} \int_{-T}^{T} g(t)e^{-iwt\pi/T}\, dt = 0.$$

PROOF: Assume first that $g := \chi_{[a,b[}$:

$$\int_{-T}^{T} g(t)e^{-iwt\pi/T}\, dt = \int_{a}^{b} 1 \cdot e^{-iwt\pi/T}\, dt$$

$$= \frac{1}{iw\pi/T} \cdot e^{-iwt\pi/T}\Big|_{a}^{b} = \frac{e^{-iwb\pi/T} - e^{-iwa\pi/T}}{iw\pi/T}$$

$$= \frac{e^{-iw(a+b)\pi/(2T)}}{w\pi/T} \cdot 2\frac{e^{-iw(b-a)\pi/(2T)} - e^{-iw(a-b)\pi/(2T)}}{2i}$$

$$= \frac{e^{-iw(a+b)\pi/(2T)}}{w} \cdot 2\sin(w[b-a]\pi/(2T)),$$

which tends to zero as w increases, because for all $a, b, w \in \mathbb{R}$,

$$|e^{-iw(a+b)\pi/(2T)}| = 1, \qquad |\sin(w[b-a]\pi/[2T])| \leq 1.$$

For any other piecewise continuous function g, approximate g uniformly on $[-T, T]$ by simple step functions, and apply the foregoing result to each such step function $\chi_{[a,b[}$. □

Remark 6.31 The Riemann–Lebesgue lemma holds because as the frequency of the complex exponential increases, its narrower and narrower oscillations average one another out. □

Definition 6.32 The **Fourier Series** of a periodic function $f \in C_{1,2T}^{0}(\mathbb{R}, \mathbb{C})$ consists of the sequence of the partial sums $(S_N(f))$. Thus, if the sequence of partial sums $(S_N(f))$ converges, then the limit of the sequence $(S_N(f))_{N=0}^{\infty}$ is the limit of the Fourier series:

$$\sum_{k \in \mathbb{Z}} c_{f,k} e^{i \cdot k \cdot t \cdot \pi/T} = \lim_{N \to \infty} S_N(f)(t). \qquad □$$

The first result shows that the Fourier series of a function f converges to f on the domain of f'.

Proposition 6.33   *For every $f \in C^0_{I,2T}(\mathbb{R}, \mathbb{C})$ and for every $t \in \mathbb{R}$ where the derivative $f'(t)$ exists, the partial sums $S_N(f)(t)$ converge to $f(t)$:*

$$\lim_{N \to \infty} S_N(f)(t) = f(t).$$

PROOF: Use the integral representation, and apply the Riemann–Lebesgue lemma:

$$S_N(f)(t) - f(t) = S_N(f)(t) - 1 \cdot f(t)$$

$$= \frac{1}{2T} \int_{-T}^{T} f(s) \cdot \frac{\sin[N + \frac{1}{2}](t - s)\pi/T}{\sin(t - s)\pi/(2T)} \, ds$$

$$- f(t) \cdot \frac{1}{2T} \int_{-T}^{T} \frac{\sin[N + \frac{1}{2}](t - s)\pi/T}{\sin(t - s)\pi/(2T)} \, ds$$

$$= \frac{1}{2T} \int_{-T}^{T} [f(s) - f(t)] \cdot \frac{\sin[N + \frac{1}{2}](t - s)\pi/T}{\sin(t - s)\pi/(2T)} \, ds$$

$$= \frac{1}{2T} \int_{-T}^{T} \frac{f(s) - f(t)}{s - t} \frac{s - t}{\sin(t - s)\pi/(2T)}$$

$$\times \sin([N + \tfrac{1}{2}](t - s)\pi/T) \, ds,$$

which tends to 0 as N increases, by the Riemann–Lebesgue lemma.   □

6.2.4   Convolutions and Dirac's "Function" δ

The foregoing proof of the convergence of the partial sums of Fourier series demonstrates the general concept of an "approximate identity" by means of Dirichlet's kernel

$$D_N(s) = \frac{\sin([N + \frac{1}{2}]s\pi/T)}{\sin(s\pi/[2T])}$$

and a combination of functions called a convolution.

Definition 6.34   For each pair of piecewise continuous functions with common period $2T$, $f : \mathbb{R} \to \mathbb{C}$ and $h : \mathbb{R} \to \mathbb{C}$, define the **convolution** of f and h as the function denoted by $f * h$ with

$$(f * h)(r) := \frac{1}{2T} \int_{T}^{T} f(s)h(r - s) \, ds.$$

With convolutions, the partial sum $S_N(f)$ admits the expression

$$S_N(f)(t) = \frac{1}{2T} \int_{-T}^{T} f(s) \cdot \frac{\sin[N + \frac{1}{2}](t - s)\pi/T}{\sin(t - s)\pi/(2T)} \, ds$$

$$= \frac{1}{2T} \int_{-T}^{T} f(s) \cdot D_N(t - s) \, ds$$

$$= (f * D_N)(t),$$

and the convergence of the partial sums to f means that

$$f(t) = \lim_{N \to \infty} \frac{1}{2T} \int_{-T}^{T} f(s) \cdot D_N(t - s) \, ds = \lim_{N \to \infty} (f * D_N)(t).$$

A plot provides a graphical interpretation of such a convergence. As N increases, the Dirichlet kernel D_N decreases in magnitude away from the origin but increases in magnitude near the origin, as in Figure 6.6. A horizontal shift by t then shows that as N increases, Dirichlet's kernel D_N decreases in magnitude away from $s := t$ but increases in magnitude near $s := t$. Consequently, as N increases, the integrand $f(s) \cdot D_N(t - s)$ weighs the values $f(s)$ near $s := t$ more heavily than the values away from $s := t$, eventually reproducing the value $f(t)$ at the limit. The method of inverting a transform—reconstructing a function from its transform—through a convolution with suitable kernels applies to situations

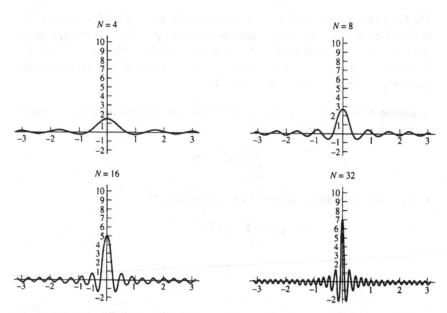

Figure 6.6   Dirichlet's kernels with $N \in \{4, 8, 16, 32\}$ and $s \mapsto D_N(s) := \{\sin([N + \frac{1}{2}]s\pi/T)\}/\{\sin(s\pi/[2T])\}$.

other than with Fourier series, for instance, with general Fourier transforms on the line, plane, and space, as explained in the following chapter.

Though not needed in the sequel, the following concept relates Dirichlet's kernel to an operator common in signal analysis.

Definition 6.35   Dirac's distribution δ is the operator such that for every function f defined at the origin,

$$\delta(f) := f(0).$$

Similarly, the shift of δ by t is the operator δ_t such that for every function f defined at t, $\delta_t(f) := f(t)$.   □

Remark 6.36   Dirac's distribution δ is *not* a function defined on the real line. Instead, δ is a function from the set of functions defined at the origin to the set of their values, $\delta : f \mapsto f(0)$. The convergence

$$\delta = \lim_{N \to \infty} D_N$$

then does not occur in the space of functions on the real line, but instead takes place in a space of functions similar to δ—called "distributions"—which associate functions to numbers. [35, p. 74].   □

6.2.5   Uniform Convergence of Fourier Series

The foregoing results established the convergence at each isolated point where the derivative exists. The following results show that the partial sums converge faster, and uniformly so, if the derivative f' is continuous by interval: $f' \in C^0_{I,2T}(\mathbb{R}, \mathbb{C})$. Because Fourier series can involve nonzero coefficients for all integral indices, the following terminology will prove convenient.

Definition 6.37   The linear space $\ell^2(\mathbb{Z}, \mathbb{C})$ consists of all sequences of complex numbers $(z_k)_{k \in \mathbb{Z}}$ such that

$$\sum_{k \in \mathbb{Z}} |z_k|^2 < \infty,$$

with the inner product $\langle \, , \, \rangle_{\ell^2}$ and norm $\| \, \|_{\ell^2}$ defined by

$$\langle z, w \rangle_{\ell^2} := \sum_{k \in \mathbb{Z}} z_k \overline{w_k},$$

$$\|z\|_{\ell^2} := \sqrt{\langle z, z \rangle_{\ell^2}} = \sqrt{\sum_{k \in \mathbb{Z}} |z_k|^2}.$$ □

The series defining the inner product converges absolutely for all sequences in $\ell^2(\mathbb{Z}, \mathbb{C})$, thanks to the Cauchy–Schwarz inequality applied to the inner product

$\langle\,,\,\rangle$ for $\mathbb{C}^N$:

$$\sum_{k=-N}^{N} |z_k \overline{w_k}| = \langle(|z_1|,\ldots,|z_N|), (|w_1|,\ldots,|w_N|)\rangle$$

$$\leq \|(|z_1|,\ldots,|z_N|)\|_2 \cdot \|(|w_1|,\ldots,|w_N|)\|_2$$

$$\leq \sqrt{\sum_{k\in\mathbb{Z}} |z_k|^2} \cdot \sqrt{\sum_{k\in\mathbb{Z}} |w_k|^2} < \infty.$$

Hence, all the theorems on orthogonal projections hold for $\ell^2(\mathbb{Z}, \mathbb{C})$, for instance, the Cauchy–Schwarz inequality:

$$|\langle z, w \rangle_{\ell^2}| \leq \|z\|_{\ell^2} \cdot \|w\|_{\ell^2}.$$

Lemma 6.38 **(Bessel Inequalities.)** *For every* $f, g \in C^0_{1,2T}(\mathbb{R}, \mathbb{C})$,

$$\sum_{k\in\mathbb{Z}} |c_{f,k}|^2 \leq \int_{-T}^{T} |f(t)|^2\, dt < \infty,$$

$$\sum_{k\in\mathbb{Z}} |c_{f,k} \cdot \overline{c_{g,k}}| \leq \|f\|_2 \cdot \|g\|_2 < \infty.$$

PROOF: The result is a particular case of a general theorem on orthogonal projections in the space $C^0_{1,2T}(\mathbb{R}, \mathbb{C})$:

$$f = S_N(f) + [f - S_N(f)],$$

$$S_N(f) \perp [f - S_N(f)],$$

$$\|f\|_2^2 = \|S_N(f) + [f - S_N(f)]\|_2^2 = \|S_N(f)\|_2^2 + \|[f - S_N(f)]\|_2^2,$$

$$\|S_N(f)\|_2^2 = \|f\|_2^2 - \|[f - S_N(f)]\|_2^2 \leq \|f\|_2^2,$$

$$\sum_{k\in\mathbb{Z}} |c_{f,k}|^2 = \|S_N(f)\|_2^2 \leq \|f\|_2^2 = \int_{-T}^{T} |f(t)|^2\, dt.$$

Hence, the Cauchy–Schwarz inequality for the space $\ell^2(\mathbb{Z}, \mathbb{C})$ gives

$$\left(\sum_{k\in\mathbb{Z}} |c_{f,k} \cdot \overline{c_{g,k}}|\right)^2 \leq \left(\sum_{k\in\mathbb{Z}} |c_{f,k}|^2\right) \cdot \left(\sum_{k\in\mathbb{Z}} |c_{g,k}|^2\right)$$

$$\leq \left(\int_{-T}^{T} |f(t)|^2\, dt\right) \cdot \left(\int_{-T}^{T} |g(t)|^2\, dt\right). \qquad \square$$

Proposition 6.39 *For every $f \in C^1_{I,2T} (\mathbb{R}, \mathbb{C})$, the partial sums $S_N(f)$ converge uniformly to f, and*

$$\sum_{k \in \mathbb{Z}} |c_{f,k}| < \infty.$$

PROOF: The proof uses the existence of the Fourier coefficients of the derivative f', an integration by parts, and the periodicity of f:

$$c_{f',k} = \frac{1}{2T} \int_{-T}^{T} f'(s) \cdot e^{-iks\pi/T} \, ds$$

$$= \frac{1}{2T} \left\{ f(s)e^{-iks\pi/T} \Big|_{-T}^{T} - \frac{1}{2T} \int_{-T}^{T} f(s)(-ik\pi/T)e^{-iks\pi/T} \, ds \right\}$$

$$= 0 + (ik\pi/T) \cdot c_{f,k}.$$

Applied to the function f', Bessel's inequalities (6.38) show that

$$\left(\frac{\pi}{T}\right)^2 \sum_{k \in \mathbb{Z}} |k|^2 \cdot |c_{f,k}|^2 = \sum_{k \in \mathbb{Z}} |c_{f',k}|^2 \leq \int_{-T}^{T} |f'(t)|^2 \, dt < \infty.$$

Moreover, the algebraic identity and inequality

$$0 \leq (p - q)^2 = p^2 - 2pq + q^2, \qquad pq \leq \frac{p^2 + q^2}{2}$$

applied to $p := |k| \cdot |c_{f,k}|$ and $q := 1/|k|$ give

$$|c_{f,k}| = (|k| \cdot |c_{f,k}|) \cdot (1/|k|) \leq \frac{(|k| \cdot |c_{f,k}|)^2 + (1/|k|)^2}{2},$$

which reveals that

$$\sum_{k \in \mathbb{Z}} |c_{f,k}| \leq \frac{\sum_{k \in \mathbb{Z}} |k|^2 \cdot |c_{f,k}|^2 + \sum_{k \in \mathbb{Z}} (1/|k|)^2}{2} < \infty.$$

Consequently, by definition of the convergence of a series, for every $\varepsilon > 0$ there exists $K_\varepsilon \in \mathbb{N}$ such that for each $N > K_\varepsilon$,

$$\sum_{|k| > N} |c_{f,k}| < \varepsilon,$$

whence for each $t \in \mathbb{R}$,

$$\sum_{|k| > K_\varepsilon} |c_{f,k} e^{ikt\pi/T}| = \sum_{|k| > K_\varepsilon} |c_{f,k}| < \varepsilon,$$

and therefore, for each $N > K_\varepsilon$,

$$|f(t) - S_N(f)(t)| = \left| \sum_{k\in\mathbb{Z}} c_{f,k} e^{ikt\pi/T} - \sum_{k=-N}^{N} c_{f,k} e^{ikt\pi/T} \right|$$

$$\leq \sum_{|k|>N} |c_{f,k} e^{ikt\pi/T}| < \varepsilon,$$

which means that the convergence occurs uniformly: with the same ε and the same K_ε for every t.   □

Remark 6.40 Induction shows that the higher the differentiability of f, the faster its Fourier series converges, in the sense that if $f \in C_{1,2T}^{(m)}$, then $\sum_{k\in\mathbb{Z}} |k^{m-1} \cdot c_{f,k}| < \infty$.   □

Corollary 6.41 (Parseval's Identity.) *For every* $f \in C_{1,2T}^1 (\mathbb{R}, \mathbb{C})$,

$$\sum_{k\in\mathbb{Z}} |c_{f,k}|^2 = \int_{-T}^{T} |f(t)|^2 \, dt.$$

PROOF: The result is a particular case of a general theorem on orthogonal projections, together with uniform convergence, which allows for the permutation of limit and integration:

$$\|S_N(f)\|_2^2 = \|f\|_2^2 - \|[f - S_N(f)]\|_2^2 \leq \|f\|_2^2,$$

$$\lim_{N\to\infty} [f - S_N(f)] = 0,$$

$$\lim_{N\to\infty} \|S_N(f)\|_2^2 = \|f\|_2^2,$$

$$\sum_{k\in\mathbb{Z}} |c_{f,k}|^2 = \lim_{N\to\infty} \|S_N(f)\|_2^2 = \|f\|_2^2 = \int_{-T}^{T} |f(t)|^2 \, dt.$$

Specifically, the equality $\lim_{N\to\infty} \|S_N(f)\|_2^2 = \|f\|_2^2$ follows from uniform convergence and orthogonality: If $N > K_\varepsilon$, then $|S_N(f)(t) - f(t)| < \varepsilon$ for every t, whence

$$0 \leq \|f\|_2^2 - \|S_N(f)\|_2^2$$

$$= \|f\|_2^2 - \left\{ \|f\|_2^2 - \|f - S_N(f)\|_2^2 \right\} = \|f - S_N(f)\|_2^2$$

$$= \int_{-T}^{T} |f(t) - S_N(f)(t)|^2 \, dt < \int_{-T}^{T} \varepsilon^2 \, dt = 2T\varepsilon^2.$$ □

Corollary 6.42 (Generalized Parseval Identity.) *For all* $f, g \in C_{1,2T}^1$ $(\mathbb{R}, \mathbb{C})$,

$$\sum_{k\in\mathbb{Z}} c_{f,k} \cdot \overline{c_{g,k}} = \int_{-T}^{T} f(t) \cdot \overline{g(t)} \, dt.$$

PROOF: Apply Parseval's identity and the Polar identity in each of the spaces $\ell^2(\mathbb{Z}, \mathbb{C})$ and $C^1_{I,2T}(\mathbb{R}, \mathbb{C})$:

$$
\begin{aligned}
\langle c_f, c_g \rangle_{\ell^2} &= \left(\frac{1}{4}\right)\left[\left(\|c_f + c_g\|^2_{\ell^2} - \|c_f - c_g\|^2_{\ell^2}\right)\right.\\
&\qquad\left. + i\left(\|c_f + ic_g\|^2_{\ell^2} - \|c_f - ic_g\|^2_{\ell^2}\right)\right]\\
&= \left(\frac{1}{4}\right)\left[\left(\|c_{f+g}\|^2_{\ell^2} - \|c_{f-g}\|^2_{\ell^2}\right) + i\left(\|c_{f+ig}\|^2_{\ell^2} - \|c_{f-ig}\|^2_{\ell^2}\right)\right]\\
&= \left(\frac{1}{4}\right)\left[\left(\|f + g\|^2_2 - \|f - g\|^2_2\right) + i\left(\|f + ig\|^2_2 - \|f - ig\|^2_2\right)\right]\\
&= \langle f, g \rangle. \hspace{5cm}\square
\end{aligned}
$$

Needed in the proof of the main inversion theorem, the following lemma relates one-sided limits to one-sided derivatives: If a function and its derivative are both piecewise continuous, then limits of one-sided difference quotients converge to one-sided limits of derivatives.

Lemma 6.43 *Assume that $f \in C^0_{I,2T}(\mathbb{R}, \mathbb{C})$ and $f' \in C^0_{I,2T}(\mathbb{R}, \mathbb{C})$. Then for every $\ell \in \{1, \ldots, m\}$ there exists a $\delta_\ell > 0$ such that*

$$
\lim_{h \searrow 0^+}\left[\frac{f(x+h) - f(x)}{h}\right] = \lim_{t \searrow x^+} f'(t)
$$

uniformly on $[s_{\ell-1}, s_{\ell-1} + 2\delta_\ell] \times [0, \delta_\ell]$.

PROOF: Apply the fundamental theorem of calculus on any interval $[x, x + h]$ where f is continuous and f' is continuous on $]x, x + h]$. Consider the auxiliary function defined in Lemma 6.21:

$$
\widetilde{f}'(t) := \begin{cases} f'(x) & \text{if } x < t \leq x + h, \\ \lim_{t \searrow x^+} f'(t) & \text{if } t := x. \end{cases}
$$

Then $\widetilde{f}'$ is continuous on $[x, x + h]$ and differs from f' at at most one point, x. Hence their integrals coincide:

$$
\begin{aligned}
f(x + h) - f(x) &= \int_x^{x+h} f'(s)\, ds = \int_x^{x+h} \widetilde{f}'(s)\, ds\\
&= \int_x^{x+h} [\widetilde{f}'(x) + \widetilde{f}'(s) - \widetilde{f}'(x)]\, ds\\
&= \int_x^{x+h} \widetilde{f}'(x)\, ds + \int_x^{x+h} [\widetilde{f}'(s) - \widetilde{f}'(x)]\, ds,
\end{aligned}
$$

where for the first integral,

$$\int_x^{x+h} \widetilde{f}'(x)\,ds = h \cdot \widetilde{f}'(x).$$

For the second integral, by uniform continuity of $\widetilde{f}'$ on the compact set $[x, x+h]$, for every $\varepsilon > 0$ there exists a $\delta > 0$ such that if $|x-s| < \delta$, then $|\widetilde{f}'(s) - \widetilde{f}'(x)| < \varepsilon$, whence

$$\left| \int_x^{x+h} [\widetilde{f}'(s) - \widetilde{f}'(x)]\,ds \right| < \int_x^{x+h} \varepsilon\,ds = h \cdot \varepsilon.$$

Consequently, if $0 < h < \delta$, then

$$\left| \{f(x+h) - f(x)\} - h\widetilde{f}'(x) \right| < h\varepsilon,$$

$$\left| \frac{f(x+h) - f(x)}{h} - \lim_{t \searrow x^+} f'(t) \right| = \frac{\left| \{f(x+h) - f(x)\} - h\widetilde{f}'(x) \right|}{|h|}$$

$$< \left| \frac{h\varepsilon}{h} \right| = \varepsilon. \qquad \square$$

The following theorem shows that at "jump" discontinuities the Fourier series converges to the average of the two one-sided limits.

Theorem 6.44 *Assume that $f \in C^0_{l,2T}(\mathbb{R}, \mathbb{C})$ and $f' \in C^0_{l,2T}(\mathbb{R}, \mathbb{C})$. Then for every $t \in \mathbb{R}$ the partial sums $S_N(f)$ converge to the average of the left-hand and right-hand limits of f:*

$$\lim_{N \to \infty} S_N(f)(t) = \frac{\lim_{s \searrow t^+} f(s) + \lim_{s \nearrow t^-} f(s)}{2}.$$

PROOF: Use the integral representation of S_N and the periodicity of f, with the change of variable $s := x + t$:

$$S_N(f)(t) = \frac{1}{2T} \int_{-T}^{T} f(s) \cdot \frac{\sin[N + \frac{1}{2}](t - s)\pi/T}{\sin(t - s)\pi/(2T)}\,ds$$

$$= \frac{1}{2T} \int_{-t-T}^{-t+T} f(x + t) \cdot \frac{\sin[N + \frac{1}{2}](x)\pi/T}{\sin(x)\pi/(2T)}\,dx$$

$$= \frac{1}{2T} \int_{-T}^{T} f(x + t) \cdot \frac{\sin[N + \frac{1}{2}](x)\pi/T}{\sin(x)\pi/(2T)}\,dx.$$

Because the integrand is a periodic even function,

$$\frac{1}{2T} \int_0^T \frac{\sin[N + \frac{1}{2}](x)\pi/T}{\sin(x)\pi/(2T)}\,dx$$

$$= \frac{1}{2} \cdot \frac{1}{2T} \int_{-T}^{T} f(x + t) \cdot \frac{\sin[N + \frac{1}{2}](x)\pi/T}{\sin(x)\pi/(2T)}\,dx = \frac{1}{2} \cdot 1 = \frac{1}{2}.$$

Consequently, for the second half-period,

$$\frac{1}{2T} \int_0^T f(x+t) \cdot \frac{\sin\left[N+\frac{1}{2}\right](x)\pi/T}{\sin(x)\pi/(2T)}\, dx - \left(\frac{1}{2}\right) \cdot \lim_{s \searrow t^+} f(s)$$

$$= \frac{1}{2T} \int_0^T \frac{\sin\left[N+\frac{1}{2}\right](x)\pi/T}{\sin(x)\pi/(2T)} \cdot \left[f(x+t) - \lim_{s \searrow t^+} f(s)\right] dx$$

$$= \frac{1}{2T} \int_0^T \frac{f(x+t) - \lim_{s \searrow t^+} f(s)}{x} \cdot \frac{x}{\sin(x)\pi/(2T)}$$

$$\times \sin\left(\left[N+\frac{1}{2}\right](x)\pi/T\right) dx.$$

By hypotheses on f, the integrand is piecewise continuous, whence the Riemann–Lebesgue lemma shows that the integral tends to 0 as N increases. The first half-period lends itself to a similar proof. □

6.3　PERIODIC FUNCTIONS

This subsection shows that every period of a periodic function equals an integral multiple of that function's smallest positive period, which is then called *the* period of that function. Recall that a function f has a **period** P, or, equivalently, is **periodic** with period P, if $f(t+P) = f(t)$ for every t. The following results demonstrate that if a function has a nonzero real period, then it also has a smallest positive period, and every period is an integral multiple of that function's smallest positive period.

Lemma 6.45　*If a function $f : \mathbb{R} \to \mathbb{C}$ has periods R and P, then for all integers $m, n \in \mathbb{Z}$ the number $mR + nP$ is also a period of f. If f has a nonzero period, then f also has a positive period.*

PROOF: If f has a period S, then the identity $t = [t - S] + S$ and the period S lead to

$$f(t) = f([t - S] + S) = f([t - S]),$$

whence $-S$ is also a period of f. In particular, if $S \neq 0$, then $P := S$ or $P := -S$ is a positive period of f.

Induction on m and then on n confirms that $f(t + mR + nT) = f(t)$. Indeed, the identity $t + (0 \cdot S) = t$ and the periodicity $f(t + mS) = f(t)$ for every $t \in \mathbb{R}$ give

$$f(t + [m+1]S) = f([t+mS] + S) = f(t + mS) = f(t),$$

whence $[m+1]S = [m+1] \cdot (\pm R) = \pm[m+1]R$ is also a period of f. The same argument applied to $S := \pm P$ then yields

$$f(t + mR + nP) = f([t + mR] + nP) = f(t + mR) = f(t). \qquad \square$$

Lemma 6.46 *If a continuous nonconstant function $f : \mathbb{R} \to \mathbb{C}$ has a nonzero real period R, then the set of all positive periods of f contains a smallest element.*

PROOF: This proof proceeds by contraposition: If a continuous function f has a nonzero period but fails to have a smallest positive period, then f is constant.

For all distinct reals $r < s$ the function f is continuous on the compact interval $[r, s]$ and hence also uniformly continuous there. This means that for every $\varepsilon > 0$ there exists a $\delta_\varepsilon > 0$ such that $|f(v) - f(w)| < \varepsilon$ for all $v, w \in [r, s]$ for which $|v - w| < \delta_\varepsilon$.

If the set of positive periods fails to have a smallest element, then for every $\varepsilon > 0$ the function f has a positive period $P_\varepsilon < \delta_\varepsilon$. Define $N_\varepsilon := \lfloor |r - s|/P_\varepsilon \rfloor$, which denotes the largest integer that does not exceed $|r - s|/P_\varepsilon$, so that $N_\varepsilon \cdot P_\varepsilon \leq |r - s|$ and

$$-(r - s) \leq N_\varepsilon \cdot P_\varepsilon \leq r - s,$$
$$s - \delta_\varepsilon < s - P_\varepsilon < r + N_\varepsilon \cdot P_\varepsilon \leq s,$$
$$-\delta_\varepsilon < -P_\varepsilon < r + N_\varepsilon \cdot P_\varepsilon - s \leq 0,$$
$$|r + N_\varepsilon \cdot P_\varepsilon - s| < \delta_\varepsilon.$$

Hence,

$$|f(r) - f(s)| = |f(r + N_\varepsilon \cdot P_\varepsilon) - f(s)| < \varepsilon.$$

Because the inequalities just proved hold for every $\varepsilon > 0$, it follows that $f(r) = f(s)$, whence it also follows that f is constant. $\quad\square$

Lemma 6.47 *If a continuous nonconstant function $f : \mathbb{R} \to \mathbb{C}$ has a nonzero real period, then every period is an integral multiple of the smallest positive period.*

PROOF: If P denotes the smallest positive period of f, and if R denotes any period of f, then define $m := \lfloor R/P \rfloor$, so that

$$m \leq R/P < m + 1,$$
$$mP \leq R < (m + 1)P,$$
$$0 \leq R - mP < P.$$

Because $R - mP$ is also a period of f, and because P is the smallest positive period of f, the inequalities $0 \leq R - mP < P$ then imply that $0 = R - mP$, whence $R = mP$. $\quad\square$

Remark 6.48 The foregoing results extend to each coordinate of functions

$$f : \mathbb{R} \to V, \quad f(t) = (f_1(t), \ldots, f_n(t))$$

from the real line $\mathbb{R}$ into any finite-dimensional vector space V over $\mathbb{R}$ or $\mathbb{C}$. In particular, such results extend to functions $f : \mathbb{R} \to \mathbb{C}^N$. $\quad\square$

PART C

Computation and Design
of Wavelets

CHAPTER 7

Fourier Transforms on the Line and in Space

7.0 INTRODUCTION

This chapter introduces a few features of the Fourier transform, mainly for its use in the design of Daubechies' continuous, compactly supported, orthogonal wavelets. The principal use of the Fourier transform in the design of wavelets will consist in transforming "convolutions" of functions into ordinary multiplications of functions, which allow for an analysis easier than with convolutions. The material presented here involves mostly calculus, with only a few results with brief statements—such as Fubini's theorem on the permutation of the order of multiple integrals [12, pp. 65–66], [21, pp. 384–388], [28, pp. 226–239]—borrowed from the literature in mathematical real analysis and complex analysis.

7.1 THE FOURIER TRANSFORM

7.1.1 Definition and Examples of the Fourier Transform

The basic Fourier transform pertains to a restricted class of functions with finite integrals over the entire real line.

Definition 7.1 A function $f : \mathbb{R} \to \mathbb{C}$ is **integrable** if the integral of its absolute value exists and is finite:

$$\int_{\mathbb{R}} |f(x)| \, dx < \infty. \qquad \Box$$

Example 7.2 The function $t \mapsto 1/(1 + t^2)$ is integrable:

$$\int_{\mathbb{R}} \left| \frac{1}{1+t^2} \right| \, dt = \int_{\mathbb{R}} \frac{1}{1+t^2} \, dt = \mathrm{Arctan}\,(t)|_{-\infty}^{\infty} = \frac{\pi}{2} - \frac{-\pi}{2} = \pi < \infty. \quad \Box$$

Counterexample 7.3   The function $t \mapsto t/(1 + t^2)$ is *not* integrable:

$$\int_{\mathbb{R}} \left| \frac{t}{1 + t^2} \right| dt = 2 \int_0^\infty \frac{t}{1 + t^2} dt = \ln(1 + t^2)\Big|_0^\infty = \infty - 0 = \infty. \qquad \square$$

Definition 7.4   The **Fourier transform** of an integrable function $f : \mathbb{R} \to \mathbb{C}$ is the function $\mathcal{F}f : \mathbb{R} \to \mathbb{C}$ defined by

$$(\mathcal{F}f)(w) := \frac{1}{\sqrt{2\pi}} \int_{\mathbb{R}} f(r) \cdot e^{-i \cdot r \cdot w} \, dr. \qquad \square$$

Example 7.5   For all distinct real numbers $r < s$, consider the characteristic function $\chi_{[r,s]} : \mathbb{R} \to \mathbb{C}$ of the closed interval $[r, s]$, defined by

$$\chi_{[r,s]}(x) := \begin{cases} 1 & \text{if } x \in [r, s], \\ 0 & \text{if } x \notin [r, s]. \end{cases}$$

A formula for its Fourier transform follows from calculus:

$$\begin{aligned}
(\mathcal{F}\chi_{[r,s]})(w) &= \frac{1}{\sqrt{2\pi}} \int_{\mathbb{R}} \chi_{[r,s]}(x) \cdot e^{-i \cdot w \cdot x} \, dx \\
&= \frac{1}{\sqrt{2\pi}} \int_r^s 1 \cdot e^{-i \cdot w \cdot x} \, dx \\
&= \frac{1}{\sqrt{2\pi}} \cdot \frac{1}{-iw} \cdot e^{-i \cdot w \cdot x} \Big|_r^s = \frac{e^{-i \cdot w \cdot s} - e^{-i \cdot w \cdot r}}{-iw\sqrt{2\pi}} \\
&= e^{-i \cdot w \cdot (r+s)/2} \cdot \frac{e^{-i \cdot w \cdot (s-r)/2} - e^{-i \cdot w \cdot (r-s)/2}}{-iw\sqrt{2\pi}} \\
&= \frac{2e^{-i \cdot w \cdot [r+s]/2}}{\sqrt{2\pi}} \cdot \frac{\sin(w \cdot [s - r]/2)}{w}.
\end{aligned}$$

In particular, for an interval $[-s, s]$ symmetric about the origin,

$$(\mathcal{F}\chi_{[-s,s]})(w) = \sqrt{\frac{2}{\pi}} \cdot \frac{\sin(w \cdot s)}{w} = \sqrt{\frac{2}{\pi}} \cdot s \cdot \text{sinc}(w \cdot s),$$

where $\text{sinc}(z) := \sin(z)/z$, as displayed for $s := 1$ in Figure 7.1. $\qquad \square$

The foregoing example also shows that while the function $\chi_{[-s,s]}$ is discontinuous and integrable, its Fourier transform $\mathcal{F}\chi_{[-s,s]}$ is differentiable but not integrable. Specific relations between the features of a function and the features of its Fourier transform will appear later.

The Fourier transform serves mainly to provide insight and steer calculation in effective directions, rather than to calculate Fourier transforms of specific func-

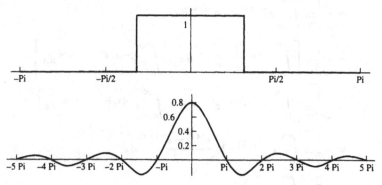

Figure 7.1 *Top.* Characteristic function χ of $[-1, 1]$. *Bottom.* The Fourier transform $\mathcal{F}\chi = \sqrt{2/\pi} \sin c$.

tions. Indeed, few easily calculable examples of Fourier transforms exist [21, p. 407]. Among the few such examples, the following one will provide a means to calculate the inverse Fourier transform [21, p. 407].

Example 7.6   (Abel's kernel A_B; see Figure 7.2.)   For each $B > 0$, define the exponential function K_B and Abel's function A_B by

$$K_B(x) := e^{-(|x|/B)},$$

$$A_B(w) := \sqrt{\frac{2}{\pi}} \cdot \frac{B}{1 + w^2 \cdot B^2}.$$

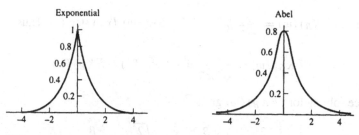

Figure 7.2   Fourier transform $\mathcal{F}K_1 = A_1$. *Left.* Exponential kernel, $K_1(x) = e^{-|x|}$. *Right.* Abel's kernel, $A_1(w) = (\sqrt{2/\pi})/(1 + w^2)$.

A direct calculation shows that $(\mathcal{F}K_B)(w) = A_B(w)$. To this end, use $|x| = x$ for $x > 0$ and $|x| = -x$ for $x < 0$:

$$(\mathcal{F}K_B)(w) = \frac{1}{\sqrt{2\pi}} \int_{-\infty}^{\infty} K_B(x) \cdot e^{-ixw} \, dx$$

$$= \frac{1}{\sqrt{2\pi}} \int_{-\infty}^{\infty} e^{-(|x|/B)} \cdot e^{-ixw} \, dx$$

$$= \frac{1}{\sqrt{2\pi}} \left[\int_{-\infty}^{0} e^{(x/B)} \cdot e^{-ixw} \, dx + \int_{0}^{\infty} e^{-(x/B)} \cdot e^{-ixw} \, dx \right]$$

$$= \frac{1}{\sqrt{2\pi}} \left[\int_{0}^{\infty} e^{-(x/B)} \cdot e^{ixw} \, dx + \int_{0}^{\infty} e^{-(x/B)} \cdot e^{-ixw} \, dx \right]$$

$$= \frac{1}{\sqrt{2\pi}} \int_{0}^{\infty} e^{-(x/B)} \cdot 2 \cdot \frac{e^{ixw} + e^{-ixw}}{2} \, dx$$

$$= \frac{2}{\sqrt{2\pi}} \int_{0}^{\infty} e^{-(x/B)} \cdot \cos(xw) \, dx \quad (\dagger)$$

$$= \frac{2}{\sqrt{2\pi}} \cdot (-B) \cdot e^{-(x/B)} \cdot \cos(xw) \Big|_{0}^{\infty}$$

$$\quad - \frac{2}{\sqrt{2\pi}} \int_{0}^{\infty} (-B) \cdot e^{-(x/B)} \cdot [-w \cdot \sin(xw)] \, dx$$

$$= \frac{2}{\sqrt{2\pi}} \cdot B + \frac{2}{\sqrt{2\pi}} B^2 \cdot e^{-(x/B)} \cdot w \cdot \sin(xw) \Big|_{0}^{\infty}$$

$$\quad + \frac{2}{\sqrt{2\pi}} \int_{0}^{\infty} (-B^2) \cdot e^{-(x/B)} \cdot [w^2 \cdot \cos(xw)] \, dx$$

$$= \frac{2}{\sqrt{2\pi}} \cdot B - (w^2 \cdot B^2) \cdot \frac{2}{\sqrt{2\pi}} \cdot \int_{0}^{\infty} e^{-(x/B)} \cdot \cos(xw) \, dx$$

$$= \frac{2}{\sqrt{2\pi}} \cdot B - (w^2 \cdot B^2) \cdot (\mathcal{F} K_B)(w),$$

thanks to $(\mathcal{F} K_B)(w) = \frac{2}{\sqrt{2\pi}} \int_{0}^{\infty} e^{-(x/B)} \cdot \cos(xw) \, dx$ in step $(\dagger)$. Thus,

$$(\mathcal{F} K_B)(w) = \frac{2}{\sqrt{2\pi}} \cdot B - (w^2 \cdot B^2) \cdot (\mathcal{F} K_B)(w),$$

whence solving for $(\mathcal{F} K_B)(w)$ gives

$$(\mathcal{F} K_B)(w) = \frac{2}{\sqrt{2\pi}} \cdot \frac{B}{1 + w^2 \cdot B^2} = \sqrt{\frac{2}{\pi}} \cdot \frac{B}{1 + w^2 \cdot B^2} = A_B(w). \quad \Box$$

EXERCISES

Exercise 7.1. Calculate $(\mathcal{F} h)(w)$ for the function h defined by

$$h(x) := \begin{cases} 1 - |x| & \text{if } -1 \le x \le 1, \\ 0 & \text{otherwise.} \end{cases}$$

Exercise 7.2. Calculate $(\mathcal{F}g)(w)$ for the function g defined by

$$g(x) := \begin{cases} 1 - 3x^2 + 2|x|^3 & \text{if } -1 \le x \le 1, \\ 0 & \text{otherwise.} \end{cases}$$

7.2 CONVOLUTIONS AND INVERSION OF THE FOURIER TRANSFORM

The mathematical concept of "convolution" constitutes a continuously weighted average of a function by another, which proves effective in solving differential and integral equations, for instance, the integral equation that arises in inverting the Fourier transform. This subsection demonstrates the use of convolutions to invert the Fourier transform with bounded functions; a subsequent subsection will handle more general functions. As for the inversion of Fourier series, the strategy presented here to invert the Fourier transform relies on suitable "approximate identities" called Abel's kernels and similar to Dirichlet's kernels for the inversion of Fourier series.

Definition 7.7 A function $f : \mathbb{R} \to \mathbb{C}$ is **bounded** if there is a $B \ge 0$ such that $|f(r)| \le B$ for every $r \in \mathbb{R}$. $\qquad\square$

Definition 7.8 For each pair of bounded and integrable functions $f : \mathbb{R} \to \mathbb{C}$ and $h : \mathbb{R} \to \mathbb{C}$, the **convolution** of f and h is the function denoted by $f * h$ with [21, p. 401]

$$(f * h)(r) := \frac{1}{\sqrt{2\pi}} \int_{\mathbb{R}} f(r - s)h(s)\, ds. \qquad\square$$

(Alternative definitions omit the factor $1/\sqrt{2\pi}$ [12, p. 260].)

Proposition 7.9 *Convolution commutes: For each pair of bounded and integrable functions* $f, h : \mathbb{R} \to \mathbb{C}$,

$$f * h = h * f.$$

PROOF: Perform the change of variable $u := r - s$, so that $s = r - u$ and $ds = -du$:

$$(f * h)(r) = \frac{1}{\sqrt{2\pi}} \int_{-\infty}^{\infty} f(r - s)h(s)\, ds$$

$$= \frac{1}{\sqrt{2\pi}} \int_{\infty}^{-\infty} f(u)h(r - u)\,(-du)$$

$$= \frac{1}{\sqrt{2\pi}} \int_{-\infty}^{\infty} h(r - u) f(u)\, du$$

$$= (h * f)(r). \qquad\square$$

Definition 7.10   For every integrable function h, let $\|h\|_1 := \int_{\mathbb{R}} |h(t)| \, dt$.   □

Proposition 7.11   *If f and g are integrable, then $f * g$ is integrable.*

PROOF: By Fubini's theorem, and with the change of variable $z := x - w$,

$$\int\int_{\mathbb{R}^2} |f(x - w)g(w)| \, dx \, dw = \int_{\mathbb{R}} |g(w)| \int_{\mathbb{R}} |f(x - w)| \, dx \, dw$$

$$\leq \int_{\mathbb{R}} |g(w)| \cdot \|f\|_1 \, dw = \|g\|_1 \cdot \|f\|_1.$$

Hence

$$\int_{\mathbb{R}} |f * g(x)| \, dx = \int_{\mathbb{R}} \left| \frac{1}{\sqrt{2\pi}} \int_{\mathbb{R}} f(x - w)g(w) \, dw \right| \, dx$$

$$\leq \int_{\mathbb{R}} \frac{1}{\sqrt{2\pi}} \int_{\mathbb{R}} |f(x - w)g(w)| \, dw \, dx \leq \frac{1}{\sqrt{2\pi}} \|g\|_1 \cdot \|f\|_1. \quad □$$

As a preliminary step toward the inversion of the Fourier transform, an integral of $\mathcal{F}f$ with K_B gives the convolution $f * A_B$.

Lemma 7.12   *For each integrable function $f : \mathbb{R} \to \mathbb{C}$ such that $\mathcal{F}f$ is also integrable,*

$$(f * A_B)(r) = \frac{1}{\sqrt{2\pi}} \int_{\mathbb{R}} (\mathcal{F}f)(w) \cdot K_B(w) \cdot e^{i \cdot w \cdot r} \, dw.$$

PROOF: Refer to the two definitions, utilize the fact that Abel's kernel is an even function, so that $A_B(r - s) = A_B(s - r)$, and swap the order of integration:

$$(f * A_B)(r) = (A_B * f)(r) = \frac{1}{\sqrt{2\pi}} \int_{\mathbb{R}} A_B(r - s) \cdot f(s) \, ds$$

$$= \frac{1}{\sqrt{2\pi}} \int_{\mathbb{R}} f(s) \cdot A_B(s - r) \, ds$$

$$= \frac{1}{\sqrt{2\pi}} \int_{\mathbb{R}} f(s) \cdot (\mathcal{F}K_B)(s - r) \, ds$$

$$= \frac{1}{\sqrt{2\pi}} \int_{\mathbb{R}} f(s) \cdot \left[\frac{1}{\sqrt{2\pi}} \int_{\mathbb{R}} K_B(w) \cdot e^{-i \cdot (s - r) \cdot w} \, dw \right] ds$$

$$= \frac{1}{\sqrt{2\pi}} \int_{\mathbb{R}} \left[\frac{1}{\sqrt{2\pi}} \int_{\mathbb{R}} f(s) \cdot e^{-i \cdot s \cdot w} \, ds \right] \cdot K_B(w) \cdot e^{i \cdot r \cdot w} \, dw$$

$$= \frac{1}{\sqrt{2\pi}} \int_{\mathbb{R}} (\mathcal{F}f)(w) \cdot K_B(w) \cdot e^{i \cdot w \cdot r} \, dw. \qquad\qquad □$$

As B increases, the convolution $f * A_B$ converges to f.

Lemma 7.13   *For each bounded integrable function $f : \mathbb{R} \to \mathbb{C}$ continuous at r,*

$$f(r) = \lim_{B \to \infty} (f * A_B)(r).$$

PROOF: This proof resembles that of the inversion of Fourier series with Dirichlet's kernels but uses Abel's kernels instead. Most of the proof consists in verifying mathematical features of Abel's kernels. Firstly, if f is continuous at r, then for each $\varepsilon > 0$ there exists some $c > 0$ such that $|f(x) - f(r)| < \varepsilon$ for each x such that $|x - r| < c$. Secondly, perform the change of variable $v := w \cdot B$ to calculate

$$\frac{1}{\sqrt{2\pi}} \int_{-c}^{c} A_B(w)\, dw = \frac{1}{\sqrt{2\pi}} \int_{-c}^{c} \sqrt{\frac{2}{\pi}} \cdot \frac{B}{1 + w^2 \cdot B^2}\, dw$$

$$= \frac{1}{\sqrt{2\pi}} \int_{-c \cdot B}^{c \cdot B} \sqrt{\frac{2}{\pi}} \cdot \frac{1}{1 + v^2}\, dv$$

$$= \frac{1}{\pi} \operatorname{Arctan}(v)\big|_{-c \cdot B}^{c \cdot B} = \frac{2}{\pi} \cdot \operatorname{Arctan}(c \cdot B).$$

However, for each such $c > 0$, there exists some $B_c > 0$ such that

$$|\operatorname{Arctan}(c \cdot B) - (\pi/2)| < \varepsilon,$$

for every $B > B_c$, because $\lim_{t \to \infty} \operatorname{Arctan}(t) = \pi/2$. Thus,

$$\left| \frac{1}{\sqrt{2\pi}} \int_{-c}^{c} A_B(w)\, dw - 1 \right| = \left| \frac{2}{\pi} \cdot \operatorname{Arctan}(c \cdot B) - 1 \right| < \frac{2\varepsilon}{\pi}.$$

Hence, with M defined such that $|f(x)| \leq M$ for every $x \in \mathbb{R}$,

$$|(f * A_B)(r) - 1 \cdot f(r)|$$

$$= \left| \frac{1}{\sqrt{2\pi}} \int_{\mathbb{R}} f(r - s) A_B(s)\, ds - \frac{1}{\sqrt{2\pi}} \int_{\mathbb{R}} A_B(s)\, ds \cdot f(r) \right|$$

$$= \left| \frac{1}{\sqrt{2\pi}} \int_{\mathbb{R}} [f(r - s) - f(r)] \cdot A_B(s)\, ds \right|$$

$$\leq \frac{1}{\sqrt{2\pi}} \int_{\mathbb{R}} |f(r - s) - f(r)| \cdot |A_B(s)|\, ds$$

$$= \frac{1}{\sqrt{2\pi}} \int_{-c}^{c} |f(r - s) - f(r)| \cdot A_B(s)\, ds$$

$$+ \frac{1}{\sqrt{2\pi}} \int_{\mathbb{R} \setminus [-c,c]} |f(r - s) - f(r)| \cdot A_B(s)\, ds$$

$$< \frac{1}{\sqrt{2\pi}} \int_{-c}^{c} \varepsilon \cdot A_B(s) \, ds$$

$$+ \frac{1}{\sqrt{2\pi}} \int_{\mathbb{R}\setminus[-c,c]} \{|f(r-s)| + |f(r)|\} \cdot A_B(s) \, ds$$

$$< \frac{1}{\sqrt{2\pi}} \int_{-c}^{c} \varepsilon \cdot A_B(s) \, ds + \frac{1}{\sqrt{2\pi}} \int_{\mathbb{R}\setminus[-c,c]} 2M \cdot A_B(s) \, ds$$

$$= \varepsilon \cdot \frac{2}{\pi} \cdot \operatorname{Arctan}(c \cdot B) + 2M \cdot \frac{2}{\pi} [(\pi/2) - \operatorname{Arctan}(c \cdot B)]$$

$$< (1 + 2M \cdot 2/\pi) \cdot \varepsilon. \qquad \square$$

The following theorem gives an **inverse Fourier transform** for some continuous, bounded, and integrable functions.

Theorem 7.14   *For each bounded and integrable function* $f : \mathbb{R} \to \mathbb{C}$ *with* f *continuous at* r *and* $\mathcal{F}f$ *also integrable, the following formula reconstructs* f *from its Fourier transform:*

$$f(r) = \frac{1}{\sqrt{2\pi}} \int_{\mathbb{R}} (\mathcal{F}f)(w) \cdot e^{i \cdot w \cdot r} \, dw.$$

PROOF: If f is bounded and continuous at r, Lemma 7.13 gives

$$f(r) = \lim_{B \to 0} (f * A_B)(r)$$

$$= \lim_{B \to 0} \frac{1}{\sqrt{2\pi}} \int_{\mathbb{R}} (\mathcal{F}f)(s) K_B(s) e^{i \cdot r \cdot s} \, ds$$

$$= \frac{1}{\sqrt{2\pi}} \int_{\mathbb{R}} (\mathcal{F}f)(s) \cdot 1 \cdot e^{i \cdot r \cdot s} \, ds.$$

The last equality follows from the integrability of $\mathcal{F}f$ and the uniform convergence of K_B to 1 near the origin. Specifically, by definition of the integrability of $\mathcal{F}f$, for each real $\varepsilon > 0$ there exists a real $R_\varepsilon > 0$ such that for each $R > R_\varepsilon$;

$$\frac{1}{\sqrt{2\pi}} \int_{\mathbb{R}\setminus[-R,R]} |(\mathcal{F}f)(s)| \, ds < \varepsilon.$$

Then by definition of K_B, for the same R_ε there exists a real $B_\varepsilon > 0$, for instance $B_\varepsilon := (R_\varepsilon + 1)/\varepsilon$, such that for each $B > B_\varepsilon$ and each t with $|t| \leq R_\varepsilon + 1$,

$$|K_B(t) - 1| = \left| e^{-|t|/B} - 1 \right| = \left| \int_0^{|t|/B} -e^{-s} \, ds \right| \leq \frac{|t|}{B} < \varepsilon.$$

Therefore, with $R := R_\varepsilon + 1$ and for every $B > B_\varepsilon$,

$$\left| \frac{1}{\sqrt{2\pi}} \int_{\mathbb{R}} (\mathcal{F}f)(s) \cdot K_B(s) \cdot e^{i \cdot s \cdot t} \, ds - \frac{1}{\sqrt{2\pi}} \int_{\mathbb{R}} (\mathcal{F}f)(s) \cdot e^{i \cdot s \cdot t} \, ds \right|$$

$$= \frac{1}{\sqrt{2\pi}} \left| \int_{\mathbb{R}} (\mathcal{F}f)(s)\{K_B(s) - 1\}e^{i \cdot s \cdot t} ds \right|$$

$$= \frac{1}{\sqrt{2\pi}} \left| \int_{[-R,R]} (\mathcal{F}f)(s)\{K_B(s) - 1\}e^{i \cdot s \cdot t} ds \right.$$

$$\left. + \int_{\mathbb{R}\setminus[-R,R]} (\mathcal{F}f)(s)\{K_B(s) - 1\}e^{i \cdot s \cdot t} ds \right|$$

$$< \frac{1}{\sqrt{2\pi}} \left\{ \int_{[-R,R]} |(\mathcal{F}f)(s)| \cdot \varepsilon \, ds + \int_{\mathbb{R}\setminus[-R,R]} |(\mathcal{F}f)(s) \cdot 1| \, ds \right\}$$

$$= \frac{1}{\sqrt{2\pi}} \{\|\mathcal{F}f\|_1 \cdot \varepsilon + \varepsilon\}. \qquad\qquad \Box$$

For functions f that are not bounded, the proof of an inverse Fourier transform will require additional considerations, as done in subsequent sections.

EXERCISES

Determine the following inverse Fourier transforms through previous examples.

Exercise 7.3. Calculate $(\mathcal{F}^{-1}A)(x)$ for the function A defined by

$$A(w) := \sqrt{\frac{2}{\pi}} \cdot \frac{1}{1 + w^2}.$$

Exercise 7.4. Calculate $(\mathcal{F}^{-1}F)(x)$ for the function F defined by

$$F(w) := \{\text{sinc } (w/2)\}^2 = \begin{cases} \{[\sin(w/2)]/(w/2)\}^2 & \text{if } w \neq 0, \\ 1 & \text{if } w = 0. \end{cases}$$

7.3   APPROXIMATE IDENTITIES

Mathematical "approximate identities" are sets or sequences of functions with integrals concentrated on smaller and smaller intervals, so that their convolutions can approximate individual values of any other function to any specified accuracy. In practice, such individual values may correspond to measurements that can suffer from inaccuracies, for instance, caused by limitations in the measuring apparatus or unforeseen external disturbances. Nevertheless, such inaccuracies may cancel one another on the average, so that the average of many measurements—a convolution with an approximate identity—can be more accurate than any single measurement.

7.3.1   Weight Functions

The usual average of a continuous function $f : [a, b] \to \mathbb{R}$ over a closed and bounded interval $[a, b]$ with $a < b$ is the integral

$$\frac{1}{b-a} \int_a^b f(t) \, dt = \int_a^b f(t) \cdot \frac{1}{b-a} \, dt.$$

Such an average "assigns" the same "weight" $1/(b - a)$ to each value $t \in [a, b]$. Yet many situations require a larger weight near a particular location of interest and smaller weights farther away. Hence arises a more general and useful concept of weighted average.

Definition 7.15   With one dimension, a **weight function** is a function $w : \mathbb{R} \to \mathbb{C}$ such that $\int_\mathbb{R} |w(t)| \, dt < \infty$ and such that

$$\int_\mathbb{R} w(t) \, dt = 1. \qquad \Box$$

Specific applications may impose further conditions upon the weight functions appropriate for the task at hand, for instance, nonnegativity, continuity, or differentiability.

Example 7.16   The function $h : \mathbb{R} \to \mathbb{C}$—with h for "hat"—defined by

$$h(t) := \begin{cases} 0 & \text{if } t < -1, \\ 1+t & \text{if } -1 \le t < 0, \\ 1-t & \text{if } 0 \le t < 1, \\ 0 & \text{if } 1 \le t, \end{cases} \qquad (7.1)$$

is a weight function, because $\int_\mathbb{R} h(t) \, dt = 1$:

$$\int_\mathbb{R} h(t) \, dt = \int_{-1}^0 (1+t) \, dt + \int_0^1 (1-t) \, dt = 2 \int_0^1 (1-t) \, dt$$

$$= 2 \left(t - \frac{t^2}{2} \right) \Big|_0^1 = 2 \left(1 - \frac{1}{2} \right) = 1.$$

Also, the weight function h is nonnegative and continuous. $\qquad \Box$

The weight function in Example 7.16 assigns the largest weight to the origin, 0, and no weight outside the interval $[-1, 1]$. Still, many situations require the possibility of adjusting the relative size of the largest weights and the interval where most of the weight lies.

Example 7.17   For each $c > 0$, the function $h_c : \mathbb{R} \to \mathbb{C}$ with

$$h_c(t) := \begin{cases} 0 & \text{if } t < -c, \\ \frac{1}{c} \cdot \left(1 + \frac{t}{c} \right) & \text{if } -c \le t < 0, \\ \frac{1}{c} \cdot \left(1 - \frac{t}{c} \right) & \text{if } 0 \le t < c, \\ 0 & \text{if } c \le t \end{cases} \qquad (7.2)$$

is a weight function, because $\int_{\mathbb{R}} h_c(t)\, dt = 1$:

$$\int_{\mathbb{R}} h_c(t)\, dt = \int_{-c}^{0} \frac{1}{c} \cdot \left(1 + \frac{t}{c}\right) dt + \int_{0}^{c} \frac{1}{c} \cdot \left(1 - \frac{t}{c}\right) dt$$

$$= 2 \int_{0}^{c} \frac{1}{c} \cdot \left(1 - \frac{t}{c}\right) dt = 2 \frac{1}{c} \cdot \left(t - \frac{t^2}{2c}\right)\Big|_{0}^{c} = 2 \frac{1}{c} \left(c - \frac{c^2}{2c}\right) = 1.$$

Thus, h_c is a weight function that assigns the largest weight to the origin, $t = 0$, with no weight outside $[-c, c]$. Moreover, the weight function h_c is also nonnegative and continuous. □

Weight functions may assign most of their weight not only near the origin, as in Example 7.17, but near any location $x \in \mathbb{R}$. Indeed, with any weight function w that assigns most of its weight near the origin, the sign reversal and the translation (shift) defined by $t \mapsto u - t$ produce by composition of functions a weight function $t \mapsto w(u - t)$ that assigns most of its weight near u.

Example 7.18 Consider the weight function h defined by equation (7.1) in Example 7.16. For each $u \in \mathbb{R}$,

$$h(u - t) = \begin{cases} 0 & \text{if } u - t < -1, \text{ that is, if } u + 1 < t, \\ 1 + (u - t) & \text{if } -1 \leq u - t < 0, \text{ that is, if } u < t \leq u + 1, \\ 1 - (u - t) & \text{if } 0 \leq u - t < 1, \text{ that is, if } u - 1 < t \leq u, \\ 0 & \text{if } 1 \leq u - t, \text{ that is, if } t \leq u - 1. \end{cases}$$

The shifted function remains a weight function, because the change of variable $s := u - t$ gives

$$\int_{\mathbb{R}} |h(u - t)|\, dt = \int_{u-1}^{u+1} h(u - t)\, dt = \int_{-1}^{1} h(s)\, ds = 1. \quad □$$

Example 7.19 For each $c > 0$, consider the function $h_c : \mathbb{R} \to \mathbb{C}$ defined in Example 7.17 by equation (7.2). For each $u \in \mathbb{R}$ the composition of the weight function h_c and the change of variable $t \mapsto u - t$ produce the shifted weight function with

$$h_c(u - t) = \begin{cases} 0 & \text{if } u - t < -c, \text{ that is, if } u + c < t, \\ \frac{1}{c} \cdot \left(1 + \frac{u-t}{c}\right) & \text{if } -c \leq u - t < 0, \text{ that is, if } u < t \leq u + c, \\ \frac{1}{c} \cdot \left(1 - \frac{u-t}{c}\right) & \text{if } 0 \leq u - t < c, \text{ that is, if } u - c < t \leq u, \\ 0 & \text{if } c \leq u - t, \text{ that is, if } t \leq u - c. \end{cases} \quad □$$

7.3.2 Approximate Identities

Any collection of such weight functions as h_c in Example 7.17 that allows for the adjustment of the interval containing most of the weight bears the name of "approximate identity" for reasons explained below.

Definition 7.20   An **approximate identity** [12, p. 236], also called an **approximate unit** [21, p. 400], is a set of integrable weight functions, $\{w_c : c > 0\}$, such that the smaller the value of c, the smaller the interval in which the weight function w_c assigns most of the weight—for each interval $[-R, R]$ with $R > 0$,

$$\lim_{c \to 0} \int_{-R}^{R} w_c(t)\, dt = 1,$$

$$\lim_{c \to 0} \int_{\mathbb{R} \setminus [-R,R]} |w_c(t)|\, dt = 0,$$

—and such that some $M > 0$ exists for which $\int_{\mathbb{R}} |w_c(t)|\, dt \leq M$ for every $c > 0$. Moreover, this exposition considers only approximate identities such that for each real $\varepsilon > 0$ and each $R > 0$, there exists a real $\delta > 0$ with

$$|w_c(t)| < \varepsilon$$

for every $c < \delta$ and every $|t| > R$. (Thus, as c tends to 0, w_c converges uniformly to 0 away from each interval containing the origin.)   □

Example 7.21   The set $\{h_c : c > 0\}$ defined by equation (7.2) in Example 7.17 forms an approximate identity. Indeed, for each $R > 0$, if $0 < c < R$, then

$$\int_{-R}^{R} |h_c(t)|\, dt = \int_{-R}^{R} h_c(t)\, dt = \int_{-c}^{c} h_c(t)\, dt = 1.$$

Thus, if $0 < c < R$, then every weight function h_c assigns the entire weight to the interval $[-R, R]$, in the sense that $\int_a^b h_c(t)\, dt = 0$ for every interval $[a, b] \subset \,]-\infty, -R[\, \cup \,]R, \infty[$.   □

The name "approximate identity" arises from the fact that averaging a function f with weight functions $t \mapsto w_c(x - t)$ approximates the value $f(x)$ with increasing accuracy as c tends to zero.

Proposition 7.22   *For each integrable function $f : \mathbb{R} \to \mathbb{C}$ continuous at $t = x$, and for each approximate identity $\{w_c : c > 0\}$,*

$$f(x) = \lim_{c \to 0} \int_{\mathbb{R}} f(t) \cdot w_c(x - t)\, dt = \lim_{c \to 0} \sqrt{2\pi}(f * w_c)(x).$$

PROOF: The definition of the continuity of f at x means that for each $\varepsilon > 0$ some $\delta > 0$ exists such that if $|t - x| < \delta$, then

$$|f(t) - f(x)| < \varepsilon.$$

Consequently, if $0 < R < \delta$ and $0 < c < \delta$, and if $x - c < t < x + c$, then $|t - x| < c < \delta$, whence

$$\left| 1 \cdot f(x) - \int_{\mathbb{R}} f(t) w_c(x - t)\, dt \right|$$

$$= \left| \int_{\mathbb{R}} f(x) w_c(t)\, dt - \int_{\mathbb{R}} f(t) w_c(x - t)\, dt \right|$$

$$= \left| \int_{\mathbb{R}} f(x) w_c(x - t)\, dt - \int_{\mathbb{R}} f(t) w_c(x - t)\, dt \right|$$

$$= \left| \int_{\mathbb{R}} \{ f(x) - f(t) \}\, w_c(x - t)\, dt \right|$$

$$\leq \int_{\mathbb{R}} |\{ f(x) - f(t) \}\, w_c(x - t)|\, dt$$

$$= \int_{[-R,R]} |\{ f(x) - f(t) \}\, w_c(x - t)|\, dt$$

$$+ \int_{\mathbb{R} \setminus [-R,R]} (|f(x)| w_c(x - t) + |f(t)| \varepsilon)\, dt$$

$$< \int_{\mathbb{R}} \varepsilon\, |w_c(x - t)|\, dt + (|f(x)| + \|f\|_1) \cdot \varepsilon$$

$$\leq \varepsilon\, (M + |f(x)| + \|f\|_1),$$

which tends to zero as ε tends to zero. Thus,

$$\lim_{c \to 0} \left| f(x) - \int_{\mathbb{R}} f(t) w_c(x - t)\, dt \right| = 0,$$

and

$$f(x) = \lim_{c \to 0} \int_{\mathbb{R}} f(t) w_c(x - t)\, dt = \lim_{c \to 0} \sqrt{2\pi}\, (f * w_c)(x). \qquad \square$$

The following results show that many weight functions can serve as building blocks for an approximate identity.

Proposition 7.23 *For each integrable weight function $w : \mathbb{R} \to \mathbb{C}$ such that $\lim_{|t| \to \infty} w(t) = 0$, the set $\{ w_c : c > 0 \}$ defined by*

$$w_c(t) := \frac{1}{c} \cdot w \left(\frac{t}{c} \right)$$

constitutes an approximate identity.

PROOF: The limit $\lim_{|t| \to \infty} w(t) = 0$ means that for each real $\varepsilon > 0$, there exists a real $T_\varepsilon > 0$ such that $|w(t)| < \varepsilon$ for every t with $|t| > T_\varepsilon$. Hence, for each $c > 0$

and each t with $|t| > c(T_{c\varepsilon})$, so that $|t/c| > T_{c\varepsilon}$,

$$|w_c(t)| = \left| \frac{1}{c} w \left(\frac{t}{c} \right) \right| < \frac{1}{c} c\varepsilon = \varepsilon.$$

Thus $\lim_{|t| \to \infty} w_c(t) = 0$. Also, the integrability of w means that

$$\|w\|_1 := \int_{\mathbb{R}} |w(t)| \, dt := \lim_{R \to \infty} \int_{-R}^{R} |w(t)| \, dt < \infty.$$

Consequently, for each real $\varepsilon > 0$, there exists a real $R_\varepsilon > 0$ such that for every real $R > R_\varepsilon$,

$$\int_{\mathbb{R} \setminus [-R,R]} |w(t)| \, dt < \varepsilon.$$

Perform the change of variables $t := c \cdot s$ and $s := t/c$. For each real $R > 0$, if $0 < c < R/R_\varepsilon$, then $R_\varepsilon < R/c$, whence

$$\int_{\mathbb{R} \setminus [-R,R]} |w_c(t)| \, dt = \int_{\mathbb{R} \setminus [-R,R]} \left| \frac{1}{c} \cdot w \left(\frac{t}{c} \right) \right| \, dt$$

$$= \int_{\mathbb{R} \setminus [-R/c, R/c]} |w(s)| \, ds < \varepsilon.$$

The same change of variable confirms that

$$\int_{\mathbb{R}} w_c(t) \, dt = \int_{\mathbb{R}} \frac{1}{c} \cdot w \left(\frac{t}{c} \right) \, dt = \int_{\mathbb{R}} w(s) \, ds = 1.$$

Similarly, $\|w_c\|_1 = \|w\|_1$ for each $c > 0$:

$$\|w_c\|_1 = \int_{\mathbb{R}} |w_c(t)| \, dt = \int_{\mathbb{R}} \left| \frac{1}{c} \cdot w \left(\frac{t}{c} \right) \right| \, dt = \int_{\mathbb{R}} |w(s)| \, ds = \|w\|_1. \quad \square$$

EXERCISES

Exercise 7.5. Verify that for each positive real $B > 0$ the multiple of Abel's kernel $A_B : \mathbb{R} \to \mathbb{C}$ defined by

$$\frac{1}{\sqrt{2\pi}} A_B(w) = \frac{1}{\pi} \cdot \frac{B}{1 + B^2 w^2}$$

is a weight function on $\mathbb{R}$, so that $(1/\sqrt{2\pi}) \int_{-\infty}^{\infty} A_B(w) \, dw = 1$.

Exercise 7.6. Verify that the Gaussian distribution $g : \mathbb{R} \to \mathbb{C}$ defined by

$$g(t) := \frac{1}{\sqrt{2\pi}} \cdot e^{-[(t^2)/2]}$$

is a weight function on $\mathbb{R}$, and consequently, that the functions defined by

$$g_c(t) := \frac{1}{c} \cdot g\left(\frac{t}{c}\right) = \frac{1}{\sqrt{2\pi c^2}} \cdot e^{-[(t^2)/(2c^2)]}$$

form an approximate identity.

Exercise 7.7. Verify that the exponential kernel $K : \mathbb{R} \to \mathbb{C}$ defined by

$$K(x) := \left(\frac{1}{2}\right) \cdot e^{-|x|}$$

is a weight function on $\mathbb{R}$, and consequently, that the functions defined by

$$K_c(x) := \frac{1}{c} \cdot K\left(\frac{x}{c}\right) = \frac{1}{2c} \cdot e^{-|x/c|}$$

form an approximate identity.

Exercise 7.8. Identify weight functions and approximate identities that have already appeared in previous chapters.

7.3.3   Dirac Delta (δ) Function

As the proof of the inversion of Fourier series for periodic functions, the foregoing proof of the inversion of the Fourier transform for continuous integrable functions by means of the limit

$$f = \lim_{B \to \infty} f * A_B$$

also shows that Abel's kernels form an approximate identity.

A plot provides a graphical interpretation of such a convergence. As B increases, Abel's kernel A_B decreases in magnitude away from the origin but increases in magnitude near the origin, as in Figure 7.3. Consequently, as B increases, the integrand $f(s) \cdot A_B(r - s)$ weighs the values $f(s)$ near $s := r$ more heavily than the values away from $s := r$, eventually reproducing the value $f(r)$ at the limit. The operation that associates a function f to its value $f(r)$ bears a special name [35, p. 19], recalled here for convenience and emphasis.

Definition 7.24   (Dirac's distribution) δ is the operator such that for every function f defined at the origin,

$$\delta(f) := f(0).$$

Similarly, the shift of δ by r is the operator δ_r such that for every function f defined at r, $\delta_r(f) := f(r)$.   □

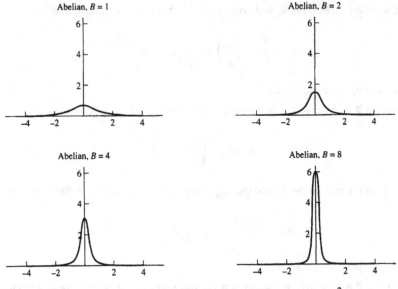

Figure 7.3   Abel's kernels: $A_B(w) = (\sqrt{2/\pi})B/[1 + (Bw)^2]$.

Remark 7.25   Dirac's distribution δ is *not* a function defined on the real line. Instead, δ is a function from the set of functions defined at the origin to the set of values of such functions:

$$\delta : f \mapsto f(0).$$

The convergence

$$\delta = \lim_{B \to \infty} A_B$$

then does not occur in the space of functions on the real line, but in a space of functions similar to δ—called "distributions"—which associate functions to numbers. [35, p. 74].   □

7.4   FURTHER FEATURES OF THE FOURIER TRANSFORM

This section establishes properties of the Fourier transform that will prove useful in deriving specific Fourier transforms not easily amenable to direct calculation, to establish the existence of Daubechies wavelets, and to interpret practical applications, for instance, Shannon's sampling theorem and Heisenberg's uncertainty principle.

7.4.1   Algebraic Features of the Fourier Transform

The following results demonstrate how such various algebraic operations as shifts and multiplications change the Fourier transforms of functions. The same results will prove useful in several subsequent applications.

Definition 7.26   For each function $f : \mathbb{R} \to \mathbb{C}$, and for each number $u \in \mathbb{R}$, denote by $T_u f$ the **shift**—or **translation**—of f by u, defined by

$$(T_u f)(x) := f(x - u).$$

Thus, the graph of $T_u f$ is the graph of f shifted by u to the right. Similarly, denote by $E_u f$ the product of f and w_u, defined by

$$(E_u f)(x) := f(x) \cdot e^{iux}. \qquad \Box$$

The following two lemmas show how a shift along the real line affects the Fourier transform.

Lemma 7.27   *For each $u \in \mathbb{R}$ and each integrable function $f : \mathbb{R} \to \mathbb{C}$, if $h(x) := f(x) \cdot e^{iux}$, then*

$$(\mathcal{F}h)(w) = (\mathcal{F}f)(w - u).$$

In other words, the Fourier transform conjugates the operator E_u of multiplication by e^{iux} and the operator T_u of translation by u:

$$\mathcal{F} \circ E_u = T_u \circ \mathcal{F}.$$

PROOF: Direct calculations confirm the result:

$$
\begin{aligned}
(\mathcal{F}h)(w) &:= \frac{1}{\sqrt{2\pi}} \int_{\mathbb{R}} h(x) \cdot e^{-i \cdot w \cdot x} \, dx \\
&= \frac{1}{\sqrt{2\pi}} \int_{\mathbb{R}} f(x) \cdot e^{iux} \cdot e^{-i \cdot w \cdot x} \, dx \\
&= \frac{1}{\sqrt{2\pi}} \int_{\mathbb{R}} f(x) \cdot e^{-i \cdot (w-u) \cdot x} \, dx \\
&= (\mathcal{F}f)(w - u). \qquad \Box
\end{aligned}
$$

Lemma 7.28   *For each $u \in \mathbb{R}$ and each integrable function $f : \mathbb{R} \to \mathbb{C}$, if $h(x) := f(x - u)$, then*

$$(\mathcal{F}h)(w) = (\mathcal{F}f)(w) \cdot e^{-iuw}.$$

In other words, the Fourier transform conjugates the operator E_{-u} of multiplication by e^{-iux} and the operator T_u of translation by u:

$$\mathcal{F} \circ T_u = E_{-u} \circ \mathcal{F}.$$

PROOF: Direct calculations confirm the result:

$$(\mathcal{F}h)(w) := \frac{1}{\sqrt{2\pi}} \int_{\mathbb{R}} h(x) \cdot e^{-i \cdot w \cdot x}\, dx = \frac{1}{\sqrt{2\pi}} \int_{\mathbb{R}} f(x-u) \cdot e^{-i \cdot w \cdot x}\, dx$$

$$= \frac{1}{\sqrt{2\pi}} \int_{\mathbb{R}} f(z) \cdot e^{-i \cdot w \cdot (z+u)}\, dx = (\mathcal{F}f)(w) \cdot e^{-iuw}. \qquad \square$$

The following lemma and theorem establish an inversion of the Fourier transform more general than Theorem 7.14.

Lemma 7.29   *For all functions $f, g : \mathbb{R} \to \mathbb{C}$ such that f and g are integrable,*

$$\int_{\mathbb{R}} (\mathcal{F}f)(s) \cdot g(s)\, ds = \int_{\mathbb{R}} f(t) \cdot (\mathcal{F}g)(t)\, dt.$$

PROOF: By Fubini's theorem on the permutation of integrations,

$$\int_{\mathbb{R}} (\mathcal{F}f)(s) \cdot g(s)\, ds = \int_{\mathbb{R}} \left(\frac{1}{\sqrt{2\pi}} \int_{\mathbb{R}} f(t) \cdot e^{-ist}\, dt \right) \cdot g(s)\, ds$$

$$= \int_{\mathbb{R}} \left(\frac{1}{\sqrt{2\pi}} \int_{\mathbb{R}} g(s) \cdot e^{-ist}\, ds \right) \cdot f(t)\, dt$$

$$= \int_{\mathbb{R}} (\mathcal{F}g)(t) \cdot f(t)\, dt. \qquad \square$$

The following theorem shows that the Fourier transform is invertible also for unbounded functions.

Theorem 7.30   *For every function $f : \mathbb{R} \to \mathbb{C}$ such that f and $\mathcal{F}f$ are integrable, if f is continuous at t, then*

$$f(t) = \frac{1}{\sqrt{2\pi}} \int_{\mathbb{R}} (\mathcal{F}f)(s) \cdot e^{i \cdot s \cdot t}\, ds.$$

PROOF: This proof follows the outline of the simpler version for bounded functions in Theorem 7.14, but uses the results just established about approximate identities. With the exponential kernels K_B defined by $K_B(t) := e^{-|t/B|}$, the Fourier transforms $A_B = \mathcal{F}K_B$ form an approximate identity as B increases, so that $c := 1/B$ tends to zero, whence

$$f(t) = \lim_{B \to \infty} (f * A_B)(t)$$

$$= \lim_{B \to \infty} \frac{1}{\sqrt{2\pi}} \int_{\mathbb{R}} f(x) \cdot (\mathcal{F}K_B)(t-x)\, dx$$

$$= \lim_{B \to \infty} \frac{1}{\sqrt{2\pi}} \int_{\mathbb{R}} f(x) \cdot (\mathcal{F}K_B)(x-t)\, dx$$

$$= \lim_{B \to \infty} \frac{1}{\sqrt{2\pi}} \int_{\mathbb{R}} f(x) \cdot (T_t \circ \mathcal{F} K_B)(x)\,dx$$

$$= \lim_{B \to \infty} \frac{1}{\sqrt{2\pi}} \int_{\mathbb{R}} f(x) \cdot (\mathcal{F} \circ E_t K_B)(x)\,dx$$

$$= \lim_{B \to \infty} \frac{1}{\sqrt{2\pi}} \int_{\mathbb{R}} (\mathcal{F} f)(s) \cdot (E_t K_B)(s)\,ds$$

$$= \lim_{B \to \infty} \frac{1}{\sqrt{2\pi}} \int_{\mathbb{R}} (\mathcal{F} f)(s) \cdot K_B(s) e^{i \cdot s \cdot t}\,ds.$$

$$= \frac{1}{\sqrt{2\pi}} \int_{\mathbb{R}} (\mathcal{F} f)(s) \cdot e^{i \cdot s \cdot t}\,ds.$$

The last equality follows from the integrability of $\mathcal{F} f$ and the uniform convergence of K_B to 1 near the origin, exactly as in the proof of Theorem 7.14.   □

7.4.2   Metric Features of the Fourier Transform

This subsection demonstrates a few equalities and inequalities between norms of functions and norms of their Fourier transforms, based in part upon the following result.

Proposition 7.31   *If $f, g : \mathbb{R} \to \mathbb{C}$ are integrable, then*

$$\mathcal{F}(f * g) = (\mathcal{F} f) \cdot (\mathcal{F} g).$$

PROOF: Fubini's theorem and direct calculations give

$$[\mathcal{F}(f * g)](w) = \frac{1}{\sqrt{2\pi}} \int_{\mathbb{R}} (f * g)(x) \cdot e^{-i \cdot w \cdot x}\,dx$$

$$= \frac{1}{\sqrt{2\pi}} \int_{\mathbb{R}} \left[\frac{1}{\sqrt{2\pi}} \int_{\mathbb{R}} f(x - v) g(v)\,dv \right] \cdot e^{-i \cdot w \cdot x}\,dx$$

$$= \frac{1}{\sqrt{2\pi}} \int_{\mathbb{R}} \left[\frac{1}{\sqrt{2\pi}} \int_{\mathbb{R}} f(x - v) e^{-i \cdot w \cdot (x-v)}\,dx \right] g(v) \cdot e^{-i \cdot w \cdot v}\,dv$$

$$= [(\mathcal{F} f)(w)] \cdot [(\mathcal{F} g)(w)].$$ □

Remark 7.32   Many results from Fourier analysis hold for integrable piecewise continuous functions. Fourier analysis also allows for integrable functions that may have vertical asymptotes, and thus are not necessarily defined at every point on the real line. In such contexts, the notation $f : \mathbb{R} \to \mathbb{C}$ then means that the domain of f is a subset, but not necessarily all, of $\mathbb{R}$, just as its range is a subset, but not necessarily all, of $\mathbb{C}$.   □

Definition 7.33   The space $\mathcal{L}^1(\mathbb{R}, \mathbb{C})$ consists of all integrable functions $f : \mathbb{R} \to \mathbb{C}$.

The space $\mathcal{L}^2(\mathbb{R}, \mathbb{C})$ consists of all functions $f : \mathbb{R} \to \mathbb{C}$ the square of which, $x \mapsto [f(x)]^2$, is integrable.

The space $\mathcal{L}^\infty(\mathbb{R}, \mathbb{C})$ consists of all *bounded* functions $f : \mathbb{R} \to \mathbb{C}$. □

The following theorem shows that the Fourier transform preserves the "energy" norm $\|f\|_2^2 := \int_{\mathbb{R}} |f(t)|^2 \, dt$.

Theorem 7.34 **(Plancherel's Identity.)** *If $f \in \mathcal{L}^1 \cap \mathcal{L}^2$, then*

$$\|\mathcal{F}f\|_2 = \|f\|_2.$$

PROOF: With the notation $f_-(x) := f(-x)$, let $h := f * [(\bar{f})_-]$. Also, $(\bar{f})_-(x) := \overline{f(-x)}$, so that $\mathcal{F}[(\bar{f})_-] = \overline{\mathcal{F}f}$. Thus, $h \in \mathcal{L}^1$, $h \in C^0$, $h \in \mathcal{L}^\infty$, and

$$\mathcal{F}h = \mathcal{F}\{f * [(\bar{f})_-]\} = \mathcal{F}f \cdot \mathcal{F}[(\bar{f})_-] = (\mathcal{F}f) \cdot (\overline{\mathcal{F}f}) = |\mathcal{F}f|^2 \geq 0.$$

The Fourier inversion theorem then gives $h = \mathcal{F}^{-1}(\mathcal{F}h)$, whence

$$\frac{1}{\sqrt{2\pi}} \int_{\mathbb{R}} f(x+w)\overline{f(w)} \, dw = \frac{1}{\sqrt{2\pi}} \int_{\mathbb{R}} f(x-w)\overline{f(-w)} \, dw$$

$$= f * [(\bar{f})_-](x) = h(x) = [\mathcal{F}^{-1}(\mathcal{F}h)](x)$$

$$= \frac{1}{\sqrt{2\pi}} \int_{\mathbb{R}} (\mathcal{F}h)(w)e^{iwx} \, dw$$

$$= \frac{1}{\sqrt{2\pi}} \int_{\mathbb{R}} (\mathcal{F}\{f * [(\bar{f})_-]\})(w)e^{iwx} \, dw$$

$$= \frac{1}{\sqrt{2\pi}} \int_{\mathbb{R}} [(\mathcal{F}f) \cdot (\overline{\mathcal{F}f})](w)e^{iwx} \, dw$$

$$= \frac{1}{\sqrt{2\pi}} \int_{\mathbb{R}} |\mathcal{F}f|^2(w)e^{iwx} \, dw.$$

Setting $x := 0$ yields the conclusion, $\|f\|_2^2 = \|\mathcal{F}f\|_2^2$. □

Corollary 7.35 *In $\mathcal{L}^1 \cap \mathcal{L}^2$, $\langle f, g \rangle = \langle \mathcal{F}f, \mathcal{F}g \rangle$.*

PROOF: Apply the polar identity, or proceed as follows. With the notation $p_-(x) := p(-x)$, set $h := f * (\bar{g}_-)$ in the proof of Plancherel's identity:

$$\frac{1}{\sqrt{2\pi}} \int_{\mathbb{R}} f(x+w)\overline{g(w)} \, dw = \frac{1}{\sqrt{2\pi}} \int_{\mathbb{R}} f(x-w)\overline{g(-w)} \, dw$$

$$= [f * (\bar{g}_-)](x) = h(x) = [\mathcal{F}^{-1}(\mathcal{F}h)](x)$$

$$= \frac{1}{\sqrt{2\pi}} \int_{\mathbb{R}} (\mathcal{F}h)(w)e^{iwx} \, dw$$

$$= \frac{1}{\sqrt{2\pi}} \int_{\mathbb{R}} (\mathcal{F}[f * (\bar{g}_-)])(w) e^{iwx} \, dw$$

$$= \frac{1}{\sqrt{2\pi}} \int_{\mathbb{R}} [(\mathcal{F}f) \cdot (\overline{\mathcal{F}g})](w) e^{iwx} \, dw.$$

Setting $x := 0$ yields the conclusion, $\langle f, g \rangle = \langle \mathcal{F}f, \mathcal{F}g \rangle$. $\qquad\square$

Remark 7.36 The foregoing result, $\langle f, g \rangle = \langle \mathcal{F}f, \mathcal{F}g \rangle$, means that the Fourier transform is a *unitary* operator

$$\mathcal{F}: (\mathcal{L}^1 \cap \mathcal{L}^2)(\mathbb{R}, \mathbb{C}) \to (\mathcal{L}^1 \cap \mathcal{L}^2)(\mathbb{R}, \mathbb{C}),$$

in other words, a *rotation* in the space $(\mathcal{L}^1 \cap \mathcal{L}^2)(\mathbb{R}, \mathbb{C})$ of integrable and square-integrable functions. $\qquad\square$

The following proposition will prove useful in deriving the Fourier transform of Gaussian distributions, and in the proof of Shannon's sampling theorem.

Proposition 7.37 *Denote by I the identity function $I : \mathbb{R} \to \mathbb{C}$ defined by $I(w) = w$. If $f : \mathbb{R} \to \mathbb{C}$ and $I \cdot f$ are integrable, then the Fourier transform $\mathcal{F}f$ is differentiable, and*

$$[D(\mathcal{F}f)](w) = -i \cdot (\mathcal{F}[I \cdot f])(w).$$

PROOF: Thanks to the integrability of $I \cdot f$, for each $\varepsilon > 0$ there exists $R > 0$ such that

$$\int_{\mathbb{R}\setminus[-R,R]} |x \cdot f(x)| \, dx < \frac{\varepsilon}{4}.$$

The following inequalities will prove useful for the interval $[-R, R]$. The fundamental theorem of calculus yields the following upper bound:

$$\left| e^{-i \cdot (w+h) \cdot x} - e^{-i \cdot w \cdot x} \right| = \left| e^{-i \cdot w \cdot x} \right| \cdot \left| \int_0^h (-i \cdot x) \cdot e^{-i \cdot t \cdot x} \, dt \right|$$

$$\leq \int_0^h \left| (-i \cdot x) \cdot e^{-i \cdot t \cdot x} \right| \, dt = \int_0^h |-i \cdot x| \, dt = |h \cdot x|.$$

Consequently,

$$\int_{\mathbb{R}\setminus[-R,R]} |f(x)| \cdot \left| \frac{e^{-i \cdot (w+h) \cdot x} - e^{-i \cdot w \cdot x}}{h} - (-i \cdot x) \cdot e^{-i \cdot w \cdot x} \right| \, dx$$

$$\leq \int_{\mathbb{R}\setminus[-R,R]} |f(x)| \cdot \left(\frac{|h \cdot x|}{h} + |x| \cdot 1 \right) \, dx$$

$$= \int_{\mathbb{R}\setminus[-R,R]} |f(x)| \cdot 2 \cdot |x|\, dx < \frac{\varepsilon}{2}.$$

Similarly,

$$\left| \frac{e^{-i\cdot(w+h)\cdot x} - e^{-i\cdot w\cdot x}}{h} - (-i\cdot x)\cdot e^{-i\cdot w\cdot x} \right|$$

$$= \left| \frac{1}{h}\int_0^h (-i\cdot x)\cdot e^{-i\cdot(w+t)\cdot x}\, dt - \frac{1}{h}\int_0^h (-i\cdot x)\cdot e^{-i\cdot w\cdot x}\, dt \right|$$

$$= \left| \frac{1}{h}\int_0^h (-i\cdot x)\cdot \left(e^{-i\cdot(w+t)\cdot x} - e^{-i\cdot w\cdot x} \right) dt \right|$$

$$\le \frac{1}{|h|}\int_0^{|h|} \left| x\cdot \left(e^{-i\cdot(w+t)\cdot x} - e^{-i\cdot w\cdot x} \right) \right| dt$$

$$= \frac{1}{|h|}\int_0^{|h|} \left| x\cdot \left(\int_0^t (-i\cdot x)\cdot e^{-i\cdot(w+s)\cdot x}\, ds \right) \right| dt$$

$$\le \frac{1}{|h|}\int_0^{|h|} |x|\cdot \left(\int_0^{|t|} \left| (-i\cdot x)\cdot e^{-i\cdot(w+s)\cdot x} \right| \right) ds\, dt$$

$$\le \frac{1}{|h|}\int_0^{|h|} |x|\cdot \left(\int_0^{|t|} |x|\, ds \right) dt$$

$$= \frac{1}{|h|}\int_0^{|h|} |x|\cdot|x|\cdot|t|\, dt = \frac{1}{|h|}\cdot|x|^2\cdot \frac{|t|^2}{2}\Bigg|_0^{|h|} = |x|^2\cdot \frac{|h|}{2}.$$

Let

$$A := \int_{-R}^{R} |x|^2\cdot|f(x)|\, dx \le R^2\|f\|_1.$$

Hence, if $|h| < \varepsilon/A$, then

$$\left| \frac{1}{\sqrt{2\pi}}\int_{-R}^{R} f(x)\cdot \frac{e^{-i\cdot(w+h)\cdot x} - e^{-i\cdot w\cdot x}}{h}\, dx \right.$$

$$\left. - \frac{1}{\sqrt{2\pi}}\int_{-R}^{R} f(x)\cdot(-ix)\cdot e^{-i\cdot w\cdot x}\, dx \right|$$

$$\le \frac{1}{\sqrt{2\pi}}\int_{-R}^{R} |f(x)|\cdot \left| \frac{e^{-i\cdot(w+h)\cdot x} - e^{-i\cdot w\cdot x}}{h} - (-i\cdot x)\cdot e^{-i\cdot w\cdot x} \right| dx$$

$$\le \frac{1}{\sqrt{2\pi}}\int_{-R}^{R} |f(x)|\cdot|x|^2\cdot \frac{|h|}{2}\, dx \le \frac{1}{\sqrt{2\pi}} A\cdot \frac{|h|}{2} < \frac{\varepsilon}{2}.$$

Finally,

$$\left| \frac{1}{\sqrt{2\pi}} \int_{\mathbb{R}} f(x) \cdot \frac{e^{-i \cdot (w+h) \cdot x} - e^{-i \cdot w \cdot x}}{h} \, dx - \frac{1}{\sqrt{2\pi}} \int_{\mathbb{R}} f(x) \cdot (-ix) \cdot e^{-i \cdot w \cdot x} \, dx \right|$$

$$= \left| \frac{1}{\sqrt{2\pi}} \int_{-R}^{R} f(x) \cdot \frac{e^{-i \cdot (w+h) \cdot x} - e^{-i \cdot w \cdot x}}{h} \, dx \right.$$

$$\left. - \frac{1}{\sqrt{2\pi}} \int_{-R}^{R} f(x) \cdot (-ix) \cdot e^{-i \cdot w \cdot x} \, dx \right|$$

$$+ \left| \frac{1}{\sqrt{2\pi}} \int_{\mathbb{R} \setminus [-R,R]} f(x) \cdot \frac{e^{-i \cdot (w+h) \cdot x} - e^{-i \cdot w \cdot x}}{h} \, dx \right.$$

$$\left. - \frac{1}{\sqrt{2\pi}} \int_{\mathbb{R} \setminus [-R,R]} f(x) \cdot (-ix) \cdot e^{-i \cdot w \cdot x} \, dx \right|$$

$$< \frac{\varepsilon}{2} + \frac{\varepsilon}{2} = \varepsilon.$$

The inequalities just established mean that

$$\{D(\mathcal{F}f)\}(w) = \lim_{h \to 0} \frac{1}{\sqrt{2\pi}} \int_{\mathbb{R}} f(x) \cdot \frac{e^{-i \cdot (w+h) \cdot x} - e^{-i \cdot w \cdot x}}{h} \, dx$$

$$= -i \cdot \{\mathcal{F}[I \cdot f]\}(w). \qquad \square$$

7.4.3 Uniform Continuity of Fourier Transforms

Besides providing information about the continuity of Fourier transforms, the following proposition will also be useful in the proof of the existence of Daubechies wavelets in a subsequent chapter.

Proposition 7.38 *For each integrable function g, the Fourier transform $\mathcal{F}g$ is uniformly continuous on $\mathbb{R}$.*

PROOF: By definition of the integrability of g, it follows that

$$\|g\|_1 := \int_{\mathbb{R}} |g(x)| \, dx = \lim_{R \to \infty} \int_{-R}^{R} |g(x)| \, dx < \infty.$$

Hence, for each real $\varepsilon > 0$ there exists a real $R_\varepsilon > 0$ such that for each $R > R_\varepsilon$,

$$\left| \int_{|x|>R} |g(x)| \, dx \right| = \left| \int_{\mathbb{R}} |g(x)| \, dx - \int_{-R}^{R} |g(x)| \, dx \right| < \frac{\varepsilon}{4}.$$

Moreover, by the continuity of the exponential function at the origin, for each real $\varepsilon > 0$ there exists a real $\delta_\varepsilon > 0$ such that for every $z \in \mathbb{C}$ with $|z| < \delta_\varepsilon$,

$$\left| e^z - e^0 \right| < \frac{\varepsilon}{2(\|g\|_1 + 1)} \leq \frac{\varepsilon}{2}.$$

Consequently, for all numbers $w \in \mathbb{R}$, $x \in \mathbb{R}$, and $r \in \mathbb{R}$ such that $|x| \leq R_\varepsilon$ and $|r| < \delta_\varepsilon / R_\varepsilon$,

$$\left| e^{-i(w+r)x} - e^{-iwx} \right| = \left| e^{-iwx} \right| \cdot \left| e^{-irx} - 1 \right| < \frac{\varepsilon}{2(\|g\|_1 + 1)} \leq \frac{\varepsilon}{2}.$$

For each $R > R_\varepsilon$, each $w \in \mathbb{R}$, and each r with $|r| < \delta_\varepsilon / R$,

$$\sqrt{2\pi} \; |(\mathcal{F}g)(w + r) - (\mathcal{F}g)(w)|$$

$$= \left| \int g(x) e^{-i(w+r)x} \, dx - \int g(x) e^{-iwx} \, dx \right|$$

$$= \left| \int g(x) \left(e^{-irx} - 1 \right) e^{-iwx} \, dx \right|$$

$$= \left| \int_{|x|>R} g(x) \left(e^{-iwx} - 1 \right) e^{-iwx} \, dx + \int_{-R}^{R} g(x) \left(e^{-iwx} - 1 \right) e^{-iwx} \, dx \right|$$

$$\leq \int_{|x|>R} |g(x)| \left| e^{-iwx} - 1 \right| dx + \int_{-R}^{R} |g(x)| \left| e^{-iwx} - 1 \right| dx$$

$$< \int_{|x|>R} |g(x)| \cdot 2 \, dx + \int_{-R}^{R} |g(x)| \frac{\varepsilon}{2(\|g\|_1 + 1)} \, dx$$

$$\leq \frac{2\varepsilon}{4} + \frac{\|g\|_1 \varepsilon}{2(\|g\|_1 + 1)}$$

$$\leq \frac{\varepsilon}{2} + \frac{\varepsilon}{2} = \varepsilon. \qquad \qquad \square$$

EXERCISES

Exercise 7.9. For each integrable function g, with $\bar{g}$ denoting the complex conjugate and with $g_-(x) := g(-x)$, verify that

$$\mathcal{F}[(\bar{g})_-] = \overline{\mathcal{F}g}.$$

Exercise 7.10. For each pair of distinct real numbers $r < s$, calculate the Fourier transform of the "hat" function $h_{r,s} : \mathbb{R} \to \mathbb{C}$ defined by

$$h_{r,s}(x) := \begin{cases} 0 & \text{if } x < r, \\ 2(x-r)/(s-r) & \text{if } r \le x < (r+s)/2, \\ 2(x-s)/(r-s) & \text{if } (r+s)/2 \le x < s, \\ 0 & \text{if } s < x. \end{cases}$$

Then verify that $h_{r,s}$ is continuous, and that its Fourier transform $\mathcal{F}h_{r,s}$ is continuous and integrable on the whole real line.

Exercise 7.11. **(a)** Verify that if f is an even real-valued function, which means that $f(-x) = f(x)$ for every x, then

$$(\mathcal{F}f)(w) = \frac{2}{\sqrt{2\pi}} \int_0^\infty f(x) \cdot \cos(x \cdot w) \, dx.$$

(b) Verify that if f is an odd real-valued function, which means that $f(-x) = -f(x)$ for every x, then

$$(\mathcal{F}f)(w) = \frac{2i}{\sqrt{2\pi}} \int_0^\infty f(x) \cdot \sin(x \cdot w) \, dx.$$

Exercise 7.12. Show that every function $f : \mathbb{R} \to \mathbb{C}$ equals the sum of an even function and an odd function.

Exercise 7.13. (Fejér's kernel F_B.) Verify that for each positive real $B > 0$ Fejér's kernel $F_B : \mathbb{R} \to \mathbb{C}$ defined by

$$F_B(w) := \frac{B}{2\pi} \cdot \left[\frac{\sin(Bw/2)}{Bw/2} \right]^2$$

for $w \ne 0$ and $F_B(0) := B/(2\pi)$ is a weight function on $\mathbb{R}$, so that $\int_{-\infty}^\infty F_B(w) \, dw = 1$.

Exercise 7.14. Verify that $\int_{\mathbb{R}} |\sin(w)/w| \, dw = \infty$ but

$$\lim_{R \to \infty} \int_{-R}^R (\sin(w)/w) \, dw$$

has a finite value.

7.5 THE FOURIER TRANSFORM WITH SEVERAL VARIABLES

This section gives a brief outline of the Fourier transform of functions defined in the plane, in space, or in a space $\mathbb{R}^n$ with any dimension $n \in \mathbb{N}^* = \{1, 2, 3, \ldots\}$.

Definition 7.39 The Fourier transform of an integrable function $f : \mathbb{R}^n \to \mathbb{C}$ is the function $(\mathcal{F}f) : \mathbb{R}^n \to \mathbb{C}$ defined by

$$(\mathcal{F}f)(\vec{w}) := \frac{1}{(\sqrt{2\pi})^n} \int_{\mathbb{R}^n} f(\vec{x}) \cdot e^{-i \cdot \vec{w} \cdot \vec{x}} \, d\vec{x},$$

where $\vec{w} \cdot \vec{x}$ represents the Euclidean inner product, and where $d\vec{x}$ denotes the volume measure in $\mathbb{R}^n$.   □

Remark 7.40   (The Fourier transform with polar coordinates.) With polar coordinates,

$$\vec{x} = r \cdot \vec{u} \text{ with } r \geq 0 \text{ and } \vec{u} \in S^{n-1} \subset \mathbb{R}^n,$$

$$\vec{w} = \rho \cdot \vec{\theta} \text{ with } \rho \geq 0 \text{ and } \vec{\theta} \in S^{n-1} \subset \mathbb{R}^n,$$

and with an abuse of notation—denoting functions in Cartesian and polar coordinates by the same symbols—the definition of the Fourier transform becomes

$$(\mathcal{F}f)(\rho, \vec{\theta}) = \frac{1}{(\sqrt{2\pi})^n} \int_{S^{n-1}} \int_{\mathbb{R}_+} f(r \cdot \vec{u}) \cdot e^{-i \cdot \rho \cdot r \cdot \vec{\theta} \cdot \vec{u}} r^{n-1} \, dr \, d\vec{u},$$

where $d\vec{u}$ denotes the rotationally invariant area measure on the unit sphere $S^{n-1} \subset \mathbb{R}^n$, and where $\mathbb{R}_+ := [0, \infty[$.   □

Polar coordinates allow for the reduction of integrals of some symmetric functions to integrals with one variable; for instance, the Fourier transform of a Gaussian distribution is still Gaussian with several variables. Hence, the same methods to invert the Fourier transform with one variable apply to the inversion with several variables.

Theorem 7.41   *For every function* $f : \mathbb{R}^n \to \mathbb{C}$ *such that* f *and* $\mathcal{F}f$ *are integrable and* f *is continuous,*

$$f(\vec{x}) = \frac{1}{(\sqrt{2\pi})^n} \int_{\mathbb{R}^n} (\mathcal{F}f)(\vec{w}) \cdot e^{i \cdot \vec{w} \cdot \vec{x}} \, d\vec{w}.$$

PROOF: With a multivariable approximate identity—for instance, a tensor product of univariate approximate identities, as outlined in the exercises—the proof proceeds *verbatim* as with one variable.   □

The Fourier transform of a multivariable Gaussian distribution is again a constant multiple of a multivariable Gaussian distribution. One method to verify this assertion uses differential equations.

Example 7.42   (Fourier transform of the univariate Gaussian.) Consider first the univariate Gaussian distribution g_c with mean 0 and variance c^2, defined by

$$g_c(x) := \frac{1}{\sqrt{2\pi \cdot c^2}} \cdot e^{-x^2/(2c^2)}.$$

With $I : \mathbb{R} \to \mathbb{C}$ denoting the identity function, g_c satisfies the ordinary differential equation

$$g_c'(x) = \frac{-x/c^2}{\sqrt{2\pi c^2}} \cdot e^{-x^2/(2c^2)} = \frac{-x}{c^2} \cdot g_c(x) = -\frac{1}{c^2} \cdot (I \cdot g_c)(x).$$

Because $I g_c$ is integrable, Proposition 7.37 shows that $(\mathcal{F} g_c)' = \mathcal{F}[-i \cdot I \cdot g_c]$, whence $\mathcal{F} g_c$ satisfies the differential equation

$$(\mathcal{F} g_c)'(w) = (\mathcal{F}[-i \cdot I \cdot g_c])(w) = (\mathcal{F}[i \cdot c^2 \cdot g_c'])(w)$$
$$= [i \cdot c^2] \cdot (\mathcal{F}[g_c'])(w) = [i \cdot c^2] \cdot (iw) \cdot (\mathcal{F} g_c)(w)$$
$$= -c^2 w \cdot (\mathcal{F} g_c)(w).$$

Thus,

$$(\mathcal{F} g_c)'(w) = -c^2 w \cdot (\mathcal{F} g_c)(w).$$

Integration gives

$$\ln[(\mathcal{F} g_c)(w)] = -c^2 w^2/2 + K,$$

and $C := e^K$ yields

$$(\mathcal{F} g_c)(w) = C \cdot e^{-c^2 \cdot w^2/2}.$$

For $w := 0$,

$$C = C \cdot e^{-c^2 \cdot 0} = (\mathcal{F} g_c)(0) = \frac{1}{\sqrt{2\pi}} \int_{\mathbb{R}} \frac{1}{\sqrt{2\pi} \cdot c^2} \cdot e^{-x^2/(2c^2)} \, dx = \frac{1}{\sqrt{2\pi}}.$$

Thus,

$$(\mathcal{F} g_c)(w) = \frac{1}{\sqrt{2\pi}} \cdot e^{-c^2 \cdot w^2/2} = \frac{(1/c)}{(1/c) \cdot \sqrt{2\pi}} \cdot e^{-c^2 \cdot w^2/2} = \frac{1}{c} \cdot g_{1/c}(w),$$

which establishes that the Fourier transform maps Gaussian distributions to Gaussian distributions:

$$(\mathcal{F} g_c)(w) = \frac{1}{c} \cdot g_{1/c}(w).$$

In particular, $\mathcal{F}(c \cdot g_c) = g_{1/c}$ is a Gaussian distribution with variance $1/c^2$, whence the set $\{\mathcal{F}(c \cdot g_c) : c \in \mathbb{R}_+\}$ forms an approximate identity: $\lim_{c \to \infty} \{f * [\mathcal{F}(c \cdot g_c)]\} = f$.

Finally, shifting a Gaussian distribution g_c by u to the right multiplies its Fourier transform by e^{-iuw}.   □

Example 7.43   The Gaussian distribution with zero mean and unit variance,

$$g_1(x) = \frac{1}{\sqrt{2\pi}} e^{-x^2/2},$$

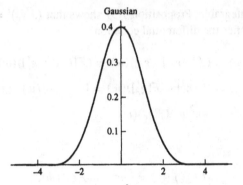

Figure 7.4   The Gaussian $g_1(x) = \frac{1}{\sqrt{2\pi}} e^{-x^2/2}$ is its own Fourier transform: $\mathcal{F}g_1 = g_1$.

equals its own Fourier transform, as in Figure 7.4:

$$\mathcal{F}g_1 = g_1. \qquad\qquad \square$$

Example 7.44   (Fourier transform of the multivariate Gaussian.) With several variables, let $S \in M_{n \times n}(\mathbb{R})$ denote any real symmetric positive definite matrix. Also, let $V = (\vec{v}_1, \ldots, \vec{v}_n)$ denote an orthonormal basis of eigenvectors of S, and let $c_1^{-2}, \ldots, c_n^{-2}$ represent the eigenvalues of the inverse matrix S^{-1}. Then the general multivariable Gaussian distribution has the form

$$g_S(\vec{x}) = \frac{1}{(\sqrt{2\pi})^n \cdot \sqrt{\det(S)}} \cdot \exp\left(\vec{x}^T \cdot S^{-1} \cdot \vec{x}/2\right).$$

Changing coordinates with $\vec{z} := V^T \vec{x}$ yields a product of univariate Gaussian distributions:

$$g_S(\vec{x}) = \frac{1}{(\sqrt{2\pi})^n \cdot \sqrt{\det(S)}} \cdot \exp\left(\left(V^T\vec{x}\right)^T \cdot V^T S^{-1} V \cdot V^T\vec{x}/2\right)$$

$$= \frac{1}{(\sqrt{2\pi})^n \cdot \sqrt{\det(S)}} \cdot \exp\left(\vec{z}^T \cdot \operatorname{diag}\left(c_1^{-2}, \ldots, c_n^{-2}\right) \cdot \vec{z}/2\right)$$

$$= \frac{1}{(\sqrt{2\pi})^n \cdot \prod_{k=1}^n c_k} \cdot \exp\left(\sum_{k=1}^n c_k^{-2} z_k^2/2\right)$$

$$= \prod_{k=1}^n \frac{1}{\sqrt{2\pi} \cdot c_k} \cdot e^{-z_k^2/(2c_k^2)} = \prod_{k=1}^n g_{c_k}(z_k).$$

Because the orthogonal matrix V of the change of coordinates preserves inner products and has determinant equal to ± 1, it follows that

$$(\mathcal{F}g_S)(\vec{\mathbf{w}}) = \frac{1}{(\sqrt{2\pi})^n} \int_{\mathbb{R}^n} g_S(\vec{\mathbf{x}}) \cdot e^{-i\vec{\mathbf{x}}\cdot\vec{\mathbf{w}}} \, d\vec{\mathbf{x}}$$

$$= \frac{1}{(\sqrt{2\pi}^n)} \int_{\mathbb{R}^n} \prod_{k=1}^{n} \left[g_{c_k}(z_k) \cdot e^{-i\cdot z_k \cdot (V^T\vec{\mathbf{w}})_k} \right] d\vec{\mathbf{z}}$$

$$= \prod_{k=1}^{n} (\mathcal{F}g_{c_k}) \left(\left(V^T\vec{\mathbf{w}}\right)_k \right) = \prod_{k=1}^{n} ((1/c_k) \cdot g_{1/c_k}) \left(\left(V^T\vec{\mathbf{w}}\right)_k \right)$$

$$= \prod_{k=1}^{n} \frac{1/c_k}{(1/c_k)\sqrt{2\pi}} \cdot e^{-[(V^T\vec{\mathbf{w}})_k]^2 c_k^2/2}$$

$$= \frac{1}{(\sqrt{2\pi})^n} \cdot \exp\left(-\left(V^T\vec{\mathbf{w}}\right)^T \cdot V^T S V \cdot V^T\vec{\mathbf{w}}/2 \right)$$

$$= \frac{\sqrt{\det(S^{-1})}}{(\sqrt{2\pi})^n \cdot \sqrt{\det(S^{-1})}} \cdot \exp\left(-\vec{\mathbf{w}}^T S\vec{\mathbf{w}}/2 \right),$$

which means that the Fourier transform of a multivariable Gaussian distribution g_S with covariance matrix S is the constant multiple $\sqrt{\det(S^{-1})}$ times the multivariable Gaussian distribution $g_{S^{-1}}$ with covariance matrix S^{-1}:

$$\mathcal{F}g_S = \sqrt{\det(S^{-1})} \cdot g_{S^{-1}}. \qquad \square$$

Definition 7.45   The **tensor product** of n functions $f_\ell : U_\ell \to \mathbb{C}$ is the function

$$\otimes_{\ell=1}^{n} f_\ell : \prod_{\ell=1}^{n} U_\ell \to \mathbb{C},$$

$$(\otimes_{\ell=1}^{n} f_\ell)(\vec{\mathbf{x}}) := \prod_{\ell=1}^{n} f_\ell(x_\ell). \qquad \square$$

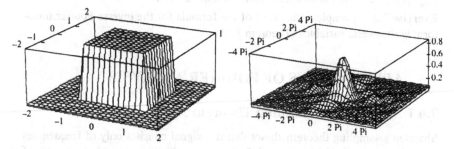

Figure 7.5   *Left.* Characteristic function χ of the square $[-1, 1] \times [-1, 1]$. *Right.* The Fourier transform $\mathcal{F}\chi = (2/\pi)\text{sinc} \otimes \text{sinc}$.

EXERCISES

Exercise 7.15. Prove that if each function f_ℓ is integrable, then

$$\mathcal{F}\left(\otimes_{\ell=1}^n f_\ell\right) = \otimes_{\ell=1}^n (\mathcal{F} f_\ell).$$

Exercise 7.16. With $\|\ \|_1$ representing the norm defined by

$$\|\vec{x}\|_1 := \sum_{\ell=1}^n |x_\ell|,$$

calculate the Fourier transform of the function $K_B : \mathbb{R}^n \to \mathbb{C}$ defined by

$$K_B(\vec{x}) := e^{-\|\vec{x}\|_1/B}.$$

Exercise 7.17. With the "hat" function $h : \mathbb{R} \to \mathbb{C}$ defined by

$$h(x) := \begin{cases} 1+x & \text{if } -1 \le x < 0, \\ 1-x & \text{if } 0 \le x \le 1, \\ 0 & \text{if } |x| > 1, \end{cases}$$

calculate the Fourier transform of the function $H_B : \mathbb{R}^n \to \mathbb{C}$ defined by

$$H_B(\vec{x}) := \prod_{\ell=1}^n [B \cdot h(B \cdot x_\ell)].$$

Exercise 7.18. With $\|\ \|_\infty$ representing the norm defined by

$$\|\vec{x}\|_\infty := \max_{\ell=1}^n |x_\ell|,$$

calculate the Fourier transform of the characteristic function of the unit cube $\chi_{[-1,1]^n} : \mathbb{R}^n \to \mathbb{C}$ defined by

$$\chi_{[-1,1]^n}(\vec{x}) := \begin{cases} 1 & \text{if } \|\vec{x}\|_\infty \le 1, \\ 0 & \text{if } \|\vec{x}\|_\infty > 1. \end{cases}$$

Exercise 7.19. Prove that if $\{g_c : c \in \mathbb{R}_+\}$ is an approximate identity on $\mathbb{R}$, then $\{\otimes_{\ell=1}^n (g_c) : c \in \mathbb{R}_+\}$ is an approximate identity on $\mathbb{R}^n$.

Exercise 7.20. Complete the proof of the formula for the inverse Fourier transform with several variables (Theorem 7.41).

7.6   APPLICATIONS OF FOURIER ANALYSIS

7.6.1   Shannon's Sampling Theorem

Shannon's sampling theorem shows that if a signal consists only of frequencies that do not exceed an upper bound T, then a sample of that signal at intervals of length π/T suffices to reconstruct the signal.

Definition 7.46 An integrable function $f : \mathbb{R} \to \mathbb{C}$ is **band-limited** if a real number T exists such that $(\mathcal{F}f)(w) = 0$ for every w for which $|w| > T$. ☐

Theorem 7.47 (**Shannon's Sampling Theorem.**) *If $f : \mathbb{R} \to \mathbb{C}$ and $I \cdot f$ are integrable and continuous, and if $(\mathcal{F}f)(w) = 0$ for every w such that $|w| > T > 0$, then*

$$f(x) = \sum_{n \in \mathbb{Z}} f(n \cdot \pi/T) \cdot \frac{\sin(xT - n\pi)}{xT - n\pi}.$$

PROOF: The Fourier inversion theorem and the hypotheses show that

$$f(x) = \frac{1}{\sqrt{2\pi}} \int_{\mathbb{R}} (\mathcal{F}f)(w) \cdot e^{i \cdot w \cdot x}\, dw = \frac{1}{\sqrt{2\pi}} \int_{-T}^{T} (\mathcal{F}f)(w) \cdot e^{i \cdot w \cdot x}\, dw.$$

Define a function $F : \mathbb{R} \to \mathbb{C}$ by

$$F(w) = (\mathcal{F}f)([w]),$$

where $[w] \equiv w$ modulo $2T$ and $[w] \in\,] - T, T]$. This means that F has period $2T$ and coincides with $\mathcal{F}f$ on the interval $] - T, T]$.

Proposition 7.37 shows that $\mathcal{F}f$, and hence F, is differentiable on $] - T, T[$; consequently, its Fourier *series* converges to F:

$$(\mathcal{F}f)(w) = F(w) = \sum_{n \in \mathbb{Z}} c_{F,n} \cdot e^{inw\pi/T},$$

where the definition of $c_{F,n}$, the hypothesis on $|w| > T$, and the inverse Fourier transform yield

$$c_{F,n} = \frac{1}{2T} \int_{-T}^{T} (\mathcal{F}f)(t) \cdot e^{-int\pi/T}\, dt$$

$$= \frac{1}{2T} \int_{\mathbb{R}} (\mathcal{F}f)(t) \cdot e^{-int\pi/T}\, dt$$

$$= \frac{1}{2T} \cdot \sqrt{2\pi} \cdot \frac{1}{\sqrt{2\pi}} \int_{\mathbb{R}} (\mathcal{F}f)(t) \cdot e^{i(-n)t\pi/T}\, dt$$

$$= \frac{1}{2T} \cdot \sqrt{2\pi} \cdot f(-n\pi/T).$$

Therefore,

$$f(x) = \frac{1}{\sqrt{2\pi}} \int_{-T}^{T} (\mathcal{F}f)(w) \cdot e^{i \cdot w \cdot x}\, dw$$

$$= \frac{1}{\sqrt{2\pi}} \int_{-T}^{T} \left(\sum_{n \in \mathbb{Z}} c_{F,n} \cdot e^{inw\pi/T} \right) \cdot e^{i \cdot w \cdot x}\, dw$$

$$= \frac{1}{\sqrt{2\pi}} \sum_{n\in\mathbb{Z}} f(-n\pi/T) \cdot \frac{1}{2T} \cdot \sqrt{2\pi} \int_{-T}^{T} \left(e^{iw(x+n\pi/T)}\right) dw$$

$$= \sum_{n\in\mathbb{Z}} f(-n\pi/T) \cdot \frac{1}{2T} \cdot \frac{1}{i(x+n\pi/T)} e^{iw(x+n\pi/T)} \Big|_{-T}^{T}$$

$$= \sum_{n\in\mathbb{Z}} f(-n\pi/T) \cdot \frac{1}{xT+n\pi} \cdot \sin(xT+n\pi)$$

$$= \sum_{n\in\mathbb{Z}} f(n\pi/T) \cdot \frac{\sin(xT-n\pi)}{xT-n\pi}. \qquad \Box$$

7.6.2   Heisenberg's Uncertainty Principle

Heisenberg's uncertainty principle has the following physical interpretation. If a function $f : \mathbb{R}^3 \times \mathbb{R} \to \mathbb{C}$ represents the de Broglie wave associated with a particle, so that $|f(\vec{x})|^2$ corresponds to the probability density that the particle passes through the point $\vec{x}$, then the Fourier transform

$$(\mathcal{F}f)(\vec{p}) = \frac{1}{\sqrt{2\pi}} \int_{\mathbb{R}^3} f(\vec{x}) e^{-i\vec{x}\cdot\vec{p}} d\vec{x}$$

decomposes the wave function f into a weighted average of waves $w_{\vec{p}}$ with $w_{\vec{p}}(\vec{x}) = e^{-i\vec{x}\cdot\vec{p}}$. The value $(\mathcal{F}f)(\vec{p})$ then represents the density of momenta at $\vec{p}$ that contributes to the wave function f. Heisenberg's uncertainty principle then states that both the location and speed of the particle cannot be specified simultaneously.

Definition 7.48   Define the **half bandwidth** W_f of each integrable function f such that $0 < \|f\|_2 < \infty$ and $\|I \cdot f\|_2 < \infty$ by

$$W_f^2 := \frac{\int_{\mathbb{R}} |x \cdot f(x)|^2 \, dx}{\int_{\mathbb{R}} |f(x)|^2 \, dx}. \qquad \Box$$

Theorem 7.49   (Heisenberg's Uncertainty Principle.)   *For band-limited functions* $f : \mathbb{R} \to \mathbb{C}$ *with* $|f|^2$, $|I \cdot f|^2$, $|\mathcal{F}f|^2$, *and* $|I \cdot \mathcal{F}f|^2$ *integrable and such that* $\lim_{|x|\to\infty} |x \cdot f(x)| = 0$,

$$W_f \cdot W_{\mathcal{F}f} \geq \frac{1}{2}.$$

PROOF: Apply the Cauchy–Schwartz inequality:

$$|\langle I \cdot f, \, f'\rangle|^2 \leq \|I \cdot f\|_2^2 \cdot \|f'\|_2^2,$$

$$\left| \int_{\mathbb{R}} x f(x) f'(x) \, dx \right|^2 \leq \int_{\mathbb{R}} |x f(x)|^2 \, dx \cdot \int_{\mathbb{R}} |f'(w)|^2 \, dw.$$

Thanks to $\lim_{|x|\to\infty} |x \cdot f(x)| = 0$, integration by parts gives

$$\int_{\mathbb{R}} x f(x) f'(x)\, dx = x \cdot \frac{[f(x)]^2}{2}\Big|_{-\infty}^{\infty} - \int_{\mathbb{R}} \frac{[f(x)]^2}{2}\, dx = -\int_{\mathbb{R}} \frac{[f(x)]^2}{2}\, dx.$$

Substitutions then yield the conclusion:

$$\left| \int_{\mathbb{R}} x f(x) f'(x)\, dx \right|^2 \leq \int_{\mathbb{R}} |x f(x)|^2\, dx \cdot \int_{\mathbb{R}} |f'(w)|^2\, dw,$$

$$\left| \int_{\mathbb{R}} \frac{[f(x)]^2}{2}\, dx \right|^2 \leq \int_{\mathbb{R}} |x f(x)|^2\, dx \cdot \int_{\mathbb{R}} |f'(w)|^2\, dw,$$

$$\frac{1}{4} \leq \frac{\int_{\mathbb{R}} |x f(x)|^2\, dx}{\int_{\mathbb{R}} |f(x)|^2\, dx} \cdot \frac{\int_{\mathbb{R}} |f'(w)|^2\, dw}{\int_{\mathbb{R}} |f(x)|^2\, dx},$$

$$\frac{1}{4} \leq \frac{\int_{\mathbb{R}} |x f(x)|^2\, dx}{\int_{\mathbb{R}} |f(x)|^2\, dx} \cdot \frac{\int_{\mathbb{R}} |(\mathcal{F} f')(w)|^2\, dw}{\int_{\mathbb{R}} |(\mathcal{F} f)(x)|^2\, dx},$$

$$\frac{1}{4} \leq \frac{\int_{\mathbb{R}} |x f(x)|^2\, dx}{\int_{\mathbb{R}} |f(x)|^2\, dx} \cdot \frac{\int_{\mathbb{R}} |i w|^2 \cdot |(\mathcal{F} f)(w)|^2\, dw}{\int_{\mathbb{R}} |(\mathcal{F} f)(x)|^2\, dx}$$

$$= W_f^2 \cdot W_{\mathcal{F} f}^2. \qquad \square$$

For additional information about Heisenberg's uncertainty principle, consult, for instance, Richard P. Feynman's [11, §37], or E. H. Wichmann's [50].

EXERCISES

Exercise 7.21. Verify that in Heisenberg's Uncertainty Principle the inequality becomes an equality for Gaussian distributions.

Exercise 7.22. Use concepts from Fourier series for periodic functions and from the Discrete Fourier transform to prove the following version of **Shannon's sampling theorem.** *If a continuous periodic function f has only finitely many nonzero Fourier coefficients, so that there exists a positive integer K such that $c_{f,k} = 0$ for every index $k > K$, then finitely many individual values of f (a finite sample from f) suffice to compute all the Fourier coefficients $c_{f,k}$ without integration.*

CHAPTER 8

Daubechies Wavelets Design

8.0 INTRODUCTION

Using the example of Daubechies wavelets, this chapter demonstrates how to use the Fourier transform to investigate the existence, the uniqueness, and the design of recursive, mutually orthogonal, compactly supported, and continuous wavelets. The proofs will reveal that the existence of such wavelets depends upon the following conditions on the recursion coefficients:

$$\varphi(x) = \sum_{k=0}^{N} h_k \varphi(2x - k), \qquad \text{(recursion)} \qquad (8.1)$$

$$\psi(x) := \sum_{k=1-N}^{1} (-1)^k h_{1-k} \varphi(2x - k), \qquad \text{(definition)} \qquad (8.2)$$

$$\sum_{k=0}^{\lfloor N/2 \rfloor} h_{2k} = 1 = \sum_{k=0}^{\lfloor (N-1)/2 \rfloor} h_{2k+1}, \qquad \text{(existence)} \qquad (8.3)$$

$$\int_{\mathbb{R}} \varphi(2x - k)\varphi(2x - \ell) \, dx = 0 \text{ for } k \neq \ell, \qquad \text{(orthogonality)} \qquad (8.4)$$

$$\sum_{k=\max\{0,2m\}}^{\min\{N,N+2m\}} h_k h_{k-2m} = \begin{cases} 2 & \text{if } m = 0, \\ 0 & \text{if } m \neq 0. \end{cases} \qquad \text{(orthogonality)} \qquad (8.5)$$

8.1 EXISTENCE, UNIQUENESS, AND CONSTRUCTION

The existence of a fast wavelet transform depends in part upon the existence of a recursive relation of the type (8.1). Consequently, one of the basic problems in the theory of mathematical wavelets consists in determining a sequence of coefficients $h_0, \ldots, h_N$ that admits a nonzero continuous building-block function φ such that $\varphi(x) = \sum_{k=0}^{N} h_k \varphi(2x - k)$ for every real number x. One method to search for such coefficients consists in extending the search to larger sets of coefficients, and then in examining whether any such coefficients satisfy the requirements.

8.1.1 The Recursion Operator and Its Adjoint

For each finite sequence of coefficients $h_0, \ldots, h_N$, consider the linear operator T that for each function $g : \mathbb{R} \to \mathbb{R}$ produces a function $Tg : \mathbb{R} \to \mathbb{R}$ defined by

$$(Tg)(x) := \sum_{k=0}^{N} h_k g(2x - k). \tag{8.6}$$

The recursion $\varphi(x) = \sum_{k=0}^{N} h_k \varphi(2x - k)$ then means that

$$\varphi = T\varphi.$$

Example 8.1 For the "box" function $g := \chi_{[0,1[}$ defined by

$$g(x) := \begin{cases} 1 & \text{if } 0 \le x < 1, \\ 0 & \text{otherwise,} \end{cases}$$

and for Daubechies' four coefficients, the operator T produces the function Tg displayed in Figure 8.1 and defined by

$$(Tg)(x) = h_0 g(2x) + h_1 g(2x - 1) + h_2 g(2x - 2) + h_3 g(2x - 3). \qquad \square$$

Several areas of mathematics address the problem of determining operators T and functions φ such that $\varphi = T\varphi$. In linear algebra and in functional analysis, such an equation means that the operator T admits φ as an eigenfunction for the eigenvalue 1, so that $T\varphi = 1 \cdot \varphi$, or, equivalently, that the operator T has a fixed point at φ. Then numerical analysis addresses the question of whether simple iterations of T will converge to such a fixed point.

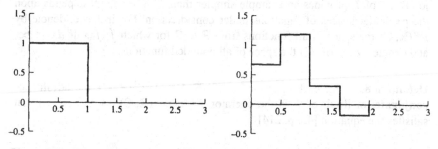

Figure 8.1 Effect of the recursion operator T. *Left*: a function g. *Right*: the function Tg.

Definition 8.2 With ∘ denoting the composition of functions, let $T^{\circ n}$ represent the nth iteration of T, defined inductively by

$$T^{\circ 0} := I,$$
$$T^{\circ 1} := T,$$
$$T^{\circ 2} := T \circ T,$$
$$T^{\circ 3} := T \circ T^{\circ 2} = T \circ T \circ T,$$
$$T^{\circ 4} := T \circ T^{\circ 3} = T \circ T \circ T \circ T,$$
$$\vdots$$
$$T^{\circ(n+1)} := T \circ T^{\circ n},$$
$$\vdots \qquad \qquad \Box$$

The notation $T^{\circ n}$ avoids ambiguities with multiplicative powers and is consistent with other texts [3], [26]. In terms of such an operator T, the basic problem leads to the following questions.

- *Existence.* Does the operator T have an eigenfunction φ for the eigenvalue 1?

- *Uniqueness.* Does the corresponding eigenspace Kernel $(T - I)$ have dimension 1, so that φ is unique?

- *Computability.* Does the power method—or other methods—produce iterations $(T^{\circ n} g)_{n=1}^{\infty}$ that converge to φ?

To demonstrate typical uses of the Fourier transform in the theory of wavelets, this section draws from Daubechies' original research article [6] and subsequent book [7] to prove that the iterations $T^{\circ n} g$ indeed converge to a function φ with the required properties, as illustrated in Figure 8.2.

As an introduction to the concept of eigenvalues of linear operators, the "adjoint" T^* of T provides an example simpler than T. The adjoint depends upon the particular spaces of functions under consideration. For instance, denote by $\mathcal{L}^p(\mathbb{R}, \mathbb{C})$ the space of all functions from $\mathbb{R}$ to $\mathbb{C}$ for which $\int |f(x)|^p\, dx < \infty$, and denote by $\mathcal{L}^\infty(\mathbb{R}, \mathbb{C})$ the space of all bounded functions.

Definition 8.3 For each operator $T : \mathcal{L}^1(\mathbb{R}, \mathbb{C}) \to \mathcal{L}^1(\mathbb{R}, \mathbb{C})$, the **adjoint**, or **conjugate**, or **dual**, of T is the operator $T^* : \mathcal{L}^\infty(\mathbb{R}, \mathbb{C}) \to \mathcal{L}^\infty(\mathbb{R}, \mathbb{C})$ that satisfies the equation [45, p. 214]

$$\langle Tg, h \rangle = \langle g, T^*h \rangle$$

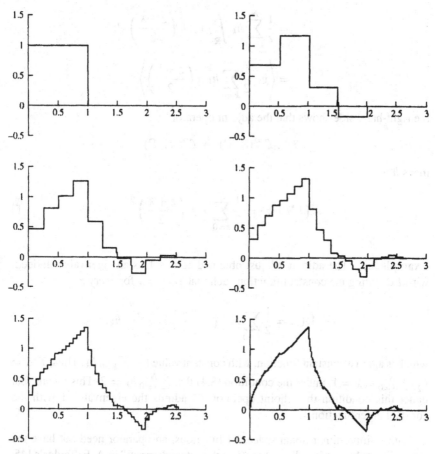

Figure 8.2 Transforms of the "box" function g. *Top:* g and Tg. *Middle:* $T^{\circ2}g$ and $T^{\circ3}g$. *Bottom:* $T^{\circ4}g$ and φ.

for every $g \in \mathcal{L}^1(\mathbb{R}, \mathbb{C})$ and every $h \in \mathcal{L}^\infty(\mathbb{R}, \mathbb{C})$, with the product $\langle \, , \, \rangle : \mathcal{L}^1 \times \mathcal{L}^\infty \to \mathbb{R}$ defined by

$$\langle g, h \rangle := \int_{\mathbb{R}} g(x)\, h(x)\, dx. \qquad \qquad \Box$$

Example 8.4 For the operator T defined by Equation (8.6), a change of variable in the integral defining the inner product leads to a formula for the adjoint T^*:

$$\langle Tg, h \rangle = \int_{\mathbb{R}} (Tg)(x)\, h(x)\, dx$$

$$= \sum_{k=0}^{N} h_k \int_{\mathbb{R}} g(2x - k)\, h(x)\, dx$$

$$= \frac{1}{2} \sum_{k=0}^{N} h_k \int_{\mathbb{R}} g(z) h\left(\frac{z+k}{2}\right) dz$$

$$= \left\langle g, \frac{1}{2} \sum_{k=0}^{N} h_k h\left(\frac{z+k}{2}\right) \right\rangle;$$

the right-hand side shows that the adjoint operator

$$T^* : \mathcal{L}^{\infty}(\mathbb{R}, \mathbb{C}) \to \mathcal{L}^{\infty}(\mathbb{R}, \mathbb{C})$$

maps h to

$$\left(T^* h\right)(z) = \frac{1}{2} \sum_{k=0}^{N} h_k h\left(\frac{z+k}{2}\right). \qquad \qquad \square$$

Example 8.5   The adjoint T^* just obtained admits 1 as an eigenvalue. Indeed, with $\mathbf{1}$ denoting the constant function such that $\mathbf{1}(x) = 1$ for every x,

$$(T^* \mathbf{1})(z) = \frac{1}{2} \sum_{k=0}^{N} h_k \mathbf{1}\left(\frac{z+k}{2}\right) = \frac{1}{2} \sum_{k=0}^{N} h_k,$$

which is again a constant function, with constant value $(\frac{1}{2}) \sum_{k=0}^{N} h_k$. Thus, $T^* \mathbf{1} = (\frac{1}{2}) \sum_{k=0}^{N} h_k \mathbf{1} = \mathbf{1}$, under the condition (8.4) that $\sum_{k=0}^{N} h_k = 2$. This means that under this condition the adjoint operator T^* admits the eigenvalue 1 with the constant eigenfunction $\mathbf{1}$. $\square$

With infinite-dimensional spaces of functions, an operator need not have the same eigenvalues as its adjoint has. (See the "state diagram" in A. E. Taylor's [45, p. 237].) Indeed, proving that a particular operator admits a specific eigenvalue often requires a substantial amount of analysis, for instance, as in the following subsections.

EXERCISES

Exercise 8.1.   For each positive integer N consider the complex linear space $\mathbb{C}^N$ with the inner product defined by $\langle \vec{z}, \vec{w} \rangle := \sum_{k=1}^{N} z_k \overline{w_k}$. Also, for each complex rectangular matrix $A \in M_{p \times q}(\mathbb{C})$ with p rows and q columns, consider the linear operator L_A defined through multiplication by the matrix A:

$$L_A : \mathbb{C}^q \to \mathbb{C}^p,$$

$$L_A(\vec{z}) := A \cdot \vec{z}.$$

Verify that the adjoint of L_A is the operator $L_{\bar{A}^T}$ defined through multiplication by the transposed complex conjugate matrix $\bar{A}^T$:

$$(L_A)^* : \mathbb{C}^p \to \mathbb{C}^q,$$

$$(L_A)^*(\vec{w}) := \bar{A}^T \cdot \vec{w}.$$

In other words, verify that for each $\vec{z} \in \mathbb{C}^q$ and each $\vec{w} \in \mathbb{C}^p$,

$$\langle L_A(\vec{z}), \vec{w} \rangle = \langle \vec{z}, L_{\bar{A}^T}(\vec{w}) \rangle.$$

Exercise 8.2. Let $C^\infty([a, b], \mathbb{R})$ denote the linear space of all real-valued functions $f : [a, b] \to \mathbb{R}$ with continuous derivatives $f', f'', \ldots : [a, b] \to \mathbb{R}$. Denote by $C^\infty_{0,0}([a, b], \mathbb{R})$ the linear subspace of all functions with boundary values $f(a) = 0 = f(b)$:

$$C^\infty_{0,0}([a, b], \mathbb{R}) = \left\{ f : f(a) = 0 = f(b) \text{ and } f \in C^\infty([a, b], \mathbb{R}) \right\}.$$

Also, consider the inner product defined for all functions f and g in

$$C^\infty_{0,0}([a, b], \mathbb{R})$$

by

$$\langle f, g \rangle := \int_a^b f(x)g(x) \, dx.$$

Moreover, consider the linear operator Δ defined by

$$\Delta : C^\infty_{0,0}([a, b], \mathbb{R}) \to C^\infty_{0,0}([a, b], \mathbb{R}),$$

$$\Delta(f) := f''.$$

(a) Verify, for instance through integration by parts, that $\Delta^* = \Delta$ on the space $C^\infty_{0,0}([a, b], \mathbb{R})$. In other words, verify that for all functions f and g in $C^\infty_{0,0}([a, b], \mathbb{R})$,

$$\langle \Delta(f), g \rangle = \langle f, \Delta(g) \rangle.$$

(b) Determine every eigenvalue λ (the Greek letter "lambda") and every eigenfunction $f \not\equiv 0$ (not identically equal to zero everywhere) such that $\Delta(f) = \lambda \cdot f$.

8.1.2   The Fourier Transform of the Recursion Operator

To study more rigorously the convolution (8.6) that defines the operator T, introduce the Fourier transform $\mathcal{F}$, defined by

$$(\mathcal{F}f)(t) := \frac{1}{\sqrt{2\pi}} \int_{\mathbb{R}} f(x)e^{-itx} \, dx.$$

The change of variable $z := 2x - k$ then leads to

$$[\mathcal{F}(Tg)](t) = \frac{1}{\sqrt{2\pi}} \int_{\mathbb{R}} \sum_{k=0}^{N} h_k g(2x - k) e^{-itx} \, dx$$

$$= \left(\frac{1}{2} \sum_{k=0}^{N} h_k e^{-ikt/2} \right) \cdot (\mathcal{F}g)(t/2),$$

whence the power method produces iterations $T^{on} g$ with transforms

$$\mathcal{F}\left(T^{on} g\right)(t) = \prod_{\ell=0}^{n} \left(\frac{1}{2} \sum_{k=0}^{N} h_k e^{-ikt/2^\ell} \right) \cdot (\mathcal{F}g)(t/2^n). \qquad (8.7)$$

Thus, for the sequence $(T^{on} g)$ to converge, it suffices that the product and the dilated transform on the right-hand side of (8.7) both converge. Consider first the sequence of dilated transforms.

Lemma 8.6 *For each function $g \in \mathcal{L}^1(\mathbb{R}, \mathbb{C})$ the sequence $(\mathcal{F}g)(\frac{t}{2^n})$ converges uniformly on compact intervals to the constant function with value $(1/\sqrt{2\pi}) \int_{\mathbb{R}} g$:*

$$\lim_{n \to \infty} [(\mathcal{F}g)(t/2^n)] = \frac{1}{\sqrt{2\pi}} \int_{\mathbb{R}} g.$$

PROOF: The continuity of $\mathcal{F}g$ at the origin, proved in the preceding chapter, means that

for each real $\varepsilon > 0$,
there exists a real $\delta > 0$,
such that $|(\mathcal{F}g)(w) - (\mathcal{F}g)(0)| < \varepsilon$
for every real w with $|w - 0| < \delta$.

Hence, for each real $R > 0$, and with $w := R/2^n$, it follows that

$$|(\mathcal{F}g)(R/2^n) - (\mathcal{F}g)(0)| < \varepsilon,$$

provided that $|R/2^n - 0| < \delta$, which means that $n > [\ln(R/\delta)]/\ln(2)$. Thus, if $n > [\ln(R/\delta)]/\ln(2)$, then for every $t \in [-R, R]$,

$$|t| \le R,$$

$$\frac{\ln(|t|/\delta)}{\ln(2)} \le \frac{\ln(R/\delta)}{\ln(2)} < n,$$

$$0 \le |t|/\delta \le R/\delta < 2^n,$$

$$0 \le |t|/2^n \le R/2^n < \delta,$$

$$\left| (\mathcal{F}g)(t/2^n) - (\mathcal{F}g)(0) \right| < \varepsilon,$$

$$\left| (\mathcal{F}g)(t/2^n) - \frac{1}{\sqrt{2\pi}} \int_{\mathbb{R}} g \right| < \varepsilon,$$

because

$$\frac{1}{\sqrt{2\pi}} \int_{\mathbb{R}} g := \frac{1}{\sqrt{2\pi}} \int_{\mathbb{R}} g(x)\, dx = \frac{1}{\sqrt{2\pi}} \int_{\mathbb{R}} g(x) \cdot e^{-i \cdot x \cdot 0}\, dx = (\mathcal{F}g)(0).$$

Thus,

for each real $\varepsilon > 0$,
for each real $R > 0$,
there exists an index $n \in \mathbb{N}$,
such that $\left| (\mathcal{F}g)(t/2^m) - (1/\sqrt{2\pi}) \int_{\mathbb{R}} g \right| < \varepsilon$
for every index $m > n$ and every $t \in [-R, R]$,

which is the definition of uniform convergence on each compact interval $[-R, R]$.
□

The following subsection verifies the convergence of the product in formula (8.7).

8.1.3 Convergence of Iterations of the Recursion Operator

Consider the product in formula (8.7), reproduced here,

$$\mathcal{F}\left(T^{\circ n}g\right)(t) = \prod_{\ell=0}^{n} \left(\frac{1}{2} \sum_{k=0}^{N} h_k e^{-ikt/2^\ell} \right) \cdot (\mathcal{F}g)(t/2^n), \qquad (8.8)$$

which involves a polynomial crucial to the forthcoming analysis.

Definition 8.7 For each sequence of coefficients $h_0, \ldots, h_N$, consider the polynomial h defined by

$$h(z) := \frac{1}{2} \sum_{k=0}^{N} h_k z^k.$$
□

The following result from complex analysis relates "infinite" products (limits of finite products) to "infinite" series (limits of finite sums) [1, p. 192], [36, pp. 279–280].

Proposition 8.8 *For each sequence of complex numbers* $(w_\ell)_{\ell=0}^{\infty}$, *the product*

$$\prod_{\ell=1}^{\infty} (1 + |w_\ell|) := \lim_{L \to \infty} \prod_{\ell=1}^{L} (1 + |w_\ell|) = P$$

converges to a limit P if, but only if, the series

$$\sum_{\ell=1}^{\infty} |w_\ell| := \lim_{L \to \infty} \sum_{\ell=1}^{L} |w_\ell| = S$$

converges to a limit S. Moreover, if the product $\prod_{\ell=1}^{\infty}(1 + |w_\ell|)$ converges, then the product $\prod_{\ell=1}^{\infty}(1 + w_\ell)$ also converges.

PROOF: If the series converges, then for each integer L,

$$0 \le \prod_{\ell=1}^{L}(1 + |w_\ell|) \le \prod_{\ell=1}^{L} \exp(|w_\ell|)$$

$$= \exp\left(\sum_{\ell=1}^{L} |w_\ell|\right) \le \exp\left(\sum_{\ell=1}^{\infty} |w_\ell|\right) = \exp(S) < \infty.$$

Consequently, the sequence of partial products $\prod_{\ell=1}^{L}(1 + |w_\ell|)$ increases but remains bounded above by the fixed number $\exp(S)$ and hence converges to a limit. Conversely, if the product converges, then for each integer L,

$$1 \le 1 + \sum_{\ell=1}^{L} |w_\ell| \le \prod_{\ell=1}^{L}(1 + |w_\ell|) \le \prod_{\ell=1}^{\infty}(1 + |w_\ell|) = P.$$

Thus, the sequence of partial sums $\sum_{\ell=1}^{L} |w_\ell|$ increases but remains bounded above by $P - 1$ and hence converges to a limit.

Finally, if the product $\prod_{\ell=1}^{\infty}(1 + |w_\ell|)$ converges to a limit P, then the series $\sum_{\ell=1}^{\infty} |w_\ell|$ converges to a limit S. Consequently, the sequence of terms tends to zero: $\lim_{\ell \to \infty} |w_\ell| = 0$. Hence, there exists an index L such that $|w_\ell| < 1$ for every $\ell > L$, which guarantees the existence of the principal branch of the logarithm $\log(1 + w_\ell)$ for every $\ell > L$. Moreover, calculus yields the limit

$$\lim_{z \to 0} \frac{\log(1 + z)}{z} = \lim_{z \to 0} \frac{\log(1 + z) - \log(1)}{z} = \log'(1) = 1.$$

Applied to $z := w_\ell$, this limit means that for every real $\varepsilon > 0$, there exists a real $\delta > 0$ such that if $|w_\ell| = |z| < \delta$, then

$$1 - \varepsilon < \frac{|\log(1 + w_\ell)|}{|w_\ell|} < 1 + \varepsilon,$$

whence

$$(1 - \varepsilon)|w_\ell| < |\log(1 + w_\ell)| < (1 + \varepsilon)|w_\ell|.$$

Therefore, if the series $\sum_{\ell=1}^{\infty} |w_\ell|$ converges, then

$$0 \le \sum_{\ell=1}^{\infty} |\log(1 + w_\ell)| < (1 + \varepsilon) \sum_{\ell=1}^{\infty} |w_\ell|,$$

whence the series $\sum_{\ell=1}^{\infty} \log(1 + w_\ell)$ also converges to some limit Z. Consequently, the continuity of the exponential function ensures that the corresponding product converges:

$$\prod_{\ell=1}^{\infty}(1 + w_\ell) = \lim_{L \to \infty} \prod_{\ell=1}^{L}(1 + w_\ell) = \lim_{L \to \infty} \prod_{\ell=1}^{L} \exp[\log(1 + w_\ell)]$$

$$= \lim_{L \to \infty} \exp\left[\sum_{\ell=1}^{L} \log(1 + w_\ell)\right] = \exp\left[\lim_{L \to \infty} \sum_{\ell=1}^{L} \log(1 + w_\ell)\right]$$

$$= \exp\left[\sum_{\ell=1}^{\infty} \log(1 + w_\ell)\right] = \exp(Z). \qquad \square$$

Thus, the infinite product

$$\prod_{\ell=1}^{\infty}\left(\frac{1}{2} \sum_{k=0}^{N} h_k e^{-ikt/2^\ell}\right) = \prod_{\ell=1}^{\infty} h\left(e^{-it/2^\ell}\right)$$

converges absolutely and uniformly on compact intervals if, but only if, so does the series $\sum_{\ell=0}^{\infty} |h(e^{-it/2^\ell}) - 1|$. Again from the condition (8.3) that $\sum_{k=0}^{N} h_k = 2$, it follows that $h(z) = 1 + (\frac{1}{2}) \sum_{k=0}^{N} h_k(z^k - 1)$, and, consequently,

$$\left|h\left(e^{-it/2^\ell}\right) - 1\right| \le \frac{1}{2} \sum_{k=0}^{N} |h_k| \cdot |e^{-ikt/2^\ell} - 1|$$

$$= \sum_{k=0}^{N} |h_k| \cdot |e^{-ikt/2^{\ell+1}}| \cdot \left|\frac{e^{-ikt/2^{\ell+1}} - e^{ikt/2^{\ell+1}}}{2i}\right|$$

$$= \sum_{k=0}^{N} |h_k| \cdot |\sin(kt/2^{\ell+1})| \le \left(\sum_{k=0}^{N} |h_k|k\right) \cdot |t| \cdot 2^{-(\ell+1)}$$

because $|\sin(x)| \le |x|$ on $\mathbb{R}$. Thus,

$$\sum_{\ell=1}^{\infty} \left|h\left(e^{-it/2^\ell}\right) - 1\right| \le |t| \cdot \left(\sum_{k=0}^{N} k|h_k|\right) \cdot \sum_{\ell=1}^{\infty} 2^{-(\ell+1)},$$

which converges uniformly on compact intervals; hence, so does the product. Therefore, the sequence of continuous transforms $(\mathcal{F}(T^{\circ n}g))$ converges uniformly on compact intervals to a continuous function $\hat{\varphi}$ defined by

$$\hat{\varphi}(t) := \lim_{n \to \infty} \left(\mathcal{F}T^{on}g \right)(t) = \prod_{\ell=1}^{\infty} \left(\frac{1}{2} \sum_{k=0}^{N} h_k e^{-ikt/2^{\ell}} \right) \int_{\mathbb{R}} g. \qquad (8.9)$$

Corollary 8.9 *There exists at most one function φ satisfying the recursion (8.1) with $\int_{\mathbb{R}} \varphi = 1$.*   □

PROOF: For all functions φ that satisfy (8.1) and such that $\int_{\mathbb{R}} \varphi = 1$, formula (8.9) shows that they all have the same Fourier transform $\hat{\varphi}$. Consequently, if φ_1 and φ_2 both satisfy (8.1) and (8.9), then the invertibility of the Fourier transform shows that

$$\varphi_1 = \mathcal{F}^{-1}[\mathcal{F}(\varphi_1)] = \mathcal{F}^{-1}[\mathcal{F}(\hat{\varphi})] = \mathcal{F}^{-1}[\mathcal{F}(\varphi_2)] = \varphi_2. \qquad □$$

Yet, the existence of the continuous function $\hat{\varphi}$ ensures only the existence of some function φ such that $\mathcal{F}\varphi = \hat{\varphi}$. The continuity of φ itself requires further conditions on $\hat{\varphi}$. For instance, it would suffice that $|\hat{\varphi}|$ and $|(\hat{\varphi})^2|$ be integrable. Under such hypotheses, the function $\varphi := \mathcal{F}^{-1}\hat{\varphi}$ defined by

$$\varphi(x) := \frac{1}{\sqrt{2\pi}} \int_{\mathbb{R}} \hat{\varphi}(t) e^{ixt} \, dt$$

is bounded and uniformly continuous on $\mathbb{R}$, $\mathcal{F}\varphi = \hat{\varphi}$, and $|\varphi^2|$ is integrable [21, pp. 401–411]. By Plancherel's theorem, $\mathcal{F}$ is an isometry on $\mathcal{L}^2(\mathbb{R}, \mathbb{C})$, and, consequently, from the convergence of $(\mathcal{F}T^{on}g)_{n=1}^{\infty}$ to $\hat{\varphi}$ in $\mathcal{L}^2(\mathbb{R}, \mathbb{C})$ follows the convergence of $(T^{on}g)_{n=1}^{\infty}$ to φ in $\mathcal{L}^2(\mathbb{R}, \mathbb{C})$. For $\hat{\varphi}$ to lie in $\mathcal{L}^1(\mathbb{R}, \mathbb{C}) \cap \mathcal{L}^2(\mathbb{R}, \mathbb{C})$, it suffices further that positive constants C and s exist for which $|\hat{\varphi}(t)| \le C/(1 + |t|)^{1+s}$ on $\mathbb{R}$, because then

$$\int_{\mathbb{R}} |\hat{\varphi}(t)| \, dt \le \int_{\mathbb{R}} \frac{C}{(1 + |t|)^{1+s}} \, dt = \frac{2}{-s} \frac{C}{(1+t)^s} \Big|_0^{\infty} = \frac{2C}{s} < \infty,$$

$$\int_{\mathbb{R}} |\hat{\varphi}(t)|^2 \, dt \le \int_{\mathbb{R}} \left[\frac{C}{(1 + |t|)^{1+s}} \right]^2 dt$$

$$= \frac{2}{-(1 + 2s)} \frac{C^2}{(1 + t)^{1+2s}} \Big|_0^{\infty} = \frac{2C^2}{1 + 2s} < \infty.$$

Such inequalities will hold under the conditions described in a subsequent lemma, which first requires a trigonometric infinite product (about which Daubechies quotes Kac, who attributes it to Vieta [7, p. 211]) and a trigonometric inequality.

Lemma 8.10 *For each $z \in \mathbb{C}$,*

$$\prod_{\ell=1}^{\infty} \cos(z/2^{\ell}) = \mathrm{sinc}\,(z) = \begin{cases} \sin(z)/z & \text{if } z \neq 0, \\ 1 & \text{if } z = 0. \end{cases}$$

PROOF: From $\sin(2w) = 2\cos(w)\sin(w)$ it follows that $\cos(w) = \frac{\sin(2w)}{2\sin(w)}$, whence the partial products simplify to one term:

$$\prod_{\ell=1}^{L} \cos(z/2^{\ell}) = \prod_{\ell=1}^{L} \frac{\sin(z/2^{\ell-1})}{2\sin(z/2^{\ell})} = \frac{\sin(z/2^0)}{2^L \sin(z/2^L)}.$$

Hence trigonometry gives $\lim_{t \to 0}[\sin(t)/t] = 1$ and finally

$$\lim_{L \to \infty} \prod_{\ell=1}^{L} \cos(z/2^{\ell}) = \lim_{L \to \infty} \frac{\sin(z/2^0)}{2^L \sin(z/2^L)}$$

$$= \frac{\sin(z)}{z} \lim_{L \to \infty} \frac{z/2^L}{\sin(z/2^L)} = \frac{\sin(z)}{z} \lim_{t \to 0} \frac{t}{\sin(t)} = \frac{\sin(z)}{z}. \quad \square$$

Lemma 8.11 *For every* $t \in \mathbb{R}$,

$$\left| \frac{\sin(t/2)}{t/2} \right| \leq \frac{4}{1 + |t|}.$$

PROOF: If $|t| \leq 1$, then from $|\sin(x)| \leq |x|$ there follows

$$\left| \frac{\sin(t/2)}{t/2} \right| \leq 1 = \frac{2}{1 + 1} \leq \frac{2}{1 + |t|}.$$

If $1 < |t|$, then from $|\sin| \leq 1$ it follows that

$$\left| \frac{\sin(t/2)}{t/2} \right| \leq \frac{1}{|t|/2} = \frac{4}{4|t|/2} = \frac{4}{|t| + |t|} < \frac{4}{1 + |t|}. \quad \square$$

There remains to prove that the infinite product converges in the formula

$$\mathcal{F}\left(T^{\text{on}}g\right)(t) = \prod_{\ell=0}^{n} \left(\frac{1}{2} \sum_{k=0}^{N} h_k e^{-ikt/2^{\ell}} \right) \cdot (\mathcal{F}g)(t/2^n). \tag{8.10}$$

Lemma 8.12 *Suppose that there exists a positive integer* K *such that* $(1 + z)^K$ *divides* $h(z)$, *so that*

$$\frac{1}{2} \sum_{k=0}^{N} h_k z^k = h(z) = \left[(\tfrac{1}{2})(1 + z) \right]^K q(z)$$

for some polynomial q,

$$q(z) = \sum_{k=0}^{N-K} q_k z^k,$$

and let $Q := \max\{|q(e^{-it})| : t \in \mathbb{R}\}$. *Then there exists a positive constant* C *for which*

$$\left| \prod_{\ell=1}^{\infty} h\left(e^{-it/2^{\ell}}\right) \right| \leq C \cdot 2^K (1 + |t|)^{-K + \ln(Q)/\ln(2)}.$$

PROOF: This proof follows Daubechies' original proof in [6, p. 949], with simplifications for the particular situation with only finitely many nonzero coefficients $h_0, \ldots, h_N$. Firstly, from the condition (8.3) that $\sum_{k=0}^{N} h_k = 2$ it follows that $q(1) = 1$:

$$1 = \frac{1}{2} \sum_{k=0}^{N} h_k = h(1) = \left[(\tfrac{1}{2})(1+1) \right]^K q(1) = q(1).$$

Secondly, from

$$\left| e^{-ir} - 1 \right| = \left| \int_0^r (-i) e^{-it} \, dt \right| \le \int_0^{|r|} \left| -i \right| \left| e^{-it} \right| dt = \int_0^{|r|} 1 \, dt = |r|,$$

it follows that

$$\left| q\left(e^{-ir} \right) - 1 \right| = \left| \sum_{k=0}^{N-K} q_k e^{-irk} - \sum_{k=0}^{N-K} q_k \right| = \left| \sum_{k=0}^{N-K} q_k \left(e^{-irk} - 1 \right) \right|$$

$$\le \sum_{k=1}^{N-K} \left| q_k \left(e^{-irk} - 1 \right) \right| \le |r| \sum_{k=1}^{N-K} k \, |q_k| \, .$$

Thus, with $r := t/2^{\ell}$ and with the constant B defined by

$$B := \sum_{k=1}^{N-K} k \, |q_k|$$

it follows that

$$\left| q\left(e^{-it/2^{\ell}} \right) - 1 \right| \le \frac{|-it|}{2^{\ell}} B$$

and hence

$$\sum_{\ell=1}^{\infty} \left| q\left(e^{-it/2^{\ell}} \right) - 1 \right| \le \sum_{\ell=1}^{\infty} \frac{|-it|}{2^{\ell}} B = |t| \, B.$$

Consequently, the infinite product

$$\prod_{\ell=1}^{\infty} q\left(e^{-it/2^{\ell}} \right) = \prod_{\ell=1}^{\infty} \left\{ 1 + \left[q\left(e^{-it/2^{\ell}} \right) - 1 \right] \right\}$$

converges uniformly for every t in any fixed compact interval, with the estimate

$$\left| \prod_{\ell=1}^{\infty} q\left(e^{-it/2^{\ell}} \right) \right| = \prod_{\ell=1}^{\infty} \left| 1 + \left[q\left(e^{-it/2^{\ell}} \right) - 1 \right] \right|$$

$$\leq \prod_{\ell=1}^{\infty} \left\{ 1 + \left| q\left(e^{-it/2^\ell}\right) - 1 \right| \right\} \leq \prod_{\ell=1}^{\infty} \exp \left| q\left(e^{-it/2^\ell}\right) - 1 \right|$$

$$= \exp \sum_{\ell=1}^{\infty} \left| q\left(e^{-it/2^\ell}\right) - 1 \right| \leq \exp\left(B|t|\right).$$

For $|t| > 1$, a smaller estimate results from splitting the infinite product at ℓ_t, defined as the smallest integer larger than or equal to $\log_2(|t|)$, specifically, at $\ell_t := \lceil \log_2(|t|) \rceil$, so that

$$\log_2(|t|) \leq \ell_t \leq 1 + \log_2(|t|),$$

$$|t/2^{\ell_t}| = 2^{\log_2(|t|)-\ell_t} \leq 1,$$

and then

$$\left| \prod_{\ell=1}^{\infty} q\left(e^{-it/2^\ell}\right) \right| = \left| \prod_{\ell=1}^{\ell_t} q\left(e^{-it/2^\ell}\right) \right| \cdot \left| \prod_{\ell=1+\ell_t}^{\infty} q\left(e^{-it/2^\ell}\right) \right|$$

$$\leq Q^{\ell_t} \cdot \left| \prod_{m=1}^{\infty} q\left(e^{-i(t/2^{\ell_t})/2^m}\right) \right|$$

$$\leq Q^{\ell_t} \cdot \exp\left(B|t/2^{\ell_t}|\right) \leq Q^{\ell_t} \cdot \exp\left(B\right)$$

$$\leq \exp\left(B + Q\right) \exp\left(\log(Q)\log_2(|t|)\right) \leq C(1 + |t|)^{\log(Q)/\log(2)}.$$

Hence, from

$$1 + e^{-it/2^\ell} = e^{-it/2^{\ell+1}} \frac{e^{it/2^{\ell+1}} + e^{-it/2^{\ell+1}}}{2} = e^{-it/2^{\ell+1}} \cos(t/2^{\ell+1})$$

and

$$\prod_{\ell=1}^{\infty} e^{-it/2^{\ell+1}} = \exp\left(\sum_{\ell=1}^{\infty} \frac{-it}{2^{\ell+1}}\right) = \exp(-it/2)$$

it follows that

$$\prod_{\ell=1}^{\infty} h\left(e^{-it/2^\ell}\right) = \prod_{\ell=1}^{\infty} \left\{ \left[(\tfrac{1}{2})\left(1 + e^{-it/2^\ell}\right) \right]^K q\left(e^{-it/2^\ell}\right) \right\}$$

$$= \prod_{\ell=1}^{\infty} \left\{ \left[\left(e^{-it/2^{\ell+1}} \cos(t/2^{\ell+1})\right) \right]^K q\left(e^{-it/2^\ell}\right) \right\}$$

$$= \left[e^{-it/2} \prod_{\ell=1}^{\infty} \cos(t/2^{\ell+1}) \right]^K \prod_{\ell=1}^{\infty} q\left(e^{-it/2^{\ell}}\right)$$

$$= e^{-iKt/2} \left(\frac{\sin(t/2)}{2t/2} \right)^K \prod_{\ell=1}^{\infty} q\left(e^{-it/2^{\ell}}\right),$$

where the infinite product converges uniformly by the foregoing argument, and hence

$$\left| \prod_{\ell=1}^{\infty} h\left(e^{-it/2^{\ell}}\right) \right| = \left| e^{-iKt/2} \left(\frac{\sin(t/2)}{2t/2} \right)^K \prod_{\ell=1}^{\infty} q\left(e^{-it/2^{\ell}}\right), \right|$$

$$\leq \frac{2^K}{(1+|t|)^K} C(1+|t|)^{\log(Q)/\log(2)}$$

$$= 2^K C(1+|t|)^{\log(Q)/\log(2)-K}. \qquad \square$$

Remark 8.13   If $K - \ln(Q)/\ln(2) > 1$, then $\varphi \in C^0(\mathbb{R})$. $\qquad \square$

Remark 8.14   The factor $(1+z)^K$ divides $h(z)$ if, but only if, h has a zero of order K at -1. In particular, $0 = h(-1) = (\frac{1}{2}) \sum_{k=0}^{N} h_k(-1)^k$, and collecting the terms of odd degrees and even degrees finally explains condition (8.3). $\qquad \square$

Remark 8.15   A proof similar to that in Example 8.4 shows that if a function g satisfies the condition of orthogonality (8.4), then so does Tg, and by induction, so do all the iterations $T^{\circ n}g$. Since the box function $g = \varphi_{\text{box}}$ satisfies (8.4), it follows that so does $\varphi = \lim_{n \to \infty} T^{\circ n} h_{\text{box}}$, as verified in the next section. $\qquad \square$

EXERCISES

Exercise 8.3.

(a) Prove that if g has its support in the interval $[0, N]$, then Tg also has its support in the interval $[0, N]$.

(b) Prove that Daubechies' building block φ has its support in the interval $[0, 3]$.

Exercise 8.4.

(a) Prove that if φ has its support in the interval $[0, N]$, then ψ has its support in the interval $[(1-N)/2, (1+N)/2]$.

(b) Prove that Daubechies wavelet ψ has its support in the interval $[-1, 2]$, so that the first shifted wavelet ψ_1 with $\psi_1(x) = \psi(x-1)$ has its support in $[0, 3]$.

Exercise 8.5. Compute and plot the "hat" function h with

$$h(x) := \begin{cases} 16\{(\frac{3}{4}) - |x - (\frac{3}{4})|\}/9 & \text{if } 0 \leq x \leq \frac{3}{2}, \\ 0 & \text{otherwise,} \end{cases}$$

and its first four transforms Th, $T^{\circ 2}h$, $T^{\circ 3}h$, and $T^{\circ 4}h$, and compare them with φ.

Exercise 8.6. Consider Daubechies' coefficients h_0, h_1, h_2, h_3.

(a) Verify that

$$h(z) = (\tfrac{1}{2})(h_0 + h_1 z + h_2 z^2 + h_3 z^3)$$
$$= [(\tfrac{1}{2})(1 + z)]^2 [(1 - \sqrt{3})z + (1 + \sqrt{3})]/2,$$

so that $K = 2$ and $q(z) = [(1 - \sqrt{3})z + (1 + \sqrt{3})]/2$.

(b) Verify that if $|z| = 1$, then $|q(z)| \leq \sqrt{3}$, so that $Q = \sqrt{3}$.

(c) Verify that $K - \log_2(Q) = 2 - \log_2(\sqrt{3}) > 1$.

Exercise 8.7. Consider Haar's coefficients $h_0 = 1 = h_1$, so that

$$h(z) = (\tfrac{1}{2})(1 + z).$$

(a) Find the exponent K, the quotient q, and its bound Q.

(b) Investigate whether $K - \log_2(Q) > 1$ for Haar's wavelets.

Exercise 8.8. Consider Haar's coefficients with $h(z) = (1 + z)/2$.

(a) Find a formula for the corresponding recursion operator T.

(b) With the box function g, find $\lim_{n \to \infty} T^{on} g$.

8.2   ORTHOGONALITY OF DAUBECHIES WAVELETS

This section verifies the orthogonality of Daubechies wavelets under the previously stated conditions

$$\varphi(x) = \sum_{k=0}^{N} h_k \varphi(2x - k), \qquad \text{(recursion)} \qquad (8.1)$$

$$\psi(x) := \sum_{k=1-N}^{1} (-1)^k h_{1-k} \varphi(2x - k), \qquad \text{(definition)} \qquad (8.2)$$

$$\sum_{k=0}^{\lfloor N/2 \rfloor} h_{2k} = 1 = \sum_{k=0}^{\lfloor (N-1)/2 \rfloor} h_{2k+1}, \qquad \text{(existence)} \qquad (8.3)$$

$$\int_{\mathbb{R}} \varphi(2x - k)\varphi(2x - \ell)\,dx = 0 \text{ for } k \neq \ell, \qquad \text{(orthogonality)} \qquad (8.4)$$

$$\sum_{k=\max\{0,2m\}}^{\min\{N,N+2m\}} h_k h_{k-2m} = \begin{cases} 2 & \text{if } m = 0, \\ 0 & \text{if } m \neq 0, \end{cases} \qquad \text{(orthogonality)} \qquad (8.5)$$

The first results verify that Daubechies' building block φ satisfies condition (8.4).

Lemma 8.16   *The recursion operator T defined by*

$$(Tg)(x) = \sum_{k=0}^{N} h_k g(2x - k)$$

preserves the "energy" norm $\|g\|_2^2 := \int_{\mathbb{R}} |g(x)|^2 \, dx$, *and preserves condition (8.4), for each compactly supported (or square-integrable) function g that satisfies condition (8.4): for all integers* $p \neq q$,

$$\|Tg\|_2 = \|g\|_2,$$

$$\int_{\mathbb{R}} (Tg)(2x - p)(Tg)(2x - q) \, dx = 0.$$

PROOF: Firstly, for each function $g : \mathbb{R} \to \mathbb{R}$ satisfying condition (8.5), apply conditions (8.4) and (8.5):

$$\|Tg\|_2^2 = \langle Tg, Tg \rangle = \int_{\mathbb{R}} \left[\sum_{k=0}^{N} h_k g(2x - k) \right] \left[\sum_{\ell=0}^{N} h_\ell g(2x - \ell) \right] dx$$

$$= \sum_{k=0}^{N} \sum_{\ell=0}^{N} h_k h_\ell \int_{\mathbb{R}} g(2x - k) g(2x - \ell) \, dx$$

$$= \sum_{k=0}^{N} h_k^2 \int_{\mathbb{R}} [g(2x - k)]^2 \, dx = 2 \int_{\mathbb{R}} [g(z)]^2 \left(\tfrac{1}{2}\right) dz = \langle g, g \rangle = \|g\|_2^2.$$

Secondly,

$$\int_{\mathbb{R}} (Tg)(2x - p)(Tg)(2x - q) \, dx$$

$$= \int_{\mathbb{R}} \left\{ \sum_{k=0}^{N} h_k g(2[2x - p] - k) \right\} \left\{ \sum_{\ell=0}^{N} h_\ell g(2[2x - q] - \ell) \right\} dx$$

$$= \sum_{k=0}^{N} \sum_{\ell=0}^{N} h_k h_\ell \int_{\mathbb{R}} g(2[2x - p] - k) g(2[2x - q] - \ell) \, dx$$

$$= \sum_{k=0}^{N} \sum_{\ell=0}^{N} h_k h_\ell \int_{\mathbb{R}} g(2[2x] - [k + 2p]) g(2[2x] - [\ell + 2q]) \, dx$$

$$= \sum_{k=0}^{N} \sum_{\ell=0}^{N} h_k h_\ell \int_{\mathbb{R}} g(2z - [k + 2p]) g(2z - [\ell + 2q]) \left(\tfrac{1}{2}\right) dz$$

$$= \sum_{k=0}^{N} h_k h_{k-2(q-p)} \int_{\mathbb{R}} g(2z - [k + 2p]) g(2z - [k + 2p]) \left(\tfrac{1}{2}\right) dz$$

$$= \left(\sum_{k=0}^{N} h_k h_{k-2(q-p)} \right) \int_{\mathbb{R}} |g(w)|^2 \left(\frac{1}{4} \right) dw$$

$$= \begin{cases} \left(\frac{1}{2} \right) \|g\|_2^2 & \text{if } m = q - p = 0, \\ 0 & \text{if } m = q - p \neq 0. \end{cases} \qquad \square$$

Lemma 8.17 *Daubechies' function satisfies condition (8.4), and its square has unit integral: for all integers $k \neq \ell$,*

$$\int_{\mathbb{R}} \varphi(2x - k)\varphi(2x - \ell)\, dx = 0,$$

$$\int_{\mathbb{R}} |\varphi(x)|^2\, dx = 1.$$

PROOF: With $g := \chi_{[0,1[}$ denoting the unit "box" function, the preceding section has established that the sequence $(T^{\circ n} g)_{n=0}^{\infty}$ converges *uniformly* to Daubechies' function φ. Therefore, the uniform convergence allows for swapping limits and integration, and the foregoing lemma gives

$$\int_{\mathbb{R}} \varphi(2x - k)\varphi(2x - \ell)\, dx = \int_{\mathbb{R}} \lim_{n \to \infty} (T^{\circ n} g)(2x - k)(T^{\circ n} g)(2x - \ell)\, dx$$

$$= \lim_{n \to \infty} \int_{\mathbb{R}} (T^{\circ n} g)(2x - k)(T^{\circ n} g)(2x - \ell)\, dx$$

$$= \lim_{n \to \infty} 0 = 0,$$

$$\int_{\mathbb{R}} |\varphi(x)|^2\, dx = \int_{\mathbb{R}} \left| \lim_{n \to \infty} (T^{\circ n} g)(2x - k) \right|^2 dx$$

$$= \lim_{n \to \infty} \int_{\mathbb{R}} |(T^{\circ n} g)(2x - k)|^2\, dx = \lim_{n \to \infty} 1 = 1. \square$$

The next results will verify that Daubechies' building blocks remain mutually orthogonal at every location and at every scale.

Definition 8.18 For all indices $k, m \in \mathbb{Z}$, define

$$\varphi_k^{(m)}(x) := \varphi(2^m x - k). \qquad \square$$

Proposition 8.19 *For all indices $k, \ell, m \in \mathbb{Z}$,*

$$\langle \varphi_k^{(m)}, \varphi_\ell^{(m)} \rangle = \begin{cases} 2^{-m} & \text{if } k = \ell, \\ 0 & \text{if } k \neq \ell. \end{cases}$$

PROOF: For $k \neq \ell$, perform the change of variable $z := 2^{m-1}x$ and apply condition (8.4):

$$\langle \varphi_k^{(m)}, \varphi_\ell^{(m)} \rangle = \int_{\mathbb{R}} \varphi_k^{(m)}(x)\, \varphi_\ell^{(m)}(x)\, dx$$

$$= \int_{\mathbb{R}} \varphi(2^m x - k)\, \varphi(2^m x - \ell)\, dx$$

$$= \int_{\mathbb{R}} \varphi(2z - k)\, \varphi(2z - \ell)\, 2^{1-m}\, dz = 0.$$

For $k = \ell$, perform the change of variable $w := 2^m x - k$ and use the condition $\int_{\mathbb{R}} \varphi(x)\, dx = 1$:

$$\langle \varphi_k^{(m)}, \varphi_k^{(m)} \rangle = \int_{\mathbb{R}} \varphi_k^{(m)}(x)\, \varphi_k^{(m)}(x)\, dx$$

$$= \int_{\mathbb{R}} \varphi(2^m x - k)\, \varphi(2^m x - k)\, dx$$

$$= \int_{\mathbb{R}} \varphi(w)\, \varphi(w)\, 2^{-m}\, dw = 2^{-m}. \qquad \square$$

Lemma 8.20   *Under conditions (8.4) and (8.5), and for each integer m and for all distinct integers $h \neq k$, the wavelets $\psi_h^{(m)}$ and $\psi_k^{(m)}$ are also mutually orthogonal:*

$$\langle \psi_h^{(m)}, \psi_k^{(m)} \rangle = \begin{cases} 2^{-m} & \text{if } h = k, \\ 0 & \text{if } h \neq k. \end{cases}$$

PROOF: Expand each wavelet $\psi_k^{(\ell)}$ into a linear combination of functions of the form $\varphi_k^{(\ell)}$, by definition (8.2), and apply conditions (8.4) and (8.5):

$$\langle \psi_h^{(m)}, \psi_k^{(m)} \rangle = \int_{\mathbb{R}} \psi(2^m x - h)\psi(2^m x - k)\, dx$$

$$= \sum_p \sum_q (-1)^p (-1)^q h_{1-p} h_{1-q}$$

$$\times \int_{\mathbb{R}} \varphi(2(2^m x - h) - p)\varphi(2(2^m x - k) - q)\, dx$$

$$= \sum_p (-1)^p (-1)^{p-2(k-h)} h_{1-p} h_{1-p+2(k-h)}$$

$$\times \int_{\mathbb{R}} \varphi(2^{m+1}x - 2h - p)^2\, dx$$

$$= \left(\sum_p (-1)^{2(p+h-k)} h_{(1-p)} h_{(1-p)-2(h-k)} \right) \int_{\mathbb{R}} \varphi(z)^2\, 2^{-(m+1)}\, dz$$

$$= \begin{cases} 2^{-m} \|\varphi\|_2^2 & \text{if } h = k, \\ 0 & \text{if } h \neq k. \end{cases} \qquad \square$$

Lemma 8.21   *Under condition (8.4), each wavelet $\psi_h^{(m)}$ is orthogonal to every function $\varphi_k^{(m)}$, for all integers h and k.*

PROOF: The present proof differs only slightly from the preceding one:

$$\int_{\mathbb{R}} \psi(x-h)\varphi(x-k)\,dx$$

$$= \int_{\mathbb{R}} \left(\sum_p (-1)^p h_{1-p} \varphi(2(x-h)-p) \right) \left(\sum_q h_q \varphi(2(x-k)-q) \right) dx$$

$$= \sum_p (-1)^p h_{1-p} h_{p-2(k-h)} \int_{\mathbb{R}} \varphi(2x-2h-p)^2\,dx$$

because the integrals for which $q \neq p - 2(k-h)$ all vanish, by condition (8.4). The sum now also vanishes, but for a different reason: Setting $m := k - h$ gives

$$\sum_p (-1)^p h_{1-p} h_{p-2(k-h)} = \sum_p (-1)^p h_{1-p} h_{p-2m},$$

and then setting $n := 2m + 1 - p$, so that $p = 2m - n + 1$, leads to

$$\sum_p (-1)^p h_{1-p} h_{p-2m} = \sum_n (-1)^{2m+1-n} h_{n-2m} h_{1-n}$$

$$= (-1)^{2m+1} \sum_n (-1)^n h_{1-n} h_{n-2m}$$

$$= -\sum_n (-1)^n h_{1-n} h_{n-2m}.$$

In the string of equalities just established, adding the last sum to the first sum now yields

$$2\sum_p (-1)^p h_{1-p} h_{p-2m} = 0.$$

Finally, the change of variable $z := 2^m x$ shows that every wavelet $\psi_h^{(m)}$ is orthogonal to every function $\varphi_k^{(m)}$.   □

A similar proposition holds for the orthogonality of $\psi_h^{(m)}$ and $\psi_k^{(m)}$.

Lemma 8.22   *Under conditions (8.4) and (8.5), every wavelet $\psi_h^{(m)}$ is orthogonal to every wavelet $\psi_k^{(n)}$ for all $(m, h) \neq (n, k)$.*

PROOF: For $m = n$ and all $h \neq k$, the proposition follows from a previous result. For $m < n$, and for all h and k,

$$\int_{\mathbb{R}} \psi(2^m x - h)\psi(2^n x - k)\,dx$$

$$= \sum_{p_1} (-1)^{p_1} h_{1-p_1} \int_{\mathbb{R}} \varphi(2(2^m x - h) - p_1) \psi(2^n x - k) \, dx$$

$$= \sum_{p_1, \ldots, p_{n-m}} (-1)^{p_1} \cdots (-1)^{p_{n-m}} h_{1-p_1} \cdots h_{1-p_{n-m}}$$

$$\times \int_{\mathbb{R}} \varphi \left(2^n x - 2^{n-m} h - 2^{n-m-1} p_1 - \cdots - 2 p_{n-m-1} - p_{n-m} \right) \psi(2^n x - k) \, dx$$

$$= 0. \qquad\qquad \square$$

A particular case not included in the previous discussion pertains to the constant function.

Lemma 8.23   *Under condition (8.3), the constant function* **1** *is also orthogonal to every wavelet.*

PROOF:

$$\langle 1, \psi_h^{(m)} \rangle = \int_{\mathbb{R}} 1 \psi(2^m x - h) \, dx = \sum_p (-1)^p h_{1-p} \int_{\mathbb{R}} 1\varphi(2(2^m x - h) - p) \, dx$$

$$= \sum_p (-1)^p h_{1-p} \int_{\mathbb{R}} \varphi(z) 2^{-(m+1)} dz = 0$$

because

$$\sum_p (-1)^p h_{1-p} = \sum_q (-1)^{1-q} h_q = -\sum_q (-1)^q h_q = 1 - 1 = 0,$$

as the difference between the sum of the coefficients with odd indices and the sum of those with even indices vanishes, by (8.3). $\square$

Theorem 8.24   *Under conditions (8.4) and (8.5), the set of all wavelets $\psi_\ell^{(m)}$ for all ℓ and all nonnegative m, together with all basic building blocks $\varphi_k^{(0)}$, is orthogonal and linearly independent. Under conditions (8.3), (8.4), and (8.5), the set of all wavelets $\psi_\ell^{(m)}$, together with the constant function* **1**, *is orthogonal and linearly independent.*

PROOF: The proof consists of the preceding considerations. $\square$

8.3   MALLAT'S FAST WAVELET ALGORITHM

The Fast Haar Wavelet Transform generalizes to wavelets and basic building blocks with nonzero integrals that satisfy the conditions assumed thus far, for example, for Daubechies wavelets. The generalization will then justify the algorithm described in a previous chapter to compute the Fast Daubechies Wavelet Transform. To generalize the Fast Haar Wavelet Transform to such wavelets, for each positive integer M, and for each function f defined on $[0, N]$, approximate

f by a linear combination $\tilde{f}$ of the basic building blocks:

$$\tilde{f} := \sum_i a_i^{(M)} \varphi_i^{(M)}.$$

For the least-squares approximation, let

$$a_i^{(M)} := \frac{\langle f, \varphi_i^{(M)} \rangle}{\langle \varphi_i^{(M)}, \varphi_i^{(M)} \rangle}.$$

In practice, however, f and $\varphi_i^{(M)}$ remain incompletely known, whence approximations of the inner products become necessary, for instance, by means of Riemann sums involving the dyadic midpoints computed by recursion in a previous section (see also [16, p. 64]):

$$a_i^{(M)} = \left\langle f, \varphi_i^{(M)} \right\rangle = \int_{\mathbb{R}} f(x) \varphi_i^{(M)}(x)\, dx$$

$$\approx \sum_{j=i}^{i+N} f((2j+1)/2^{M+1}) \varphi_i^{(M)}((2j+1)/2^{M+1}) 2^{-M},$$

where the indices $j \in \{i, \dots, i+N\}$ correspond to the support $[i/2^M, (i+N)/2^M]$ of $\varphi_i^{(M)}$.

Next, denote by

$$V_M := \text{Span}\left\{\varphi_h^{(M)} : h \in \mathbb{Z}\right\},$$

$$V_M^{\perp} := \text{Span}\left\{\psi_h^{(M)} : h \in \mathbb{Z}\right\}$$

the linear subspaces spanned by the building blocks $\varphi_h^{(M)}$ and by the wavelets $\psi_h^{(M)}$, which are mutually orthogonal by virtue of the preceding sections. By the basic recursion and definition, $V_M \subset V_{M+1}$, $V_M^{\perp} \subset V_{M+1}$, and $V_M \oplus V_M^{\perp} = V_{M+1}$. In particular, $V_{M-1} \oplus V_{M-1}^{\perp} = V_M$, which means that $\tilde{f} = \sum_i a_i^{(M)} \varphi_i^{(M)}$ may also be expressed as a linear combination of the slower blocks $\varphi_\ell^{(M-1)}$ and slower wavelets $\psi_\ell^{(M-1)}$:

$$\tilde{f} = \sum_i a_i^{(M)} \varphi_i^{(M)} = \sum_i a_i^{(M-1)} \varphi_i^{(M-1)} + \sum_j c_j^{(M-1)} \psi_j^{(M-1)}.$$

Again by orthogonality of all the $\varphi_i^{(M)}$ and $\psi_h^{(M)}$, inner products yield the coefficients $a_\ell^{(M-1)}$ and $c_\ell^{(M-1)}$:

$$\left\langle \tilde{f}, \varphi_\ell^{(M-1)} \right\rangle = \left\langle \sum_i a_i^{(M)} \varphi_i^{(M)}, \sum_{k=0}^N h_k \varphi_{2\ell+k}^{(M)} \right\rangle$$

$$= \sum_i \sum_{k=0}^N a_i^{(M)} h_k \left\langle \varphi_i^{(M)}, \varphi_{2\ell+k}^{(M)} \right\rangle$$

$$= \sum_{k=0}^N a_{2\ell+k}^{(M)} h_k \left\langle \varphi_0^{(M)}, \varphi_0^{(M)} \right\rangle.$$

Consequently, since $\left\langle \varphi_\ell^{(M-1)}, \varphi_\ell^{(M-1)} \right\rangle = 2 \left\langle \varphi_k^{(\ell)}, \varphi_k^{(\ell)} \right\rangle$, it follows that

$$a_\ell^{(M-1)} = \frac{\left\langle \tilde{f}, \varphi_\ell^{(M-1)} \right\rangle}{\left\langle \varphi_\ell^{(M-1)}, \varphi_\ell^{(M-1)} \right\rangle} = \frac{1}{2} \sum_{k=0}^N h_k a_{2\ell+k}^{(M)}, \qquad (8.11)$$

which proves the formulae for the averages in the Fast Daubechies Wavelet Transform. Similarly,

$$\left\langle \tilde{f}, \psi_\ell^{(M-1)} \right\rangle = \left\langle \sum_i a_i^{(M)} \varphi_i^{(M)}, \sum_{k=1-N}^1 (-1)^k h_{1-k} \varphi_{2\ell+k}^{(M)} \right\rangle$$

$$= \sum_i \sum_{k=1-N}^1 a_i^{(M)} (-1)^k h_{1-k} \left\langle \varphi_k^{(\ell)}, \varphi_{2\ell+k}^{(M)} \right\rangle$$

$$= \sum_{k=-N+1}^1 (-1)^k h_{1-k} a_{2\ell+k}^{(M)} \left\langle \varphi_0^{(M)}, \varphi_0^{(M)} \right\rangle,$$

whence

$$c_\ell^{(M-1)} = \frac{\left\langle \tilde{f}, \psi_\ell^{(M-1)} \right\rangle}{\left\langle \psi_\ell^{(M-1)}, \psi_\ell^{(M-1)} \right\rangle} = \frac{1}{2} \sum_{k=1-N}^1 (-1)^k h_{1-k} a_{2\ell+k}^{(M)}, \qquad (8.12)$$

which proves the formulae for the wavelet coefficients in the Fast Daubechies Wavelet Transform.

The algorithm described by equations (8.11) and (8.12) applies to, instead of $\tilde{f}$, the new function

$$\tilde{f}^{(M-1)} = \sum_i a_i^{(M-1)} \varphi_i^{(M-1)},$$

which yields new coefficients $a_i^{(M-2)}$ and $c_i^{(M-2)}$, and so forth. The iterations stop with $\tilde{f}^{(0)} = \sum_i a_i^{(0)} \varphi_i^{(0)} + \sum_j c_j^{(0)} \psi_j^{(0)}$.

For further details about the numerical computations of Daubechies wavelet coefficients and wavelet transforms with applications to partial differential equations, see [16]. For the design of other types of wavelets and their application to geodesy, see [13].

CHAPTER 9

Signal Representations with Wavelets

9.0 INTRODUCTION

Using the example of Daubechies wavelets, this chapter demonstrates how to investigate the accuracy with which recursive, mutually orthogonal, compactly supported, and continuous wavelets can represent signals (mathematical functions). The proofs will depend upon the following conditions on the recursion coefficients:

$$\varphi(x) = \sum_{k=0}^{N} h_k \varphi(2x - k), \qquad \text{(recursion)} \qquad (9.1)$$

$$\psi(x) := \sum_{k=1-N}^{1} (-1)^k h_{1-k} \varphi(2x - k), \qquad \text{(definition)} \qquad (9.2)$$

$$\sum_{k=0}^{\lfloor N/2 \rfloor} h_{2k} = 1 = \sum_{k=0}^{\lfloor (N-1)/2 \rfloor} h_{2k+1}, \qquad \text{(existence)} \qquad (9.3)$$

$$\int_{\mathbb{R}} \varphi(2x - k)\varphi(2x - \ell)\, dx = 0 \text{ for } k \neq \ell, \qquad \text{(orthogonality)} \qquad (9.4)$$

$$\sum_{k=\max\{0,2m\}}^{\min\{N,N+2m\}} h_k h_{k-2m} = \begin{cases} 2 & \text{if } m = 0, \\ 0 & \text{if } m \neq 0. \end{cases} \qquad \text{(orthogonality)} \qquad (9.5)$$

9.1 COMPUTATIONAL FEATURES OF DAUBECHIES WAVELETS

Based on the recursion that defines Daubechies wavelets, this section derives further computational features of wavelets, which will help in assessing how accurately wavelets can represent signals.

9.1.1 Initial Values of Daubechies' Scaling Function

The preceding chapter has established the existence of exactly one continuous function φ that satisfies the recursion (9.1) with unit integral $\int_{\mathbb{R}} \varphi = 1$. The recursion (9.1) revealed that $\varphi(x) = 0$ for every x outside the interval $[0, N]$, but the proof of the existence and uniqueness did not yield any specific values for the function φ inside the interval $[0, N]$. The calculation of such values forms the object of this section, beginning with the values $\varphi(n)$ at the integers $n \in \{1, \ldots, N-1\}$.

Writing the recursion (9.1) for each $n \in \{1, \ldots, N-1\}$ gives

$$\varphi(n) = \sum_{k=0}^{N} h_k \varphi(2n - k),$$

and listing all such equations for $n \in \{1, \ldots, N-1\}$ produces a linear system of $N - 1$ equations:

$$\varphi(1) = h_1 \varphi(1) + h_0 \varphi(2),$$

$$\varphi(2) = h_3 \varphi(1) + h_2 \varphi(2) + h_1 \varphi(3) + h_0 \varphi(4),$$

$$\vdots$$

$$\varphi(N - 2) = h_N \varphi(N - 4) + h_{N-1} \varphi(N - 3)$$
$$+ h_{N-2} \varphi(N - 2) + h_{N-3} \varphi(N - 1),$$

$$\varphi(N - 1) = h_N \varphi(N - 2) + h_{N-1} \varphi(N - 1).$$

This system is equivalent to the statement that the vector $\vec{x}$ of values of φ at the integers,

$$\vec{x} := (\varphi(1), \varphi(2), \ldots, \varphi(N - 2), \varphi(N - 1))^T,$$

is an eigenvector with eigenvalue 1 for the matrix

$$A := \begin{pmatrix} h_1 & h_0 & & & & & \\ h_3 & h_2 & h_1 & h_0 & & & \\ h_5 & h_4 & h_3 & h_2 & h_1 & h_0 & \\ \vdots & & & & \ddots & & \\ & & & & h_N & h_{N-1} & h_{N-2} & h_{N-3} \\ & & & & & & h_N & h_{N-1} \end{pmatrix},$$

which has entries $a_{i,j} = h_{2i-j}$ if $2i - j \in \{0, \ldots, N\}$ and $a_{i,j} = 0$ if $2i - j \notin \{0, \ldots, N\}$. The columns of A with odd indices contain the coefficients with odd indices, and similarly, the columns of A with even indices contain the coefficients with even indices. In particular, if condition (9.3) holds,

$$\sum_{k=0}^{\lfloor N/2 \rfloor} h_{2k} = 1 = \sum_{k=0}^{\lfloor (N-1)/2 \rfloor} h_{2k+1},$$

then the sum of all the entries down each column of A equals one: $\sum_{i=1}^{N-1} a_{i,j} = 1$ for every $j \in \{1, \ldots, N-1\}$. This means that the vector $\vec{1} := (1, 1, \ldots, 1, 1)^T$ is an eigenvector with eigenvalue 1 for the transposed matrix A^T, because $A^T \vec{1}$ has all entries equal to

$$\left(A^T \vec{1}\right)_j = \sum_{i=1}^{N-1} (A^T)_{j,i} \vec{1}_i = \sum_{i=1}^{N-1} a_{i,j} 1 = 1 = \vec{1}_j.$$

Thus, $A^T \vec{1} = \vec{1}$. Since A^T and A have the same eigenvalues, A also has the eigenvalue 1 for some eigenvector $\vec{x}$. Defining

$$\varphi(n) := x_n$$

then yields the values of φ at the integers, and then the recursion gives the values of φ at all the "dyadic" points, of the form $\varphi(k/2^m)$ for each positive integer m and all integers $k \in \{0, \ldots, 2^m N\}$.

Example 9.1   For the particular example of Daubechies wavelets demonstrated here, $N = 3$, and the coefficients have the values

$$h_0 := \frac{1 + \sqrt{3}}{4},$$

$$h_1 := \frac{3 + \sqrt{3}}{4},$$

$$h_2 := \frac{3 - \sqrt{3}}{4},$$

$$h_3 := \frac{1 - \sqrt{3}}{4}. \tag{9.6}$$

Thus, $N = 3$, $N - 1 = 2$, and $h_0 + h_2 = 1 = h_1 + h_3$. Also, with A_D denoting the matrix A in the context of Daubechies wavelets,

$$A_D = \begin{pmatrix} h_1 & h_0 \\ h_3 & h_2 \end{pmatrix} = \begin{pmatrix} (3 + \sqrt{3})/4 & (1 + \sqrt{3})/4 \\ (1 - \sqrt{3})/4 & (3 - \sqrt{3})/4 \end{pmatrix}$$

has eigenvalues 1 and $\frac{1}{2}$. Because the corresponding building block function φ equals zero at the endpoints, φ has values $\varphi(0) = 0 = \varphi(3)$, and solving the system

$$(A_D - I)(\varphi(1), \varphi(2))^T = \vec{0},$$

$$\begin{pmatrix} (3 + \sqrt{3})/4 - 1 & (1 + \sqrt{3})/4 \\ (1 - \sqrt{3})/4 & (3 - \sqrt{3})/4 - 1 \end{pmatrix} \begin{pmatrix} \varphi(1) \\ \varphi(2) \end{pmatrix} = \begin{pmatrix} 0 \\ 0 \end{pmatrix}$$

yields scalar multiples of

$$\varphi(1) := \frac{1+\sqrt{3}}{2},$$

$$\varphi(2) := \frac{1-\sqrt{3}}{2}, \tag{9.7}$$

normalized here so that $\varphi(1)+\varphi(2) = 1$ approximates $\int_{\mathbf{R}} \varphi = 1$ by the composite midpoint rule. □

Not only for Daubechies' wavelets, but also for all wavelets satisfying conditions (9.1)–(9.5), the additional condition that the sum of the values at the integers equals one guarantees that the integral also equals one.

Proposition 9.2 *For each sequence of coefficients $h_0, \ldots, h_N$ and each continuous function φ satisfying conditions (9.1)–(9.5), if the additional condition*

$$\sum_{n \in \mathbf{Z}} \varphi(n) = 1 \tag{9.8}$$

holds, then

$$\int_{\mathbf{R}} \varphi = 1.$$

PROOF: This proof determines the integral inductively through a limit of Riemann sums at the midpoints. To this end, assume that for some nonnegative integer $m \in \mathbf{N}$, the following equation holds (which it does for $m := 0$ by hypothesis):

$$\sum_{n \in \mathbf{Z}} \varphi(n/2^m) 2^{-m} = 1. \tag{9.9}$$

Then the same equation holds for $m + 1$ instead of m:

$$\sum_{n \in \mathbf{Z}} \varphi(n/2^{m+1}) 2^{-(m+1)} = \sum_{n \in \mathbf{Z}} \sum_{k=0}^{N} h_k \varphi(2[n/2^{m+1}] - k) 2^{-(m+1)}$$

$$= \sum_{n \in \mathbf{Z}} \sum_{k=0}^{N} h_k \varphi(n/2^m - k) 2^{-(m+1)}$$

$$= \sum_{n \in \mathbf{Z}} \sum_{k=0}^{N} h_k \varphi([n - k2^m]/2^m) 2^{-(m+1)}$$

$$= \sum_{k=0}^{N} h_k \sum_{n \in \mathbf{Z}} \varphi([n - k2^m]/2^m) 2^{-(m+1)}$$

$$= \frac{1}{2} \sum_{k=0}^{N} h_k \sum_{\ell \in \mathbb{Z}} \varphi(\ell/2^m) \, 2^{-m}$$

$$= \frac{1}{2} \sum_{k=0}^{N} h_k = 1.$$

Consequently, equation (9.9) holds for every nonnegative integer $m \in \mathbb{N}$. Moreover, the continuity of φ ensures that Riemann sums converge to its integrals. Therefore, Riemann sums with midpoints separated by intervals of length 2^{-m} give

$$\int_{\mathbb{R}} \varphi = \int_0^N \varphi = \lim_{m \to \infty} \sum_{n \in \mathbb{Z}} \varphi(n/2^m) \, 2^{-m} = \lim_{m \to \infty} 1 = 1. \qquad \square$$

9.1.2   Computational Features of Daubechies' Function

Daubechies wavelets have algebraic features that allow for calculations with fewer operations than described earlier, which may not only speed up such calculations but also increase the accuracy of the results. For instance, this subsection will verify the following algebraic identities, valid for each dyadic number $r \in \mathbb{D}$:

$$\varphi(r) \in \mathbb{D}[\sqrt{3}],$$

$$\overline{\varphi(r)} = \varphi(3 - r),$$

$$1 = \sum_{k \in \mathbb{Z}} \varphi(r - k),$$

$$r = \sum_{k \in \mathbb{Z}} \left(\frac{3 - \sqrt{3}}{2} + k \right) \varphi(r - k).$$

Completing David Pollen's outline in [4, pp. 5–6], the following proofs illustrate the type of argument that establishes the foregoing formulae from the initial values, recurrence relations, and induction on the index n of $\mathbb{D}_n$. The first two propositions show how it suffices to calculate values $\varphi(r)$ only for $0 \le r \le \frac{3}{2}$, because the values $\varphi(r)$ then follow through conjugation.

Proposition 9.3   *For each dyadic number $r \in \mathbb{D}$, $\varphi(r) \in \mathbb{D}[\sqrt{3}]$.*

PROOF: The verification that $\varphi(r) \in \mathbb{D}[\sqrt{3}]$ for each dyadic number $r \in \mathbb{D}$ proceeds by induction on the index n of $\mathbb{D}_n$.

For $n = 0$, the set $\mathbb{D}_0$ consists of all the integers: $\mathbb{D}_0 = \mathbb{Z}$. For integral values r, however, the definition of the initial values $\varphi(r)$ already confirms that $\varphi(r) \in \mathbb{D}[\sqrt{3}]$:

$$\varphi(0) := 0 \in \mathbb{D}[\sqrt{3}],$$

$$\varphi(1) := \frac{1 + \sqrt{3}}{2} \in \mathbb{D}[\sqrt{3}],$$

$$\varphi(2) := \frac{1 - \sqrt{3}}{2} \in \mathbb{D}[\sqrt{3}],$$

$$\varphi(3) := 0 \in \mathbb{D}[\sqrt{3}].$$

To proceed by induction, for some integer k assume that $\varphi(r) \in \mathbb{D}[\sqrt{3}]$ for each dyadic number $r \in \mathbb{D}_k$. To verify that $\varphi(r) \in \mathbb{D}[\sqrt{3}]$ also holds for every $r \in \mathbb{D}_{k+1}$, observe that if $r \in \mathbb{D}_{k+1}$, then

$$2r, 2r - 1, 2r - 2, 2r - 3 \in \mathbb{D}_k,$$

whence the induction hypotheses ensure that

$$\varphi(2r), \varphi(2r - 1), \varphi(2r - 2), \varphi(2r - 3) \in \mathbb{D}[\sqrt{3}],$$

and all multiples of $\varphi(2r), \varphi(2r - 1), \varphi(2r - 2), \varphi(2r - 3)$ by coefficients in $\mathbb{D}[\sqrt{3}]$ remain in $\mathbb{D}[\sqrt{3}]$. Then apply the recurrence relation:

$$\varphi(r) = \frac{1 + \sqrt{3}}{4} \varphi(2r) + \frac{3 + \sqrt{3}}{4} \varphi(2r - 1) + \frac{3 - \sqrt{3}}{4} \varphi(2r - 2)$$

$$+ \frac{1 - \sqrt{3}}{4} \varphi(2r - 3) \in \mathbb{D}[\sqrt{3}]. \qquad \square$$

Proposition 9.4   *For each dyadic number $r \in \mathbb{D}$,*

$$\varphi(3 - r) = \overline{\varphi(r)}.$$

PROOF: The verification that $\varphi(3 - r) = \overline{\varphi(r)}$ for each dyadic number $r \in \mathbb{D}$ proceeds by induction on the index n of $\mathbb{D}_n$.

For $n = 0$, the set $\mathbb{D}_0$ consists of all the integers: $\mathbb{D}_0 = \mathbb{Z}$. For integral values r, however, the definition of the initial values $\varphi(r)$ already confirms that the proposition holds:

$$\varphi(1) = \frac{1 + \sqrt{3}}{2}, \quad \varphi(2) = \frac{1 - \sqrt{3}}{2} = \overline{(1 + \sqrt{3})/2} = \overline{\varphi(1)} = \varphi(3 - 1).$$

To proceed by induction, for some integer k assume that $\varphi(3 - r) = \overline{\varphi(r)}$ for each dyadic number $r \in \mathbb{D}_k$. To verify that the same formula holds for every $r \in \mathbb{D}_{k+1}$, observe that if $r \in \mathbb{D}_{k+1}$, then $2r \in \mathbb{D}_k$, whence the induction hypotheses ensure that $\varphi(2r) \in \mathbb{D}[\sqrt{3}]$, and $\varphi(3 - 2r) = \overline{\varphi(2r)}$. Then apply the recurrence relation:

$\varphi(3 - r)$

$$= \frac{1 + \sqrt{3}}{4} \varphi(2[3 - r]) + \frac{3 + \sqrt{3}}{4} \varphi(2[3 - r] - 1)$$

$$+ \frac{3 - \sqrt{3}}{4} \varphi(2[3 - r] - 2) + \frac{1 - \sqrt{3}}{4} \varphi(2[3 - r] - 3)$$

$$= \frac{1 + \sqrt{3}}{4} \varphi(3 - [2r - 3]) + \frac{3 + \sqrt{3}}{4} \varphi(3 - [2r - 2])$$

$$+ \frac{3 - \sqrt{3}}{4} \varphi(3 - [2r - 1]) + \frac{1 - \sqrt{3}}{4} \varphi(3 - [2r - 0])$$

$$= \overline{(1 - \sqrt{3})/4} \, \overline{\varphi(2r - 3)} + \overline{(3 - \sqrt{3})/4} \, \overline{\varphi(2r - 2)}$$

$$+ \overline{(3 + \sqrt{3})/4} \, \overline{\varphi(2r - 1)} + \overline{(1 + \sqrt{3})/4} \, \overline{\varphi(2r - 0)}$$

$$= \overline{\tfrac{1+\sqrt{3}}{4} \varphi(2r) + \tfrac{3+\sqrt{3}}{4} \varphi(2r - 1) + \tfrac{3-\sqrt{3}}{4} \varphi(2r - 2) + \tfrac{1-\sqrt{3}}{4} \varphi(2r - 3)}$$

$$= \overline{\varphi(r)}. \qquad\qquad \square$$

Example 9.5 The following numerical calculations illustrate the identity $\varphi(3 - r) = \overline{\varphi(r)}$.

1. From $\varphi(1) = \frac{1+\sqrt{3}}{2}$ follows

$$\varphi(2) = \varphi(3 - 1) = \overline{\varphi(1)} = \overline{(1 + \sqrt{3})/2} = \frac{1 - \sqrt{3}}{2}.$$

2. From $\varphi(\tfrac{3}{4}) = \frac{9+5\sqrt{3}}{16}$ follows

$$\varphi\left(\frac{9}{4}\right) = \varphi\left(3 - \frac{3}{4}\right) = \overline{\varphi\left(\frac{3}{4}\right)} = \overline{(9 + 5\sqrt{3})/16} = \frac{9 - 5\sqrt{3}}{16}. \qquad \square$$

The following proposition shows that the values of φ at any points at integral distances from one another add up to 1, a feature called **partition of unity**.

Proposition 9.6   *For each dyadic number $r \in \mathbb{D}$,*

$$1 = \sum_{m \in \mathbb{Z}} \varphi(r - m)$$

$$= \cdots + \varphi(r - 2) + \varphi(r - 1) + \varphi(r) + \varphi(r + 1) + \varphi(r + 2) + \cdots.$$

PROOF: This proof proceeds by induction on the index n of $\mathbb{D}_n$. For $n := 0$, the dyadics $\mathbb{D}_0$ are the integers $\mathbb{Z}$. Thus, if r represents an integer, then every shift $r \pm m$ of r by an integer m is also an integer, but only two such integers give a nonzero value of $\varphi(r \pm m)$, whence

$$\cdots + \varphi(r - 2) + \varphi(r - 1) + \varphi(r) + \varphi(r + 1) + \varphi(r + 2) + \cdots$$

$$= \cdots + 0 + \varphi(1) + \varphi(2) + 0 + \cdots = \frac{1 + \sqrt{3}}{2} + \frac{1 - \sqrt{3}}{2} = 1.$$

Thus the proposition holds for $n = 0$. Hence, assume the proposition for some n and consider $r \in \mathbb{D}_{n+1}$. Then $r + m \in \mathbb{D}_{n+1}$ and $2[r + m] \in \mathbb{D}_n$ for every integer m. The recursion

$$\varphi(r + m) = h_0 \varphi(2[r + m]) + h_1 \varphi(2[r + m] - 1)$$

$$+ h_2 \varphi(2[r + m] - 2) + h_3 \varphi(2[r + m] - 3)$$

$$= h_0 \varphi(2r + 2m) + h_1 \varphi(2r + 2m - 1)$$

$$+ h_2 \varphi(2r + 2m - 2) + h_3 \varphi(2r + 2m - 3)$$

and the hypothesis of induction both applied to every m then confirm that the proposition also holds for $r \in \mathbb{D}_n$:

$$\sum_{m \in \mathbb{Z}} \varphi(r - m) = \sum_{m \in \mathbb{Z}} [h_0 \varphi(2r + 2m) + h_1 \varphi(2r + 2m - 1)$$

$$+ h_2 \varphi(2r + 2m - 2) + h_3 \varphi(2r + 2m - 3)]$$

$$= (h_0 + h_2) \sum_{m \in \mathbb{Z}} [\varphi(2r + 2m) + \varphi(2r + 2m - 2)]$$

$$+ (h_1 + h_3) \sum_{m \in \mathbb{Z}} [\varphi(2r + 2m - 1) + \varphi(2r + 2m - 3)]$$

$$= (1) \sum_{k \in \mathbb{Z}} [\varphi(2r + 2k)] + (1) \sum_{k \in \mathbb{Z}} [\varphi(2r + 2k - 1)]$$

$$= (1) \sum_{k \in \mathbb{Z}} [\varphi(2r + 2k) + \varphi(2r + 2k - 1)]$$

$$= \sum_{m \in \mathbb{Z}} \varphi(2r + m)$$

$$= 1. \qquad \qquad \square$$

With the foregoing result, the next lemma and proposition will show in the next subsection that linear combinations of Daubechies wavelets can reproduce any affine function (polynomial of degree at most one) exactly.

Lemma 9.7 *For Daubechies' coefficients*

$$h_0 := \frac{1 + \sqrt{3}}{4},$$

$$h_1 := \frac{3 + \sqrt{3}}{4},$$

$$h_2 := \frac{3 - \sqrt{3}}{4},$$

$$h_3 := \frac{1 - \sqrt{3}}{4},$$

and with $\mu_1 := \sum_{k=0}^{N} (-1)^k k h_k$, the following relation holds:

$$0 = \mu_1 = \sum_{k=0}^{N} (-1)^k k h_k = 0 h_0 - 1 h_1 + 2 h_2 - 3 h_3;$$

or, equivalently, with even indices and odd indices separated,

$$\sum_{\ell=0}^{\lfloor N/2 \rfloor} (2\ell) h_{2\ell} = \sum_{\ell=0}^{\lfloor (N-1)/2 \rfloor} (2\ell + 1) h_{2\ell+1},$$

$$0 h_0 + 2 h_2 = 1 h_1 + 3 h_3.$$

Moreover,

$$\sum_{k=0}^{N} k \varphi(k) = \sum_{k=0}^{\lfloor N/2 \rfloor} (2k) h_{2k}.$$

PROOF: Straightforward calculations confirm that

$$\sum_{k=0}^{N} (-1)^k k h_k = 0 \frac{1 + \sqrt{3}}{4} - \frac{3 + \sqrt{3}}{4} + 2 \frac{3 - \sqrt{3}}{4} - 3 \frac{1 - \sqrt{3}}{4} = 0;$$

$$\sum_{k=0}^{N} k \varphi(k) = 1 \frac{1 + \sqrt{3}}{2} + 2 \frac{1 - \sqrt{3}}{2} = \frac{3 - \sqrt{3}}{2} = 0 \frac{1 + \sqrt{3}}{4} + 2 \frac{3 - \sqrt{3}}{4}$$

$$= \sum_{k=0}^{\lfloor N/2 \rfloor} 2k h_{2k}. \qquad \square$$

Definition 9.8   For each sequence of coefficients $h_0, \ldots, h_N$, define

$$S_1 := \sum_{\ell=0}^{N} (2\ell) h_{2\ell}. \qquad \square$$

The next proposition shows how to recover a dyadic number from values of φ, a feature also called **partition of the identity,** because a sum of values of φ adds up to the **identity** function, which maps each number to the identical number: $r \mapsto r$.

Proposition 9.9 *For each sequence of coefficients $h_0, \ldots, h_N$ with*

$$\sum_{k=0}^{N}(-1)^k k h_k = 0,$$

$$\sum_{k=0}^{N} k\varphi(k) = \sum_{k=0}^{N}(2k)h_{2k},$$

the following equation holds for each dyadic number $r \in \mathbb{D}$:

$$r = \sum_{k\in\mathbb{Z}} [S_1 + k]\,\varphi(r - k).$$

PROOF: This proof uses both the basic recursion and the preceding result, and proceeds by induction on the index n of $\mathbb{D}_n$. Thanks to the preceding result,

$$\sum_{m\in\mathbb{Z}} \frac{3 - \sqrt{3}}{2}\varphi(r - m) = \frac{3 - \sqrt{3}}{2}\sum_{m\in\mathbb{Z}}\varphi(r - m) = \frac{3 - \sqrt{3}}{2} = S_1,$$

and for the present result it suffices to prove the equivalent formula

$$r - S_1 = \sum_{k\in\mathbb{Z}} k\varphi(r - k).$$

For each integer $m \in \mathbb{Z}$,

$$\sum_{k\in\mathbb{Z}} k\varphi(m - k) = \sum_{k=m-N}^{m} k\varphi(m - k) = \sum_{\ell=0}^{N}(m - \ell)\varphi(m - [m - \ell])$$

$$= m\sum_{\ell=0}^{N}\varphi(\ell) - \sum_{\ell=0}^{N}\ell\varphi(\ell) = m - S_1.$$

Thus, the proposition holds for every integer. Proceeding by induction, assume that the proposition holds for some nonnegative index $n \in \mathbb{N}$ and for every $r \in \mathbb{D}_n$. Then for each $r \in \mathbb{D}_{n+1}$ it follows that $2r \in \mathbb{D}_n$, and the basic recursion gives

$$\sum_{k\in\mathbb{Z}} k\varphi(r - k)$$

$$= \sum_{k\in\mathbb{Z}} k \sum_{\ell\in\mathbb{Z}} h_\ell\varphi(2[r - k] - \ell)$$

$$= \sum_{\ell\in\mathbb{Z}} h_\ell \sum_{k\in\mathbb{Z}} k\varphi(2r - [2k + \ell])$$

$$= \sum_{\ell\in\mathbb{Z}} h_\ell \frac{1}{2}\sum_{k\in\mathbb{Z}}[2k + \ell - \ell]\varphi(2r - [2k + \ell])$$

$$= \sum_{m\in\mathbb{Z}} h_{2m} \frac{1}{2} \sum_{k\in\mathbb{Z}} [2k + 2m - 2m]\varphi(2r - [2k + 2m])$$

$$+ \sum_{m\in\mathbb{Z}} h_{2m+1} \frac{1}{2} \sum_{k\in\mathbb{Z}} [2k + 2m + 1 - (2m + 1)]\varphi(2r - [2k + 2m + 1])$$

$$= \sum_{m\in\mathbb{Z}} h_{2m} \frac{1}{2} \sum_{n\in\mathbb{Z}} [2n - 2m]\varphi(2r - 2n)$$

$$+ \sum_{m\in\mathbb{Z}} h_{2m+1} \frac{1}{2} \sum_{n\in\mathbb{Z}} [2n + 1 - (2m + 1)]\varphi(2r - [2n + 1])$$

$$= \frac{1}{2} \left(\sum_{m\in\mathbb{Z}} h_{2m} \right) \sum_{n\in\mathbb{Z}} [2n]\varphi(2r - 2n)$$

$$+ \frac{1}{2} \left(\sum_{m\in\mathbb{Z}} h_{2m+1} \right) \sum_{n\in\mathbb{Z}} [2n + 1]\varphi(2r - [2n + 1])$$

$$- \frac{1}{2} \left(\sum_{m\in\mathbb{Z}} [2m] h_{2m} \right) \sum_{n\in\mathbb{Z}} \varphi(2r - 2n)$$

$$- \frac{1}{2} \left(\sum_{m\in\mathbb{Z}} [2m + 1] h_{2m+1} \right) \sum_{n\in\mathbb{Z}} \varphi(2r - [2n + 1])$$

$$= \frac{1}{2} \sum_{k\in\mathbb{Z}} [k]\varphi(2r - k) - \left(\sum_{m\in\mathbb{Z}} [2m] h_{2m} \right) \sum_{k\in\mathbb{Z}} \varphi(2r - k)$$

$$= \frac{1}{2} 2r - \sum_{m\in\mathbb{Z}} [2m] h_{2m}$$

$$= r - S_1 = r - \frac{3-\sqrt{3}}{2}. \qquad \square$$

EXERCISES

Exercise 9.1. Write and program an algorithm to calculate the values $\varphi(r)$ at all the dyadic numbers $r \in \mathbb{D}_n$ for each positive integer $n \in \mathbb{N}$ by means of the formulae just presented. Test the results by plotting them and comparing them with a graph of φ.

Exercise 9.2. Prove that if $r \in \mathbb{D}_n$, then $\varphi(r) \in \mathbb{D}_{n+2}[\sqrt{3}]$. Such a feature may help in determining in advance the accuracy required for computations.

Exercise 9.3. Prove the converse of Proposition 9.2: *For each sequence of coefficients $h_0, \ldots, h_N$ and function φ satisfying conditions (9.1)–(9.5), if the addi-*

tional condition

$$\int_{\mathbb{R}} \varphi(x)\,dx = 1,$$

holds, then

$$\sum_{n \in \mathbb{Z}} \varphi(n) = 1.$$

Exercise 9.4. Determine which of the foregoing results hold, and which of the foregoing results do not hold, for Haar's wavelets.

9.1.3 Exact Representation of Polynomials by Wavelets

This section shows how the particular example of Daubechies wavelets demonstrated here can reproduce polynomials of degree at most one (straight segments) exactly. More sophisticated examples of Daubechies wavelets can reproduce polynomials of degrees greater than one. The proof will depend upon the partition of unity and the partition of identity just established:

$$1 = \sum_{k \in \mathbb{Z}} \varphi(r - k),$$

$$r = \sum_{k \in \mathbb{Z}} \left(\frac{3 - \sqrt{3}}{2} + k \right) \varphi(r - k),$$

for each dyadic number $r \in \mathbb{D}$. By continuity of φ the same equations also hold for every $r \in \mathbb{R}$, because every real number is the limit of a sequence of dyadic numbers, for instance, rounded binary expansions. The first equation means that the constant function **1** equals a linear combination of Daubechies' building blocks with all coefficients equal to one:

$$\mathbf{1}(r) = 1 = \sum_{k \in \mathbb{Z}} \varphi(r - k).$$

The second equation means that the "identity" function I equals another linear combination of Daubechies' building blocks:

$$I(r) = r = \sum_{k \in \mathbb{Z}} \left(\frac{3 - \sqrt{3}}{2} + k \right) \varphi(r - k).$$

Consequently, for each affine polynomial (of degree at most one)

$$p(r) = c_0 + c_1 r,$$

whence adding multiples of **1** and I gives

$$p(r) = c_0 + c_1 r$$

$$= c_0 \mathbf{1}(r) + c_1 I(r)$$

$$= c_0 \sum_{k \in \mathbb{Z}} \varphi(r - k) + c_1 \sum_{k \in \mathbb{Z}} \left(\frac{3 - \sqrt{3}}{2} + k \right) \varphi(r - k)$$

$$= \sum_{k \in \mathbb{Z}} \left[c_0 + c_1 \left(\frac{3 - \sqrt{3}}{2} + k \right) \right] \varphi(r - k).$$

For finite sums, the same equality holds on an interval instead of on the whole real line:

$$p(r) = c_0 + c_1 r = \sum_{k \in \mathbb{Z}} \left[c_0 + c_1 \left(\frac{3 - \sqrt{3}}{2} + k \right) \right] \varphi(r - k)$$

$$= \sum_{k=K}^{L} \left[c_0 + c_1 \left(\frac{3 - \sqrt{3}}{2} + k \right) \right] \varphi(r - k).$$

for every $r \in \mathbb{R}$ such that

$$K + N \le r \le L,$$

because if $k < K$, or $L < k$, then $r - k > (K + N) - K = N$ or $r - k < L - L = 0$, whence $\varphi(r - k) = 0$, so that the infinite sum reduces to the finite sum for such r. The results just obtained mean that if a signal happens to sample an affine function f (slanting or horizontal straight line), then the approximation $\tilde{f}$ by Daubechies's wavelets reproduces the initial signal f exactly.

9.2   ACCURACY OF SIGNAL APPROXIMATION BY WAVELETS

For signals that need not lie on a straight line, the present section derives estimates of the accuracy with which Daubechies wavelets can approximate such signals.

9.2.1   Accuracy of Taylor Polynomials

This subsection merely reviews the derivation and statement of estimates of accuracy for Taylor polynomials. Here, the notation $p \equiv 0$ means that the function p has the value zero everywhere: $p(x) = 0$ for every x.

Lemma 9.10   *For each nonnegative continuous function* $p : [a, b] \rightarrow \mathbb{R}$ *with* $a < b$, *if* $\int_a^b p = 0$, *then* $p \equiv 0$.

PROOF: Proceed by contraposition: assume that $p \ne 0$, and conclude that $\int_a^b p \ne 0$. If $p \ne 0$, then some $z \in [a, b]$ exists where $p(z) \ne 0$, whence

$p(z) > 0$, by the hypothesis of nonnegativity of p. Hence, by continuity of p, for $\varepsilon := p(z)/2$, some $\delta \in \mathbb{R}$ exists such that $0 < \delta < \max\{z - a, b - z\}$ and $|p(x) - p(z)| < \varepsilon$ for every $x \in [a, b]$ such that $|x - z| < \delta$. Consequently,

$$|p(x) - p(z)| < \varepsilon = p(z)/2,$$

$$-p(z)/2 < p(x) - p(z) < p(z)/2,$$

$$p(z) - p(z)/2 < p(x) < p(z) + p(z)/2,$$

$$0 < p(z)/2 < p(x) < 3p(z)/2,$$

$$0 < \varepsilon < p(x) < 3\varepsilon$$

for every $x \in [a, b]$ such that $|x - z| < \delta$.

The condition that $\delta < \max\{z - a, b - z\}$ ensures that either $[z - \delta, z] \subseteq [a, b]$ or $[z, z + \delta] \subseteq [a, b]$. Thus, in either case, $[a, b]$ contains an interval, $I := [z - \delta, z]$ or $I := [z, z + \delta]$, with length $\delta > 0$ and such that $|p(x) - p(z)| < \varepsilon$ for every $x \in I$. Consequently,

$$\int_a^b p \geq \int_I p > \int_I \varepsilon = \delta \cdot \varepsilon > 0.$$

Theorem 9.11 **(Mean-Value Theorem for Integrals.)** *For all continuous functions $p, q : [a, b] \to \mathbb{R}$ with $p \geq 0$ everywhere on $[a, b]$, or with $p \leq 0$ everywhere on $[a, b]$, a number $c \in [a, b]$ exists such that*

$$\int_a^b pq = q(c) \cdot \int_a^b p.$$

PROOF: Consider first the general situation where $p \geq 0$ and $p \not\equiv 0$, so that $\int_a^b p > 0$, by Lemma 9.10. By the maximum value theorem from calculus [37, p. 108 and 123–124], the continuous function q reaches its minimum and maximum values, m and M, on the compact interval $[a, b]$:

$$m := \min\{q(x) : x \in [a, b]\},$$

$$M := \max\{q(x) : x \in [a, b]\},$$

so that $m \leq q(x) \leq M$ for every $x \in [a, b]$. Multiplying the inequalities $m \leq q \leq M$ by $p \geq 0$ then gives

$$m \leq q \leq M,$$

$$mp \leq qp \leq Mp,$$

$$\int_a^b mp \leq \int_a^b qp \leq \int_a^b Mp,$$

$$m \leq \frac{\int_a^b qp}{\int_a^b p} \leq M.$$

By the intermediate value theorem from calculus [37, pp. 108, 110, 121], the continuous function q reaches all the values from its minimum through its maximum, in particular, the value $y := \int_a^b qp / \int_a^b p$. Thus, a number $c \in [a, b]$ exists such that

$$q(c) = y = \frac{\int_a^b qp}{\int_a^b p},$$

$$q(c) \cdot \int_a^b p = \int_a^b qp.$$

Consider second the special situation where $p \equiv 0$. Then for any $c \in [a, b]$, for instance $c = (a + b)/2$,

$$\int_a^b qp = \int_a^b 0 \cdot p = 0 = q(c) \cdot 0 = q(c) \cdot \int_a^b p.$$

If $p(x) \leq 0$ for every $x \in [a, b]$, the result just obtained holds for the function $-p \geq 0$, and multiplication by -1 in the result yields the conclusion for $p \leq 0$. $\square$

Definition 9.12   A function $f : I \subseteq \mathbb{R} \to \mathbb{R}$ is **of class** C^k, a fact denoted by $f \in C^k([a, b], \mathbb{R})$, if the derivatives $f', f'', \ldots, f^{(k)}$ exist and are continuous on the open interval I. $\square$

Proposition 9.13   (Integration by Parts.)   *For all functions* $u, v : I \subseteq \mathbb{R} \to \mathbb{R}$ *of class* C^k *on a nonempty open interval* I,

$$\int u \cdot dv = u \cdot v - \int v \cdot du.$$

PROOF: Apply the product rule for derivatives, $(u \cdot v)' = u' \cdot v + u \cdot v'$, rearrange terms in the form $(u \cdot v)' - u' \cdot v = u \cdot v'$, and integrate:

$$(u \cdot v)' - u' \cdot v = u \cdot v',$$

$$\int d(u \cdot v) - \int du \cdot v = \int u \cdot dv. \qquad \square$$

Theorem 9.14   (Taylor Polynomial with Integral Remainder.)   *For each function* $f : I \subseteq \mathbb{R} \to \mathbb{R}$ *of class* C^k *on a nonempty open interval* I, *and for all reals* x *and* h *such that* $x, x + h \in I$,

$$f(x + h) = f(x) + hf'(x) + \frac{h^2}{2} f''(x) + \cdots + \frac{h^{k-1}}{(k-1)!} f^{(k-1)}(x)$$

$$+ (-1)^{k+1} \int_x^{x+h} \frac{[t - (x+h)]^{k-1}}{(k-1)!} f^{(k)}(t) \, dt.$$

PROOF: Proceed by induction on k. By the fundamental theorem of calculus,

$$f(x + h) - f(x) = \int_x^{x+h} f'(t)\, dt.$$

Hence, integration by parts with

$$u(t) := f'(t), \qquad du = f'' dt,$$
$$dv := dt, \qquad v(t) := t - (x + h),$$

produces

$$\int_x^{x+h} f'(t)\, dt = [t - (x + h)] \cdot f'(t) \Big|_x^{x+h} - \int_x^{x+h} [t - (x + h)] \cdot f''(t)\, dt$$

$$= hf'(x) - \int_x^{x+h} [t - (x + h)] \cdot f''(t)\, dt,$$

whence

$$f(x + h) = f(x) + hf'(x) - \int_x^{x+h} [t - (x + h)] \cdot f''(t)\, dt.$$

Proceeding by induction, assume that for some $\ell \in \{1, \ldots, k - 1\}$,

$$f(x + h) = f(x) + hf'(x) + \frac{h^2}{2} f''(x) + \cdots + \frac{h^{\ell-1}}{(\ell - 1)!} f^{(\ell-1)}(x)$$

$$+ (-1)^{\ell+1} \int_x^{x+h} \frac{[t - (x + h)]^{\ell-1}}{(\ell - 1)!} f^{(\ell)}(t)\, dt.$$

Integration by parts with

$$u(t) := f^{(\ell)}(t), \qquad\qquad du = f^{(\ell+1)} dt,$$
$$dv := [t - (x + h)]^{\ell-1}/(\ell - 1)!\, dt, \qquad v(t) := [t - (x + h)]^{\ell}/\ell!,$$

gives

$$\int_x^{x+h} \frac{[t - (x + h)]^{\ell-1}}{(\ell - 1)!} f^{(\ell)}(t)\, dt$$

$$= \frac{[t - (x + h)]^{\ell}}{\ell!} f^{(\ell)}(t) \Big|_x^{x+h} - \int_x^{x+h} \frac{[t - (x + h)]^{\ell}}{\ell!} f^{(\ell+1)}(t)\, dt$$

$$= (-1)^{\ell-1} \frac{h^{\ell}}{\ell!} f^{(\ell)}(x) - \int_x^{x+h} \frac{[t - (x + h)]^{\ell}}{\ell!} f^{(\ell+1)}(t)\, dt,$$

and a substitution in the induction hypothesis yields the final result for $\ell = k$. $\square$

Corollary 9.15   (Taylor Polynomial with Differential Remainder.)   *For all x and h, a number $s \in [0, 1]$ exists for which*

$$f(x + h) = f(x) + hf'(x) + \frac{h^2}{2} f''(x) + \cdots + \frac{h^{k-1}}{(k-1)!} f^{(k-1)}(x)$$

$$+ \frac{h^k}{k!} f^{(k)}(x + sh).$$

PROOF: Apply the mean value theorem for integrals, with $p(t) := [t - (x + h)]^{k-1}/(k - 1)!$, $q(t) := f^{(k)}(t)$, and $c := x + sh$ lying between x and $x + h$. □

9.2.2   Accuracy of Signal Representations by Wavelets

With the particular example of Daubechies wavelets, this subsection demonstrates how to derive estimates of accuracy for the wavelet representation of a function (signal). Such estimates of accuracy will rely on the ability of Daubechies wavelets to reproduce exactly a polynomial approximating the signal, for instance, a Taylor polynomial, so that the discrepancy between the signal and its wavelet representation remains proportional to the discrepancy between the signal and its Taylor polynomial.

Theorem 9.16   *With Daubechies wavelets, for each continuous function $g :$ $[0, N] \to \mathbb{R}$, here with $N = 3$ and*

$$\|g\|_\infty := \max_{x \in [0, N]} |g(x)|,$$

the least-squares $2N$-periodic wavelet representation $\sum_{k \in \mathbb{Z}} a_{g,k} \varphi(x - k)$ remains below the upper bounds

$$|a_{g,k}| \le \|g\|_\infty \|\varphi\|_1,$$

$$\left\| \sum_{k=0}^{2N} a_{g,k} \varphi_k^{(0)} \right\|_2^2 := \int_{\mathbb{R}} \left| \sum_{k=0}^{2N} a_{g,k} \varphi_k^{(0)}(x) \right|^2 dx \le 2N \|g\|_\infty^2 \|\varphi\|_1^2,$$

$$\left\| \sum_{k \in \mathbb{Z}} a_{g,k} \varphi_k^{(0)} \right\|_\infty := \max_{x \in [0, N]} \left| \sum_{k \in \mathbb{Z}} a_{g,k} \varphi_k^{(0)}(x) \right| \le 2N \|g\|_\infty \|\varphi\|_1 \|\varphi\|_\infty.$$

PROOF: The wavelet representation constitutes the orthogonal projection on the linear space spanned by the orthonormal functions $\varphi_k^{(0)} = \varphi(x - k)$, whence

$$|a_{g,k}| = |\langle g, \varphi_k^{(0)} \rangle| = \left| \int_k^{N+k} g(x) \varphi(x - k) dx \right|$$

$$\le \int_k^{N+k} \|g\|_\infty |\varphi(x - k)| dx = \|g\|_\infty \|\varphi\|_1.$$

The functions $\varphi_k^{(0)}$ are mutually orthonormal, whence

$$\left\| \sum_{k=0}^{2N} a_{g,k}\varphi_k^{(0)} \right\|_2^2 = \sum_{k=0}^{2N} |a_{g,k}|^2 \|\varphi_k^{(0)}\|_2^2$$

$$\leq \sum_{k=0}^{2N} \|g\|_\infty^2 \|\varphi\|_1^2 \cdot 1 = 2N\|g\|_\infty^2 \|\varphi\|_1^2.$$

Similarly, at most $2N$ functions $\varphi_k^{(0)}$ have nonzero values in $[0, N]$, whence

$$\left\| \sum_{k\in\mathbb{Z}} a_{g,k}\varphi_k^{(0)} \right\|_\infty = \max_{x\in[0,N]} \left| \sum_{k\in\mathbb{Z}} a_{g,k}\varphi_k^{(0)}(x) \right|$$

$$\leq \max_{x\in[0,N]} \sum_{k\in\mathbb{Z}} |a_{g,k}| \cdot \left|\varphi_k^{(0)}(x)\right|$$

$$\leq 2N \left(\|g\|_\infty \|\varphi\|_1\right) \|\varphi\|_\infty. \qquad \square$$

Theorem 9.17 *For each twice continuously differentiable function $f \in C^1(\mathbb{R}, \mathbb{R})$, and with $P_{f,r,1}$ denoting the Taylor polynomial of f at r with degree at most 1, so that*

$$P_{f,r,1}(x) = f(r) + (x - r)f'(r),$$

and for each of the norms $\|\ \|_1$, $\|\ \|_2$, and $\|\ \|_\infty$, there exists a real constant $C_p \geq 0$ such that the Daubechies wavelet representation $\tilde{f} = \sum_{k\in\mathbb{Z}} a_{f,k}\varphi_k^{(0)}$ of f satisfies the inequality

$$\|f - \tilde{f}\|_p \leq C_p \|f''\|_\infty = C_p \max_{u\in[0,N]} |f''(u)|$$

with

$$C_1 = \frac{N^3}{8\sqrt{5}},$$

$$C_2 = \frac{N^{\frac{5}{2}}}{8\sqrt{5}},$$

$$C_\infty = \frac{N^3}{4} \left(\frac{1}{2N} + \|\varphi\|_1 \|\varphi\|_\infty\right).$$

PROOF: For any norm $\|\ \|$, and with $\tilde{\ }$ representing the approximation by Daubechies wavelets, the exact reproduction of affine polynomials by Daubechies wavelets guarantees that $P_{f,r,1} = \tilde{P}_{f,r,1}$, whence

$$\|f - \tilde{f}\| = \|(f - P_{f,r,1}) + (P_{f,r,1} - \tilde{P}_{f,r,1}) + (\tilde{P}_{f,r,1} - \tilde{f})\|$$
$$= \|(f - P_{f,r,1}) + 0 + (\tilde{P}_{f,r,1} - \tilde{f})\|$$
$$= \|(f - P_{f,r,1}) - (\tilde{f} - \tilde{P}_{f,r,1})\|.$$

With $r := N/2$, the differential remainder of Taylor's approximation applied to the function $g := f - P_{f,r,1}$ and to each $x \in [0, N]$ yields

$$|g(x)| = |f(x) - P_{f,r,1}(x)|$$

$$\leq \frac{|x - r|^2}{2} \max_{u \in [0,N]} |f''(u)| \leq \frac{(N/2)^2}{2} \max_{u \in [0,N]} |f''(u)|.$$

For the energy norm $\| \ \|_2$,

$$\|g\|_2^2 := \int_0^N |g(x)|^2 \, dx \leq \int_0^N \left| \frac{|x - r|^2}{2} \|f''\|_\infty \right|^2 dx = \frac{N^5}{320} \|f''\|_\infty^2.$$

Also, $\tilde{g}$ represents the orthogonal projection of g on a finite-dimensional linear subspace of wavelets, whence $(g - \tilde{g}) \perp \tilde{g}$ and the Pythagorean theorem yields

$$\|g\|_2^2 = \|(g - \tilde{g}) + \tilde{g}\|_2^2 = \|g - \tilde{g}\|_2^2 + \|\tilde{g}\|_2^2,$$
$$\|g - \tilde{g}\|_2^2 = \|g\|_2^2 - \|\tilde{g}\|_2^2 \leq \|g\|_2^2,$$

with equality if, and only if, g is orthogonal to the subspace of wavelets. Consequently,

$$\|f - \tilde{f}\|_2 = \|g - \tilde{g}\|_2 \leq \|g\|_2 \leq \frac{N^{\frac{5}{2}}}{8\sqrt{5}} \cdot \|f''\|_\infty.$$

For the area norm $\| \ \|_1$, the Cauchy–Schwarz inequality gives

$$\|h\|_1 := \int_0^N |h(x)| \, dx = \int_0^N 1 \cdot |h(x)| \, dx = \langle 1, |h| \rangle$$

$$\leq \|1\|_2 \|h\|_2 = \sqrt{\int_0^N 1^2 \, dx} \sqrt{\int_0^N |h(x)|^2 \, dx} = \sqrt{N} \|h\|_2.$$

Applied to $h := f - \tilde{f}$, the inequality just obtained gives

$$\|f - \tilde{f}\|_1 \leq \sqrt{N} \|f - \tilde{f}\|_2 \leq \sqrt{N} \frac{N^{\frac{5}{2}}}{8\sqrt{5}} \cdot \|f''\|_\infty = \frac{N^3}{8\sqrt{5}} \cdot \|f''\|_\infty.$$

For the maximum norm $\| \ \|_\infty$, the triangle inequality and Theorem 9.16 yield

$$\|f - \tilde{f}\|_\infty = \|g - \tilde{g}\|_\infty \leq \|g\|_\infty + \|\tilde{g}\|_\infty$$

$$\leq \frac{N^2}{8} \|f''\|_\infty + 2N \frac{N^2}{8} \|f''\|_\infty \|\varphi\|_1 \|\varphi\|_\infty$$

$$= \frac{N^2}{8} (1 + 2N \|\varphi\|_1 \|\varphi\|_\infty) \|f''\|_\infty. \qquad \square$$

Similar results hold with Taylor polynomials of degrees greater than 1, and hence with greater accuracy, but only with more sophisticated examples of Daubechies wavelets, which require different and longer sequences of coefficients $h_0, \ldots, h_N$ for which additional "moments" μ_m equal zero:

$$0 = \mu_1 := \sum_k (-1)^k k h_k,$$

$$0 = \mu_2 := \sum_k (-1)^k k^2 h_k,$$

$$0 = \mu_3 := \sum_k (-1)^k k^3 h_k,$$

$$\vdots$$

$$0 = \mu_p := \sum_k (-1)^k k^p h_k.$$

Sharper inequalities can result from more sophisticated approximating polynomials, for instance, Hermite's interpolating polynomials [27, Ch. 6, §6.3, pp. 463–374], least squares approximation [27, Ch. 6, §6.8, pp. 421–434], or Chebyshev's "best uniform approximation" [27, Ch. 6, §6.9, pp. 434–449].

9.2.3 Approximate *Interpolation* by Daubechies' Function

A simple and common choice of the coefficients a_k consists in setting, for each $k \in \{0, \ldots, 2^n - 1\}$,

$$a_k := s_k.$$

The choice $a_k = s_k$ admits several technical interpretations. The first interpretation shows that the approximation

$$\tilde{f}(r) = \sum_{k=0}^{2^n-1} s_k \varphi(r - k)$$

nearly interpolates f at the sample points $s_k = f(k)$. Indeed, consider the three points circled in Figure 9.1,

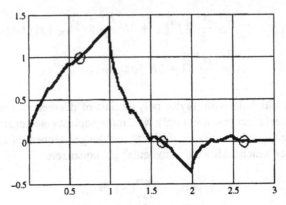

Figure 9.1   Circles mark points where φ takes the values 1, 0, and 0.

$$r_0 \approx \frac{3 - \sqrt{3}}{2} \approx 1.634, \qquad \varphi(r_0) = 1,$$

$$r_1 \approx \frac{5 - \sqrt{3}}{2} \approx 1.634, \qquad \varphi(r_1) = 0,$$

$$r_2 \approx \frac{7 - \sqrt{3}}{2} \approx 2.634, \qquad \varphi(r_2) = 0.$$

More precisely, Table 9.1 locates the points r_0, r_1, and r_2 with an absolute accuracy of 0.001. The preceding chapter has established the continuity of φ, which guarantees the existence of the points r_0, r_1, and r_2. Moreover, the values in Table 9.1 result from automated symbolic and numerical computations based on the recursion (9.1). Consequently, barring any computational errors,

$$\frac{650}{1024} < r_0 < \frac{651}{1024}, \quad \frac{1673}{1024} < r_1 < \frac{1674}{1024}, \quad \frac{2695}{1024} < r_2 < \frac{2696}{1024}.$$

Table 9.1   Locations of r_0, r_1, and r_2 within 0.001

r	$\varphi(r)$		
650/1024	$\varphi\,(650/1024)$	$\approx$	+0.999 985
r_0	$\varphi(r_0)$	=	1
651/1024	$\varphi\,(651/1024)$	$\approx$	+1.001 000
1673/1024	$\varphi\,(1673/1024)$	$\approx$	+0.000 285
r_1	$\varphi(r_1)$	=	0
1674/1024	$\varphi\,(1674/1024)$	$\approx$	−0.000 155
2695/1024	$\varphi\,(2695/1024)$	$\approx$	−0.000 248
r_2	$\varphi(r_2)$	=	0
2696/1024	$\varphi\,(2696/1024)$	$\approx$	+0.000 137
2698/1024	$\varphi\,(2698/r1024)$	$\approx$	+0.000 170

Therefore,

$$f(0)\varphi(r_0) = f(0),$$
$$f(1)\varphi(r_1) = 0,$$
$$f(2)\varphi(r_2) = 0,$$

so that for each integer ℓ,

$$\tilde{f}(r_0 + \ell) = \sum_{k=0}^{2^n-1} f(k)\varphi(r_0 + \ell - k) \approx \cdots + 0 + f(\ell) + 0 + \cdots +$$

nearly interpolates each $f(\ell)$ at $r_0 + \ell$.

EXERCISES

Exercise 9.5.

(a) Identify the computational features of Daubechies' scaling function φ that yield all the potential triples of numbers (r_0, r_1, r_2) separated by intervals of unit length, so that $r_2 - r_1 = 1 = r_1 - r_0$, and where φ might assume the values $\varphi(r_0) = 1$, $\varphi(r_1) = 0 = \varphi(r_2)$.

(b) Prove that no such triple exists.

Exercise 9.6. Determine an upper bound on the accuracy of representations of signals by Haar's wavelets.

PART D

Directories

Acknowledgments

Included here by permission, ©COMAP, Inc.,
the parts of Chapter 4 on 3-D graphics are excerpts from

[a] Yves Nievergelt, "3-D Graphics in Calculus and Linear Algebra," *Tools for Teaching 1991*, pp. 125–169; also reprinted as *UMAP Module 717, 3-D Graphics in Calculus and Linear Algebra,* both by COMAP, Lexington, MA, 1992. [This work was supported in part by a Seed Grant (Grant Number 143150-92-02) from the Washington Center for Improving the Quality of Undergraduate Education, at Evergreen State College, in Olympia, WA.]
Parts of Chapters 4, 5, and 6 are excerpts from

[b] Yves Nievergelt, "Orthogonal Projections and Applications in Linear Algebra," *UMAP Journal*, Vol. 18, No. 4 (winter 1997) pp. 403–432; also reprinted as *UMAP Module 756, Orthogonal Projections and Applications in Linear Algebra,* COMAP, Lexington, MA: 1997.
Parts of Chapter 7 are excerpts from

[c] Yves Nievergelt, "Computed Tomography in Multivariable Calculus," *UMAP Modules 1996: Tools for Teaching*, COMAP, Lexington, MA, 1997, pp. 135–191; also reprinted as *UMAP Module 753, Computed Tomography in Multivariable Calculus,* COMAP, Lexington, MA: 1997. [This work was supported in part by the National Science Foundation's grant DUE-9455061.]

The author acknowledges the use of Donald Knuth's TeX mathematical typesetting language and the American Mathematical Society's fonts with Blue Sky Research's TEXTURES™, Wolfram Research's *Mathematica*™ software for graphics, and Symantec Corporation's C and Absoft Corporation's FORTRAN 77 compilers for the computations, with an Apple Computer, Inc.'s, Macintosh® IIcx.

Collection of Symbols

	Item	Section	
$[\#]$		1.0	reference number in the bibliography
$[u, w[$	Def. 1.1	1.1	half-open interval
$\varphi_{[u,w[}$	Fig. 1.2	1.1	Haar's step function
$\sum$		1.1	summation
$\psi_{[0,1[}$	Exercise 1.4	1.1	Haar's unit wavelet
$\psi_{[u,w[}$		1.2.3	Haar wavelet
$\varphi_j^{(n)}, \psi_j^{(n)}$		1.3	step functions, wavelets
$\vec{\mathbf{s}}^{(n-\ell)}$		1.3	array of coefficients after ℓ steps
$\tilde{f}, \hat{f}$		1.3.2	coarse and fine approximations
$\approx$		1.6.2	approximately
$\otimes$	Def. 2.4	2.1.1	tensor product of functions
$\Phi_{(k,\ell)}^{a,(j)}, \Psi_{(k,\ell)}^{a,(j)}$	Def. 2.8	2.1.1	2-D Haar step functions and wavelets
Ω		2.1.3	matrix of change of Haar basis
$\langle \vec{\mathbf{v}}, \vec{\mathbf{w}} \rangle$	Def. 3.1	3.0	inner product of two vectors
φ	(3.2)	3.1	Daubechies' building block function
h_0, h_1, h_2, h_3	(3.3)	3.1	Daubechies' coefficients
ψ	(3.4)	3.1	Daubechies wavelet
$\mathbb{D}$	Def. 3.5	3.1	the set of dyadic numbers
$\bar{r}$	Def. 3.7	3.1	conjugate
$_D\Omega$		3.4	matrix of change of Daubechies' basis
I		3.4	identity matrix
$\Phi_{(k,\ell)}^{a,(j)}, \Psi_{(k,\ell)}^{a,(j)}$	Fig. 3.9	3.6	2-D Daubechies' functions and wavelets
$\mathbb{F}$	Def. 4.1	4.1	number field
$\mathbb{Q}$	Ex. 4.2	4.1	field of rational numbers
$\mathbb{R}$	Ex. 4.3	4.1	field of real numbers
$\mathbb{C}$	Ex. 4.4	4.1	field of complex numbers
$\mathbb{Z}$	Counterex 4.7	4.1	set of integers
$\langle f, g \rangle$	Ex. 4.21	4.2	inner product of functions
$\| \ \|$	Def. 4.23	4.2	norm on a linear space
$\| \ \|_2$	Def. 4.26	4.2	Euclidean norm on a linear space
$\| \ \|_\infty$	Ex. 4.28	4.2	"maximum" norm on a linear space
$\| \ \|_1$	Ex. 4.29	4.2	"taxicab" norm on a linear space

289

Bibliography

[1] Lars Valerian Ahlfors, *Complex Analysis,* McGraw-Hill, New York, NY, 1979. ISBN 0-07-000657-1. QA 331.A45 1979. LCCC No. 78-17078. 515'.93.

[2] Garrett Birkhoff and Saunders Mac Lane, *A Survey of Modern Algebra,* 4th ed., Macmillan, New York, NY, 1977 (50th Anniversary Printing 1991). ISBN 0-02-310070-2. QA162.B57 1977. LCCC No. 75-42402. 512.

[3] Bodil Branner, "The Mandelbrot Set," in reference [8], pp. 75–105.

[4] Charles K. Chui, *Wavelets: A Tutorial in Theory and Applications,* Academic Press, Boston, MA, 1992. LCCC No. 91-58833.

[5] James W. Cooley and John W. Tukey, "An Algorithm for the Machine Calculations of Complex Fourier Series," *Mathematics of Computation,* Vol. 19 (1965), pp. 297–301.

[6] Ingrid Daubechies, "Orthonormal Bases of Compactly Supported Wavelets," *Communications on Pure and Applied Mathematics,* Vol. XLI, No. 8 (December 1988), pp. 909–996.

[7] Ingrid Daubechies, *Ten Lectures on Wavelets,* SIAM, Philadelphia, PA, 1992. ISBN 0-89871-274-2. QA403.3.D38 1992. LCCC No. 92-13201.

[8] Robert L. Devaney and Linda Keen, editors, *Chaos and Fractals: The Mathematics Behind the Computer Graphics,* American Mathematical Society, Providence, RI, 1989. ISBN 0-8218-0137-6. T385.C454 1989. LCCC No. 89-7003. 006.6–dc20.

[9] Mark F. Dubach, "Distribution of Intracerebrally Injected Dopamine as Studied by a Punch-Scintillation Modeling Technique," *Neuroscience,* Vol. 45, No. 1 (1991), pp. 103–115.

[10] Heinz W. Engl, Alfred K. Louis, and William Rundell, *Inverse Problems in Geophysical Applications,* Society for Industrial and Applied Mathematics (SIAM), Philadelphia, PA, 1997. ISBN 0-89871-381-1. LCCC No. 96-71040.

[11] Richard Phillips Feynman, Robert B. Leighton, and Matthew L. Sands, *The Feynman Lectures on Physics, Vol. I: Mechanics, Radiation, and Heat,* Commemorative Issue, Addison-Wesley, Redwood City, CA, 1989. ISBN 0-201-51003-0. LCCC No. 89-433. QC21.2.F49 1989. 530–dc19.

[12] Gerald B. Folland, *Real Analysis: Modern Techniques and Their Applications,* Wiley, New York, NY, 1984. ISBN 0-471-80958-6. QA300.F67 1984. LCCC No. 84-10435. 515.

[13] W. Freeden, F. Schneider, and M. Schreiner, "Gradiometry—An Inverse Problem in Modern Satellite Geodesy," in reference [10], pp. 179–239.

[14] Stephen H. Friedberg, Arnold J. Insel, and Lawrence E. Spence, *Linear Algebra*, 2nd ed., Prentice Hall, Englewood Cliffs, NJ, 1989. ISBN 0-13-537102-3. QA184.F75 1989. LCCC No. 88-28568. 512'.5–dc19.

[15] Felix R. Gantmacher, *Matrizentheorie*, Springer-Verlag, Berlin–Heidelberg–New York–Toronto, 1986. ISBN 3-540-16582-7.

[16] Roland Glowinski, Wayne Lawton, Michel Ravachol, and Eric Tenenbaum, "Wavelet Solutions of Linear and Nonlinear Elliptic, Parabolic and Hyperbolic Problems in One Space Dimension," in reference [17], pp. 55–120.

[17] Roland Glowinski and Alain Lichnewsky, eds., *Computing Methods in Applied Sciences and Engineering*, Society for Industrial and Applied Mathematics, Philadelphia, PA, 1990. ISBN 0-89871-264-5. QC39.I49 1990. LCCC No. 90-23772. 530–dc20.

[18] David Gottlieb and Chi-Wang Shu, "On the Gibbs Phenomenon and Its Resolution," Society for Industrial and Applied Mathematics' *SIAM Review*, Vol. 39, No. 4 (December 1997), pp. 644–668.

[19] Alfred Haar, "Zur Theorie der orthogonalen Funktionen-Systeme," *Mathematische Annalen*, Vol. 69 (1910), pp. 331–371.

[20] Eugenio Hernández and Guido Weiss, *A First Course on Wavelets*, CRC Press, Boca Raton, FL, 1996. ISBN 0-8493-8274-2. QA403.3.H47 1996. LCCC No. 96-27111. 515'.2433–dc20.

[21] Edwin Hewitt and Karl Stromberg, *Real and Abstract Analysis*, Springer-Verlag, Berlin–Heidelberg, 1965. LCCC No. 65-26609.

[22] Edwin Hewitt and Robert E. Hewitt, "The Gibbs-Wilbraham Phenomenon: An Episode in Fourier Analysis," *Archive for History of Exact Sciences*, Vol. 21, No. 2 (1979–1980), pp. 129–160.

[23] Thomas W. Hungerford, *Algebra*, Holt, Rinehart and Winston, New York, NY, 1974. ISBN 0-03-086078-4. QA155.H83. LCCC No. 73-15693. 512.

[24] Bill Jacob, *Linear Algebra*, W. H. Freeman, New York, NY, 1990. ISBN 0-7167-2031-0. QA184.J33 1990. LCCC No. 89-11801. 512'.5–dc20.

[25] Alan H. Karp, "Bit Reversal on Uniprocessors," *[Society for Industrial and Applied Mathematics] SIAM Review*, Vol. 38, No. 1 (March 1996), pp. 1–26.

[26] Linda Keen, "Julia Sets," in reference [8], pp. 57–74.

[27] David R. Kincaid and E. Ward Cheney, *Numerical Analysis: The Mathematics of Scientific Computing*, 2nd ed., Brooks/Cole, Pacific Grove, CA, 1996. ISBN 0-534-33892-5. QA297.K563 1996. LCCC No. 95-53860. 519.4–dc20.

[28] Thomas William Körner, *Fourier Analysis*, Cambridge University Press, Cambridge, UK, 1988. ISBN 0-521-38991-7. QA403.5.K67 1986. LCCC No. 85-17410. 515'.2433.

[29] Thomas William Körner, *Exercises for* Fourier Analysis, Cambridge University Press, Cambridge, UK, 1993. ISBN 0-521-43849-7. QA403.5.K66 1993. LCCC No. 92-30757. 515'.723–dc20.

[30] Catherine C. McGeoch, "Data Compression," *American Mathematical Monthly*, Vol. 100, No. 5 (May 1993), pp. 493–497.

[31] Yves Meyer, *Ondelettes et Opérateurs I: Ondelettes*, Hermann, Paris, France, 1990. ISBN 2-7056-6125-0.

[32] Yves Meyer, *Ondelettes et Opérateurs II: Opérateurs de Calderón-Zygmund,* Hermann, Paris, France, 1990. ISBN 2-7056-6126-7.

[33] David Pollen, "Daubechies' Scaling Function on [0, 3]," in reference [4], pp. 3–13.

[34] Richard J. Pulskamp and James A. Delaney, "Computer and Calculator Computation of Elementary Functions," *UMAP Journal,* Vol. 12, No. 4 (Winter 1991), pp. 315–348, and *UMAP Modules 1991: Tools For Teaching,* pp. 1–34. Also reprinted as *Computer and Calculator Computation of Elementary Functions,* UMAP Module 708, COMAP, Lexington, MA, 1992.

[35] Laurent Schwartz, *Théorie des distributions,* corrected ed., Hermann, Paris, F, 1973. ISBN 2-7056-5551-4.

[36] Murray R. Spiegel, *Complex Variables,* Schaum's Outlines Series, McGraw-Hill, New York, NY, 1964. ISBN 07-060230-1.

[37] Michael Spivak, *Calculus,* Publish or Perish, Inc., Wilmington, DE, 1980. ISBN 0-914098-77-2. LCCC No. 80-82517.

[38] James Stewart, *Calculus,* 3rd ed., Brooks/Cole, Pacific Grove, CA, 1995. ISBN 0-534-21798-2. QA303.S8825. LCCC No. 94-29764. 515'.15–dc20.

[39] Josef Stoer and Roland Bulirsch, *Introduction to Numerical Analysis,* 2nd ed., Springer-Verlag, New York, NY, 1993. ISBN 0-387-97878-X. QA297.S8213 1992. LCCC No. 92-20536. 519.4–dc20.

[40] Gilbert Strang, *Introduction To Applied Mathematics,* Wellesley-Cambridge Press, Wellesley, MA, 1986. ISBN 0-9614088-0-4. QA37.S87 1986. LCCC No. 84-52450. 510.

[41] Gilbert Strang, "Wavelets and Dilation Equations: A Brief Introduction," *SIAM Review,* Vol. 31, No. 4 (December 1989), pp. 614–627.

[42] Gilbert Strang, *Calculus,* Wellesley-Cambridge Press, Wellesley, MA, 1991. ISBN 0-13-032946-0. QA303.S8839 1991. LCCC No. 90-49977. 515.20.

[43] Gilbert Strang and Truong Nguyen, *Wavelets and Filter Banks,* Wellesley-Cambridge Press, Wellesley, MA, 1996. ISBN 0-9614088-7-1. QA403.3.S87.

[44] Robert S. Strichartz, "How To Make Wavelets," *American Mathematical Monthly,* Vol. 100, No. 6 (June–July 1993), pp. 539–556.

[45] Angus E. Taylor, *Introduction to Functional Analysis,* Wiley, New York, NY, 1958. LCCC 58-12704.

[46] Bartel Leenert van der Waerden, *Modern Algebra,* Vol. I, Frederick Ungar, New York, NY, 1953.

[47] Bartel Leenert van der Waerden, *Modern Algebra,* Vol. II, Frederick Ungar, New York, NY, 1950.

[48] Gilbert G. Walter, "Wavelets: A New Tool in Applied Mathematics," *UMAP Journal,* Vol. 14, No. 2 (Summer 1993), pp. 155–178.

[49] Gilbert G. Walter, *Wavelets and Other Orthonormal Systems With Applications,* CRC Press, Boca Raton, FL, 1994. ISBN 0-8493-7878-8.

[50] Eyvind H. Wichmann, *Quantum Physics,* Berkeley Physics Course, Vol. 4, McGraw-Hill, New York, NY, 1971. LCCC 64-66016.

[51] Antoni Zygmund, *Trigonometrical Series* (reprint of 1935 ed.), Dover, New York, NY, 1955.

[52] Antoni Zygmund, *Trigonometric Series,* 2nd ed., Cambridge University Press, Cambridge, UK, 1990. ISBN 0-521-35885-X.

Index

Page numbers in *italics* locate definitions.